Laboratory Manual
to accompany

Chemistry:
The Practical Science

Laboratory Manual
to accompany

Chemistry: The Practical Science

Paul Kelter
University of Illinois at Urbana-Champaign
Michael Mosher
University of Nebraska at Kearney
Andrew Scott
Perth College, UHI Millennium Institute

James Almy
Cornell University

HOUGHTON MIFFLIN COMPANY   **Boston   New York**

Vice President and Executive Publisher: George Hoffman
Vice President and Publisher: Charles Hartford
Senior Marketing Manager: Laura McGinn
Development Editor: Rebecca Berardy Schwartz
Assistant Editor: Amy Galvin
Editorial Associate: Henry Cheek
Senior Project Editor: Charline Lake
Manufacturing Coordinator: Susan Brooks
Marketing Assistant: Kris Bishop

Printed in the U.S.A.

ISBN-10: 0-618-00092-5
ISBN-13: 978-0-618-00092-0

1 2 3 4 5 6 7 8 9-POO-11-10 09 08 07

CONTENTS

* Indicates an experiment in the guided-inquiry format.

* Indicates an experiment in the guided-inquiry format.

To the Student

WHY STUDY CHEMISTRY?

Large numbers of students take chemistry at most schools, relatively few of which intend to be chemists. Why, then, this high enrollment? Simply put: Chemistry teaches things of value. To some extent, it is the chemistry content itself that is valuable. Many fields other than chemistry deal with chemicals. Professionals in these fields need a basic understanding of chemical principles to perform well at their chosen work. But it is also because of the methods chemists employ to solve problems, methods that can be applied usefully outside of chemistry. It was with these needs of students in mind, both the chemistry-specific and the broadly applicable, that the laboratory exercises in this Lab Manual were designed.

WHAT WILL I LEARN IN THE LAB?

The general skills emphasized by the laboratory exercises in this Lab Manual are

A Better Understanding of Chemistry Concepts

Learning is an active, participatory endeavor—and there is no better way to actively engage with the subject matter of chemistry than in the laboratory. Performing the laboratory exercises in this Lab Manual will deepen and broaden your understanding of the concepts that will be presented to you during your course lectures.

Laboratory Techniques

Performing the laboratory exercises in this Lab Manual will help you to develop an understanding of and competence with a number of laboratory techniques used in chemistry. Examples include titrations, crystallizations, and spectroscopy. You also will become skilled at using the equipment common to the chemistry laboratory: burets, pipets, flasks, graduated cylinders, and more. Many of the laboratory techniques and pieces of equipment you will learn about in the chemistry lab are applicable to other scientific disciplines as well.

Data Assessment and Interpretation

Once you've collected data in the lab, it is necessary to assess its quality. Are the data as you expected? Are the data from multiple trials of the same experiment consistent with each other? Should the experiment be repeated? Questions such as these are pertinent in every experiment, whether the experiment involves chemistry or some other discipline.

The logical follow-up to assessing data is their interpretation. Correctly interpreting data is of critical importance in science. After all, our current understanding of science is based on an interpretation of *what actually happened* in various experiments, not *what was expected* to happen. The laboratory exercises in this Lab Manual will provide you with numerous opportunities to interpret data. Various techniques will be used to interpret the data: calculations, graphs, and qualitative comparisons, just to name a few. These are techniques, again, that can be applied to fields other than chemistry.

Experimental Design

Perhaps the most valuable skill you have an opportunity to learn from the exercises in this Lab Manual is that of designing an experimental procedure, a skill applicable across the sciences. The exercises in this Lab Manual have either one of two formats: *traditional* or *guided-inquiry*. The fundamental difference between these two formats is the degree of procedural detail provided. Exercises in the traditional format are typified by complete, detailed, easy-to-follow procedures. Guided-inquiry format exercises, however, require the student, to some extent, to design the experimental procedure. In some instances the degree of experimental design is slight, perhaps just the modification of a given procedure, whereas in others it is quite large, the complete determination of how to accomplish the experimental objective using the available reagents and equipment. Your instructor has selected laboratory exercises in either one or both formats for you to perform. The format of each exercise is indicated in the Table of Contents.

PREPARING FOR LAB

Read the Laboratory Exercise

Each laboratory exercise begins with an *Objective* section. The Objective section provides a concise statement of the purpose of the exercise. Use the Objective to orient your thinking about the exercise. Following the Objective is the *Introduction* section. The Introduction provides the background necessary for understanding the exercise. It reviews the theory and methods pertinent to the exercise. You should supplement the material in the Introduction by reading the sections of your textbook that deal with the same concepts. Next comes the *Additional Reading* section. The Additional Reading indicates which sections from the *Laboratory Techniques* part of this Lab Manual are pertinent to the exercise. If you have not already done so, you should read each of these sections. Each section explains how to safely and effectively perform a particular experimental technique. The *Safety Precautions* necessary for each exercise follow the Additional Reading section. Make sure that you understand the hazards associated with each exercise. The Safety Precautions are boxed to make finding them easier. The *Procedure* or *Experiment* section follows next. Complete, detailed procedures are provided for all exercises in the traditional format. These should be followed explicitly and without alteration unless you are told otherwise by your instructor. Lab exercises in the guided-inquiry format have Experiment sections. Each Experiment section explains what needs to be accomplished and any restrictions you are to work under. The Experiment section does not tell you how to accomplish your given task. The different name is used to emphasize that you are responsible for determining at least part of the experimental procedure. For lab exercises in the guided-inquiry format, an *Equipment and Reagents* section follows. This section lists the chemicals and equipment available for your use during the experiment. It is vital that you account for these limitations when designing your experimental procedures in guided-inquiry exercises.

Answer the Prelaboratory Questions

Each laboratory exercise has a number of *Prelaboratory Questions* that accompany it. The Prelaboratory Questions are listed with the exercise. The purpose of these questions is to prepare you for the upcoming laboratory exercise. In some instances, the answers to these questions are necessary for developing and applying the experimental procedure. These questions will have you do such things as calculate the amounts of reagents to be used in the exercise and research the safety hazards of the chemicals to be used, make you aware of procedural necessities, solve questions similar to those required when analyzing the data you will collect during the exercise, and more. Think carefully about the answers to these Prelaboratory Questions; they will have an impact on the effectiveness and efficiency of your work in the laboratory.

Develop a Flowchart of the Experimental Procedure

The flowchart should indicate any major procedural steps and their sequence. For the lab exercises in the traditional format, the flowchart will be a summary of the given Procedure. The main value of the flowchart in these exercises is organizational: The flowchart helps to ensure that you perform the correct procedural step at the correct time. Moreover, it also makes your lab work more efficient by revealing what tasks can be done simultaneously.

A flowchart of the experimental procedure is even more helpful for the lab exercises in the guided-inquiry format. In these exercises, the flowchart of the experimental procedure has all of the same advantages as in the traditional format experiments, but in addition, it causes you to define, at least in overview, the experimental procedure before arriving in lab. After reading the experiment in this Lab Manual and answering the Prelaboratory Questions, you will have a reasonable idea of how to proceed experimentally once in lab. Writing your notions about the experimental procedure in the form a flowchart will further develop them. Not only will the flowchart make your lab work more efficient, but it also will make it safer. Indeed, creating a flowchart of the experimental procedure is so beneficial that your instructor might require it of you in these guided-inquiry experiments.

WORKING IN LAB

Listen to the Instructor's Experiment, Equipment, and Safety Instructions at the Start of Lab

At the beginning of each lab period, your instructor will state the purpose of the experiment, remind you of any procedural restrictions you are under, instruct you in the proper use of any equipment unfamiliar to you, and stress the safe handling and disposal of all chemicals.

Finalize Your Experimental Procedure

For exercises in the traditional format, your instructor might alter the given experimental procedure. For example, your instructor might have you perform only one part of a multipart exercise. The finalized form of your experimental procedure should include any changes of this sort.

For exercises in the guided-inquiry format, you might decide after listening to your instructor's introductory comments that the flowchart of your experimental procedure needs to be changed. If so, make the necessary changes. If you are performing the experiment as part of a group, as is often the case with guided-inquiry experiments, you and your group members should meet, share your ideas about the experimental procedure, and come to a consensus on a finalized experimental procedure.

Perform Your Experimental Procedure, Recording Data and Observations

Keep the safety precautions associated with each experiment in mind as you perform the experimental procedure. When performing experiments in the guided-inquiry format, if you have any doubts as to the safety of your experimental procedure, ask your instructor.

Record your data and observations in the *Results/Observations* section of the exercise, which is located just after the Prelaboratory Questions section.

Analyze Your Data

As you collect your data, ask yourself whether they appear imprecise, inaccurate, or inconclusive. If so, and you are performing an exercise in the traditional format, you will want to repeat the procedure, time permitting. If you are performing an exercise in the guided-inquiry format, however, you will need to decide if it is your procedure that is flawed or simply your execution of it. If it is the latter, repeat the exercise. If you conclude that your procedure is unsound, use your results to help you revise your experimental procedure before beginning the exercise again. Do not be overly concerned if your initial experimental procedure is flawed. It is assumed that this will happen occasionally. The guided-inquiry format exercises have been devised so that time for revisions is available. More important, realizing that you have made a mistake is a sign of learning. All scientists make mistakes. The point is to learn from them.

AFTER THE LAB

Perform Any Necessary Calculations or Graphs

Perform any necessary calculations and graphs in the spaces provided in the Results/Observations section. (*Note:* If it is your preference and your instructor allows it, you may instead prepare your graphs using a software program such as Excel.) Use the results of these calculations and graphs to answer any questions asked in the Results/Observations section. Performing the calculations, constructing the graphs, and answering the questions asked in the Results/Observations section will allow you to achieve the objective of the exercise.

Answer the Postlaboratory Questions

Each exercise has a number of *Postlaboratory Questions* that accompany it. The Postlaboratory Questions are listed with the exercise, just after the Results/Observations section. Answering these questions will further solidify your understanding of the objectives of the exercise and how the procedure and subsequent calculations and graphs allow the objective to be met. Moreover, it will further your understanding of the chemistry concepts pertinent to the exercise.

To the Instructor

This Lab Manual contains a mixture of *traditional* and *guided-inquiry* laboratory exercises. The exercises in this Lab Manual are grouped in categories (kinetics, equilibrium, the mole and stoichiometry, etc.). Each category contains at least one experiment in each format. The format of each exercise is indicated in the Table of Contents. The fundamental difference between these two formats is the degree of procedural detail provided. Exercises in the traditional format are typified by complete, detailed, easy-to-follow procedures. Guided-inquiry format exercises, however, require the student, to some extent, to design the experimental procedure. In some instances the degree of experimental design is slight, perhaps just the modification of a given procedure; in others it is quite large, the complete determination of how to accomplish the experimental objective. When it comes to procedural aspects that are considered too subtle for students to realize given their limited experience, experimental caveats are given. Restrictions are placed on the chemicals and equipment available for use in each exercise. The prelab questions also help students to devise their experimental procedure by having them answer questions that make them aware of procedural necessities.

By combining lab exercises in these two formats, this Lab Manual strives to provide you the utmost in course design flexibility. For example, it is possible to use this Lab Manual to teach an entire year-long course of general chemistry in the guided-inquiry format. Alternatively, you could couple traditional and guided-inquiry format exercises during the course: The traditional exercise is performed first. It serves to develop lab skills and student confidence. The following guided-inquiry exercise then serves as an application of the newly developed skills and concept understanding. A third option is to have students perform traditional exercises for most of the term, so as to build skills and confidence, and to end the term with one or more guided-inquiry exercises, a capstone event that enables students to apply the skills and understanding they have developed throughout the term. A fourth option is to use this Lab Manual to slowly phase guided-inquiry exercises into your laboratory program. In this scheme, one or two guided-inquiry exercises would be performed the first year, most likely as capstone events at the end of the course. During the second year, one or two more guided-inquiry exercises would be added to the curriculum. In the third and subsequent years, ever more guided-inquiry exercises would be added to the course, until the desired proportion of guided-inquiry to traditional exercises has been reached. Of course, it is even possible to use this Lab Manual for a year-long course in the traditional format. Each of these approaches has its advantages. You best determine the approach most suited to your and your students' circumstances. The goal of this Lab Manual is to provide you with the tools necessary for applying the approach you deem best.

One advantage of guided-inquiry exercises over those in the traditional format, the reason for their ever increasing popularity, is their ability to better promote student understanding of chemistry. Because students proceed through the material under their own direction, guided by their own developing understanding of particular chemistry topics, students better learn the material. Another advantage of the guided-inquiry format is that it better represents to students what it is chemists in particular, and scientists in general, do. Chemists do not spend all of their time following instructions to answer questions. Chemists devise experiments that allow them to answer questions. Thirdly, guided-inquiry exercises better serve the educational needs of most students enrolled in general chemistry courses. Over ninety percent of students enrolled in general chemistry courses are not chemistry majors. While it is true that these students need a basic understanding of chemistry for their future professions in some science related endeavor, it can be readily argued that they need to learn to design experiments even more. The ability to design an experiment and analyze the validity of the results is a skill that all students can benefit from learning, whatever their professional ambitions.

My experiences with guided-inquiry format exercises have been overwhelmingly positive. As an educator, there are few sights more satisfying than that of students talking animatedly about chemistry while in the lab: students explaining to each other why something will (or won't) work, students arguing the merits of various experimental procedures, students talking to one another about their laboratory observations and their meaning, students teaching one another chemistry. It sounds almost too good to be true. But I've seen it. Not once or twice, but many times over a number of years. It's pleasing to be a part of such liveliness, such an obvious demonstration of students in the act of learning. When you try it yourself, I know you'll agree.

James Almy
jba26@cornell.edu

Common Laboratory Equipment

The items listed below are the likely contents of your lab drawer.

pinch clamp	screw clamp	medicine dropper with bulb
Pasteur pipet	amber rubber bulb	rubber policeman
magnetic stirbar	glass stirring rod	split rubber stopper with hole
13 × 100 mm testtubes	16 × 150 mm testtubes	25 × 200 mm testtubes
110°C thermometer	scintillation vial with cap	wash bottle
crucible with lid	crucible tongs	evaporating dish
Florence flasks	funnel	spatula
testtube holder	clay triangle	watchglass
beakers	testtube brush	ceramic tile
Erlenmeyer flasks	graduated cylinder	testtube rack
wire gauze	funnel rack	evaporating dish

There are many other pieces of laboratory equipment you will be using in this course that, owing to their size, cost, or infrequency of use, are not included in your lab drawer. Such items include

Büchner funnel	gas collecting bottle	buret
volumetric pipet	Mohr pipet	mortar and pestle
water aspirator	pipet bulb	pneumatic trough
beaker tongs	dish tongs	Bunsen burner
sidearm flask	volumetric flask	buret clamp
utility clamp	ringstand	support ring
extension clamp holder	extension clamp	

Although some of these items already will be familiar to you, many will not. Use the drawings on this and the following pages to identify the items unknown to you.

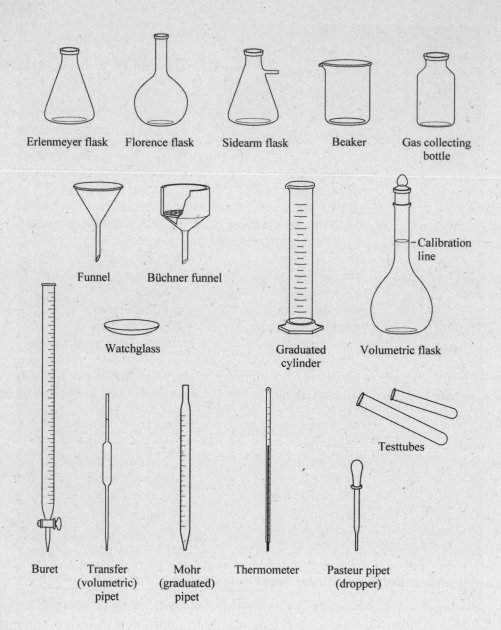

Figure 1

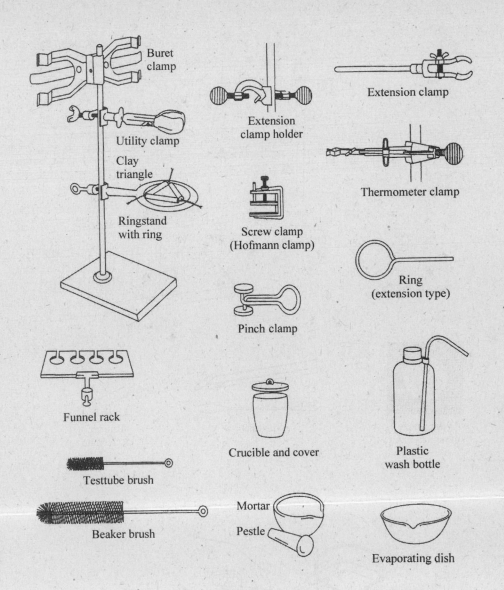

Figure 2

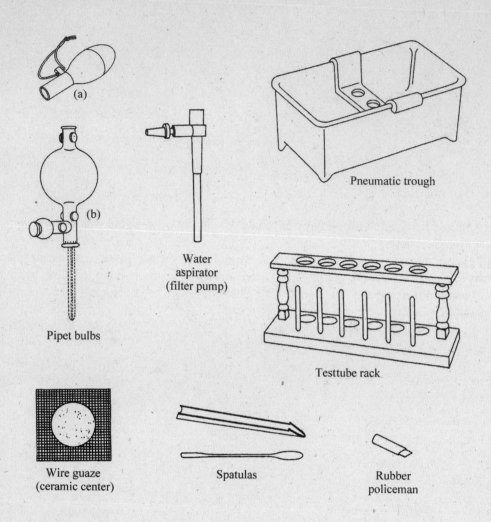

(a)

(b)

Pipet bulbs

Water
aspirator
(filter pump)

Pneumatic trough

Testtube rack

Wire guaze
(ceramic center)

Spatulas

Rubber
policeman

Figure 3

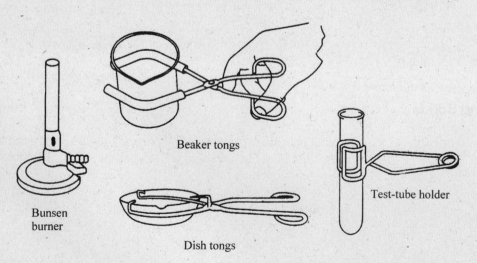

Beaker tongs

Bunsen
burner

Dish tongs

Test-tube holder

Figure 4

Laboratory Techniques

CLEANING GLASSWARE

All glassware should be cleaned properly prior to use. Using dirty glassware can lead to imprecise data and inaccurate results.

Clean all glassware with soap or a detergent solution and *tap* water. (Deionized or distilled water is too expensive to use for washing glassware.) Use a sponge or testtube, pipet, or buret brush as appropriate. Be sure not to use too much soap. Once the glassware has been cleaned properly, rinse several times with tap water and once or twice with small amounts of deionized or distilled water from a squirt bottle. Make sure that each rinse reaches the entire inner surface of the vessel before discarding.

Do not wipe or blow-dry the glassware; these acts could lead to contamination. Instead, invert the clean glassware on a paper towel or rubber mat to dry. The glassware is clean if no water droplets adhere to the inside of the vessel.

HANDLING CHEMICALS

When handled properly, the chemicals you will use in this course pose no great hazard to you. Use the following guidelines when handling reagents to maximize your safety and your degree of successful experimentation:

- Read the label on a reagent bottle at least twice before removing any chemicals. Many reagents have similar names and formulas but vastly different properties. Using the wrong reagent could lead to serious accidents.

- Avoid dispensing much more of the reagent than is necessary.

Transferring Solids

Grasp the reagent bottle so that the label is against the palm of your hand; always following this procedure will minimize the likelihood of contact between your flesh and any chemical that might dribble down the outside of the reagent bottle. Remove the cap from the reagent bottle, and place it top-side down on the lab bench; placing the lid top-side down will avoid contamination of the reagent bottle. Dispense the solid into a labeled clean beaker by tilting the bottle and rolling the solid back and forth [see Figure 1(a)]. *Do not* put a spatula inside a reagent bottle, an act that could lead to the reagent bottle becoming contaminated. Transfer solid from the beaker to the desired container using a spatula [see Figure 1(b)]. Do not return any excess solid to the reagent bottle, which could lead to contamination. Share any excess with another student. Recap the reagent bottle when you are finished.

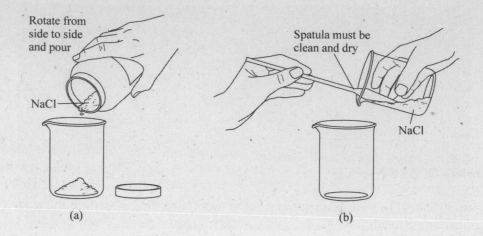

Figure 1
Technique for transferring a solid from (a) a reagent bottle and (b) from a beaker.

Transferring Liquids and Solutions

Inspect the outside of the bottle for evidence of liquids on the bottle. You might need to wash the outside of the bottle. If so, consult your instructor before handling.

Grasp the reagent bottle so that the label is against the palm of your hand; always following this procedure will minimize the likelihood of contact between your flesh and any chemical that might dribble down the outside of the reagent bottle. Remove the cap or glass stopper from the reagent bottle. If the lid is a screw cap, place it top-side down on the lab bench; placing the lid top-side down will avoid contamination of the reagent bottle. If the lid is a glass stopper, hold it between the fingers of the hand that is holding the reagent bottle; the glass stopper is never laid on the lab bench lest it lead to impurities being introduced into the reagent bottle. To dispense the liquid or solution, hold a glass stirring rod against the lip of the reagent bottle, and pour the liquid down the stirring rod, which also should touch the inner wall of the receiving vessel; pouring down the glass stirring rod will minimize the amount of liquid that dribbles down the outside of the reagent bottle (see Figure 2). Replace the lid on the reagent bottle when finished.

Figure 2
Technique for transferring a liquid or solution from a reagent bottle.

- Never smell, touch, or taste chemicals unless directed to do so. If a chemical accidentally spills on your skin, begin rinsing with water, an act that should last for 15 minutes. Ask a classmate to notify your instructor.

- Familiarize yourself with the specific safety precautions provided as part of each experiment.

WEIGHING

Mass determinations are made most commonly in the laboratory using an electronic balance. Two methods are used frequently to weigh chemicals: weighing by difference and weighing by taring.

Weighing by Difference

First, weigh a piece of weighing paper or a receiving vessel such as a beaker on the pan of the balance. Then the chemical to be weighed is added, and a second mass reading is taken. The mass of the chemical is the difference between these two masses.

Weighing by Taring

Place a piece of weighing paper or a receiving vessel on the pan of the balance. Press the *tare* button. This will cause the display to read "0.000 g." Pressing the tare button causes the balance to ignore the mass of the weighing paper or receiving vessel. This is called *taring* the balance. The chemical to be weighed is added, and its mass is read directly from the balance.

Both methods of weighing are equally effective. The choice between the weighing methods usually is decided on the basis of convenience.

Observe the following guidelines when using the balance:

- Check that the balanced is leveled. An unleveled balance will yield erroneous masses. If the balance is not leveled, inform your instructor.

- Always use weighing paper or a receiving vessel when weighing a chemical; never place chemicals directly on the balance pan. This protects the balance from corrosion and also allows you to recover the entirety of the chemical being weighed.

- Be sure that the material to be weighed is at room temperature. Hot objects create updrafts that cause the balance to give lower readings

- If a glass draft shield surrounds the balance pan, close all the sliding doors when weighing. This prevents changes in air pressure, which result in changes in the measured mass.

- After placing a container on the balance pan and closing the draft shield doors, wait a few seconds for the digital display to read a constant mass. Record all the displayed digits, even trailing zeros; all these digits are significant.

- Keep the balance clean. Remove any spilled chemicals promptly. Use a brush to remove any solids. Clean liquid spills with paper towels.

BURETS

A buret is a glass tube marked with gradations that allows variable volumes of liquid to be delivered and measured (see Figure 3). The buret is open at one end and fitted with a stopcock valve at the other. The stopcock valve controls the flow of liquid out of the buret.

Figure 3
A buret.

Begin by rinsing the buret. Use a buret clamp attached to a ringstand to support the buret. Close the stopcock valve. Place a funnel in the open end of the buret, and add 7–10 mL of the liquid to be delivered. Remove the funnel from the buret and the buret from the buret clamp. Roll the liquid around the buret until the entire inner surface has been rinsed. Return the buret to the buret clamp. Position a waste beaker under the buret tip, and open the stopcock to allow the liquid to drain from the buret. Repeat the rinsing twice more.

Fill the buret above the 0.00 mL mark at the top. Drain a few milliliters of liquid through the stopcock into a waste beaker to remove any air bubbles trapped in the tip of the buret, just below the stopcock. Air bubbles trapped in the tip of a buret at the beginning of an experiment could fill during the experiment, leading to an erroneous determination of the volume delivered. Drain liquid through the stopcock until the meniscus is at or below the 0.00 mL mark. It isn't necessary to adjust the meniscus to precisely 0.00 mL. In fact, doing so is usually a waste of time because all volumes delivered by buret are measured as the difference between an initial and final reading. Touch the side of the waste beaker to the tip of the buret to remove any drop that might be adhered to it.

Measure and record the initial volume of liquid in the buret. When measuring liquid levels, it is important that your eye be at the same level as the liquid (see Figure 4). This avoids a parallax error: If your eye is above the level of the liquid when you make the measurement, the liquid will seem to be higher than it actually is. If your eye is below the level of the liquid when you make the measurement, the liquid will seem lower than it actually is. The volume measurement should be made using the meniscus of the liquid. The *meniscus* refers to the curvature of the liquid in a narrow container. If the surface of the liquid is concave (downward curving), the liquid volume is measured using the bottom of the meniscus. If the surface of the liquid is convex (upward curving), the liquid volume is measured using the top of the meniscus. For very dark solutions with a difficult-to-observe meniscus, the volume should be read at the edge of the liquid.

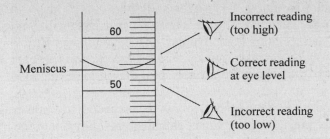

Figure 4
To properly read a meniscus, your eye must be at the same level as the liquid.

After delivering the desired amount of liquid, the final volume is measured and recorded. The volume of liquid delivered is the difference between the two volume measurements.

Before returning the buret to your instructor, rinse it several times with deionized water, being sure to drain each rinse through the tip.

PIPETS

Pipets are instruments used to deliver a known volume of liquid. The volume delivered by a given pipet and the precision with which it delivers the liquid are stated on the pipet.

To use the pipet, grasp it near the top with the fingers of one hand (see Figure 5). Adjust your grasp so that your index finger can be positioned over the top of the pipet. Place the pipet tip well below the surface of the liquid to be transferred. Squeeze a suction bulb, place it snugly atop the pipet, and slowly release the pressure on the suction bulb to draw liquid into the pipet.

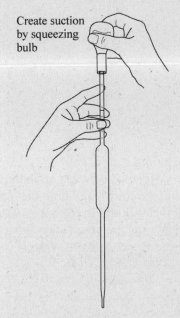

Create suction
by squeezing
bulb

Figure 5
Technique for filling a pipet with a suction bulb.

Chemicals are *never* pipeted by mouth. Once a few milliliters of liquid have been pulled up into the pipet, remove the suction bulb and quickly cover the top of the pipet with your index finger to keep the liquid from draining out. Roll this liquid around the pipet until the entire inner surface has been rinsed. Position the pipet tip over a waste beaker, and release the pressure exerted by your index finger to allow the liquid to drain from the pipet. Repeat the rinsing twice more.

To deliver the volume stated on the pipet, follow the preceding procedure using the suction bulb for drawing liquid into the pipet, only this time continue to pull liquid into the pipet until the liquid is 2–3 cm above the marking on the pipet. Depending on the size of the pipet and one's skill, this might require covering the top of the pipet with an index finger, squeezing the suction bulb, and replacing the suction bulb on the pipet. Remove the pipet from the liquid, and wipe the tip off with a Kimwipe. Hold the pipet above a clean beaker, and use finger pressure to drain liquid out of the pipet until the bottom of the meniscus is at the level of the marking on the pipet [see Figure 6(a)]. Touch the tip of the pipet to the wall of the beaker to remove any drop suspended from the pipet tip. If too much is allowed to leave the pipet, you can recover this from your beaker. Otherwise, share the excess with another student or dispose of it as directed by your instructor.

Place the pipet in the receiving vessel such that the tip of the pipet touches the inner wall of the vessel. Allow the liquid to drain [see Figure 6(b)]. Keep the pipet against the wall of the container for a count of 20 seconds after it appears that all the liquid has drained. Do not blow out the last bit of liquid that remains in the tip of the pipet; this liquid has been included in the calibration of the pipet [see Figure 6(c)].

Before returning the pipet to your instructor, rinse it several times with deionized water, being sure to drain each rinse through the tip.

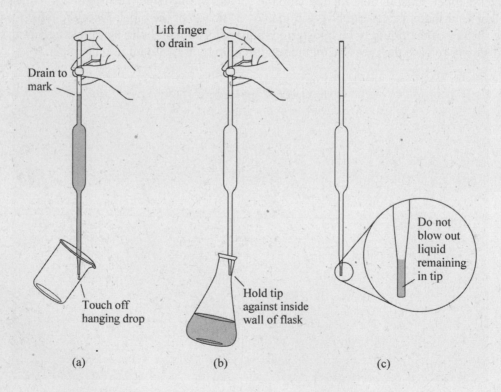

Drain to mark

Touch off hanging drop

(a)

Lift finger to drain

Hold tip against inside wall of flask

(b)

Do not blow out liquid remaining in tip

(c)

Figure 6
Technique for measuring and delivering liquid with a pipet.

VOLUMETRIC FLASKS

Volumetric flasks are designed to contain a particular volume of liquid. The volume contained by a given volumetric flask is etched on the flask. The volumetric flask contains the specified volume when the liquid level is adjusted so that the bottom of the meniscus is aligned with the single graduation mark on the neck.

Volumetric flasks are used commonly to prepare solutions of known concentration. The amount of solute necessary to prepare a solution of the desired concentration is calculated. The calculated amount of solute then is measured. Solvent is added to the volumetric flask until it is one-third to one-half full. Solute then is added to the volumetric flask. The contents of the volumetric flask then are swirled until dissolution is complete. Solvent is added to the volumetric flask until the meniscus is 1–2 mL below the calibration mark on the neck of the flask. A dropper is used to add solvent until the bottom of the meniscus is even with the calibration mark. The volumetric flask then is stoppered and, while securely holding the stopper, slowly inverted 10–15 times to ensure complete mixing.

Volumetric flasks are not used for storage. Transfer the contents of a volumetric flask to another container after preparation. Verify that the receiving vessel is dry, or else the solution concentration will be altered. Before returning the volumetric flask to your instructor, rinse it several times with deionized water.

DECANTATION

Decantation is the technique of separating a liquid from a solid by pouring. The solid is first allowed to settle to the bottom of the vessel. Place a glass stirring rod under the bottom of the vessel on the side opposite the spout. This will cause the solid to settle below the spout, leading to less solid being transferred during the pouring [see Figure 7(a)]. Pour the liquid (called the *supernatant*) down a glass stirring rod into another vessel. Lift and decant the supernatant slowly so as to disturb the solid as little as possible [see Figure 7(b)].

Decantation has the advantages of requiring no specialized equipment and being easy to perform. However, it suffers the disadvantage of leading to an incomplete separation, no matter how careful you are.

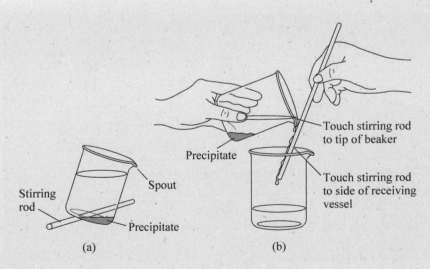

Stirring rod

Spout

Precipitate

Precipitate

Touch stirring rod to tip of beaker

Touch stirring rod to side of receiving vessel

(a)   (b)

Figure 7
Technique for decantation: (a) place a glass stirring rod under the bottom of the vessel on the side opposite the spout, causing the solid to settle below the spout, before (b) decanting with the aid of a glass stirring rod.

GRAVITY FILTRATION

Gravity filtration is a technique for separating a liquid from a solid. Gravity filtration is used when the desired substance is the liquid; the solid typically is discarded afterward.

   Obtain a piece of filter paper. First, fold the filter paper in half. Then fold it in half again. Tear off a corner; the tear allows the filter paper to fit more snugly in the funnel. Open the side of the filter paper that has not been torn to make a cone [see Figure 8(a)]. Fit this cone of filter paper snugly into the funnel. Wet the filter paper with the solvent of the liquid–solid mixture being filtered, and press the filter paper against the top wall of the funnel to form a seal.

   Attach a support ring to a ringstand. Place a funnel in the support ring and a vessel to receive the filtrate under the support ring. The tip of the funnel should touch the wall of the receiving vessel to reduce splashing of the filtrate [see Figure 8(b)].

   Decant the liquid–solid mixture into the funnel by pouring down a glass stirring rod. Fill the funnel less than two-thirds full with the mixture; otherwise, the solid might creep up and over the edge of the filter paper. Add the liquid–solid mixture to the funnel at a rate that keeps the funnel stem full of filtrate; the weight of the liquid in the stem creates a slight suction on the filter paper, speeding the filtration. Solid adhering to the walls of the container or glass stirring rod should be dislodged with a rubber policeman and flushed into the funnel with liquid from a squirt bottle. The liquid in the squirt bottle should be the same as the solvent in the liquid–solid mixture.

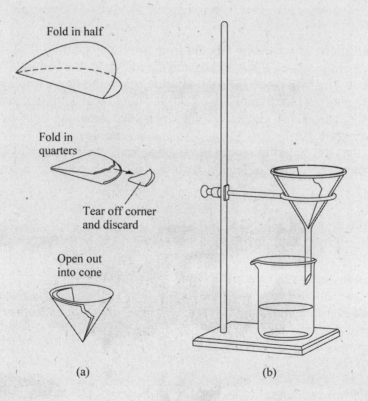

Fold in half

Fold in quarters

Tear off corner and discard

Open out into cone

(a)   (b)

Figure 8
Gravity filtration: (a) preparing filter paper for gravity filtration, and (b) the gravity filtration setup.

VACUUM FILTRATION

Vacuum filtration is the liquid–solid separation technique used when the desired substance is the solid; the liquid typically is discarded afterward.

Attach a sidearm flask to a filter trap via vacuum tubing (see Figure 9). The tubing slides on easily if the sidearm is wetted first. The filter trap should be attached to an aspirator with a second length of vacuum tubing. A support ring should be placed around the trap to prevent it from tipping over. Use a clamp attached to a ringstand to support the sidearm flask. The stem of a Büchner funnel should be fitted through a one-holed rubber stopper, which then is inserted in the neck of the sidearm flask.

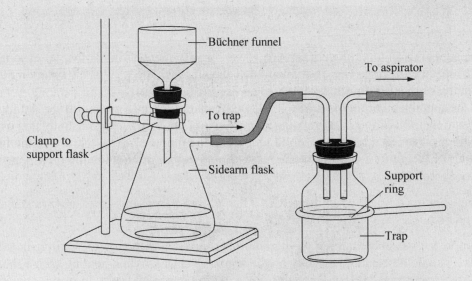

Figure 9
Vacuum filtration setup

The filter paper should be large enough that the holes in the Büchner funnel are just covered. If filter paper larger than the diameter of the funnel is used, a tight seal cannot be obtained, which could lead to some of the solid being lost over the unsealed edge. The aspirator should be turned on fully to create a vacuum, and a minimal amount of the solvent of the liquid–solid mixture being filtered should be used to seal the filter paper to the funnel. The aspirator creates a vacuum by pulling air into a fast-moving stream of cold water.

The mixture to be filtered is transferred to the Büchner funnel by pouring down a glass stirring rod. Solid adhering to the walls of the container or glass stirring rod should be dislodged with a rubber policeman and flushed into the funnel with liquid from a squirt bottle. The liquid in the squirt bottle should be the same as the solvent in the liquid–solid mixture.

After the liquid has been drawn through the funnel, the solid typically is washed with a small amount of solvent to remove any unwanted ions or other molecules still adsorbed on the solid. If the solvent is high boiling, the trace amounts remaining after washing are rinsed away by washing with a small amount of a more volatile solvent. Air then is drawn through the filter for a few minutes to further dry the crystals. It is sometimes necessary to press out the last remnants of solvent by pressing down on the solid with a clean spatula.

The suction is removed by first disconnecting the vacuum hose from the sidearm flask and then turning off the aspirator. Reversing this order could lead to a backup of water into the filtrate.

The dry solid and filter paper should be removed from the Büchner funnel using a spatula and placed in an appropriate container. Use a spatula to remove the solid from the filter paper, being careful not to contaminate the solid with pieces of filter paper.

HEATING LIQUIDS AND SOLUTIONS

Heating liquids and solutions is a commonly performed act in the laboratory. Despite the apparent simplicity of this act, it can be dangerous if performed without regard for proper technique. Adhere to the following guidelines when heating liquids and solutions in the lab:

- *Never* heat flammable liquids with a flame; always use a stirrer-hotplate.

- *Never* heat a liquid or solution in a closed container. The large resulting pressure could lead to the container exploding.

- When heating a liquid or solution on a stirrer-hotplate, always add a few boiling stones or a magnetic stirbar to prevent "bumping," which is the sudden, vigorous formation of bubbles from a superheated liquid. (Boiling stones are small, porous pieces of ceramic. When heated, the air contained within the pores is released, causing a gentle mixing of the liquid. The speed of the magnetic stirbar is adjusted using the control knob on the stirrer-hotplate. Both the stirbar and boiling stones minimize bumping by mixing the liquid to bring about a more uniform heating. The vessel containing the liquid can be placed directly on the surface of the stirrer-hotplate.)

- When heating a liquid or solution with a flame from a Bunsen burner, always add a few boiling stones.

- Support the vessel containing the liquid properly. When heating with a stirrer-hotplate, the vessel should be centered directly on the hotplate surface. When heating with a flame, attach a support ring to a ringstand, center a piece of wire gauze on the support ring, place the vessel on the wire gauze, and position the flame under the support ring (see Figure 10).

- Place a support ring around the top of the vessel containing the liquid or solution to prevent it from being accidentally knocked over.

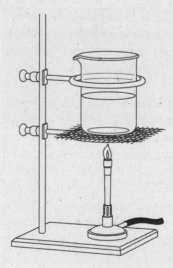

Figure 10
Technique for heating a liquid or solution with a flame.

CRYSTALLIZATION

Crystallization is the process of formation of solid crystals from solution. The shape of the compound creates a stacking pattern that continues and favors adding only that compound, and no others, to this

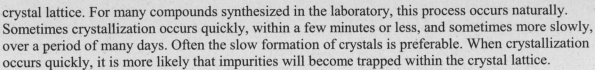

crystal lattice. For many compounds synthesized in the laboratory, this process occurs naturally. Sometimes crystallization occurs quickly, within a few minutes or less, and sometimes more slowly, over a period of many days. Often the slow formation of crystals is preferable. When crystallization occurs quickly, it is more likely that impurities will become trapped within the crystal lattice.

Sometimes, however, the expected crystallization does not occur. In these instances, there are number of techniques that can be used in an attempt to induce crystallization.

- Change the temperature of the solution. All compound solubilities vary with temperature, albeit some more greatly than others. Cooling the solution in an ice-water bath or heating often will induce crystals to form.

- Add a seed crystal of the desired compound to the solution. The seed crystal provides a surface for the solute particles to adsorb onto, increasing the likelihood of many solute particles coming together and forming a crystal.

- Scratch the bottom of the vessel that contains the solution with a glass stirring rod. Scratching produces microscopic crystals of glass that can act as seed crystals.

- Heat the solution gently to evaporate some solvent. Compound solubility decreases as the total volume of solvent decreases. Crystallization often occurs after the solution of reduced volume cools.

MELTING-POINT DETERMINATIONS

Melting points are used to identify unknown compounds and gauge the purity of known ones. For a pure substance, melting typically occurs at a specific, characteristic temperature. For an impure substance, however, melting will occur over a range of a few degrees, the so-called melting range. Not only will the presence of an impurity broaden the melting point of the primary component, but it also will lower the temperature at which melting begins. The nearer the experimentally determined melting point is to the accepted value and the smaller the melting range, the higher the purity of the compound.

Place 2–4 mg of the solid (about as much as you could pile on the eraser atop a pencil) on a watchglass. Use a tapping motion to push the open end of a melting-point capillary tube into the sample. Repeat until the solid has a height of about 0.5 cm in the capillary tube. Invert the capillary tube. Gently tap the bottom of the capillary tube on the benchtop until the solid is packed at the bottom.

Mount the capillary tube beside the bulb of a thermometer with a rubber band (see Figure 11). Place the thermometer in a split one-holed rubber stopper. Use a clamp attached to a ringstand to support the thermometer; tighten the clamp about the split rubber stopper, not the bare thermometer. Assemble a water bath using a beaker, a magnetic stirbar, and a stirrer-hotplate. Insert the thermometer so that the solid is below the level of the water, but the top of the capillary tube is above the level of the water. Slowly heat the water in the beaker. Record the temperature range over which the solid melts. Allow the water bath to cool to just below this approximate melting point. Place another sample of solid in another capillary tube. Again, heat the water bath until the solid melts, but at a slower rate this time. Record this approximate melting-point range. Repeat the heating–cooling cycle until reproducibility is achieved. This is the melting point of the sample.

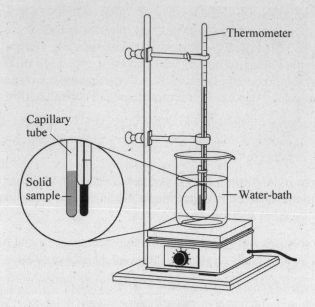

Figure 11
Melting point setup.

INSERTING GLASS TUBING THROUGH A RUBBER STOPPER

Warning: Serious cuts to the hand can occur quite easily if this task is performed incorrectly.

Lubricate the glass tubing and the hole in the rubber stopper with glycerol. With a small towel between your hand and the glass tubing for protection, grasp the glass tubing about 1 inch above the end to be inserted in the rubber stopper (holding the tubing further back than this usually causes the glass to break). Hold the rubber stopper with your other hand while simultaneously twisting and pushing the glass tubing into the stopper (see Figure 12). Pull your hand back so that about 1 inch of glass tubing is again between the end of your hand and the rubber stopper, and repeat the simultaneous twisting and pushing of the glass tubing into the stopper. These incremental insertions are less likely to cause the glass tubing to break in your grasp and lead to a serious cut. When the glass tubing has been inserted as far as desired, wash away any excess glycerol with water.

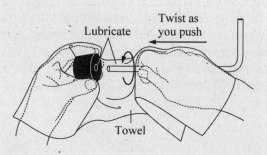

Figure 12
Inserting glass tubing through a rubber stopper.

BUNSEN BURNERS

The Bunsen burner is a device used for heating.

Caution: Never use a Bunsen burner in the presence of flammable compounds because of the danger of an unexpected and perhaps uncontrollable fire. Even if the flammable compound is across the lab, the vapors can reach the flame, ignite, and travel back to the flammable compound. In the event of a fire, notify your instructor. A fire in a small container can be smothered by covering the vessel with a watchglass. Use a fire extinguisher to eliminate a larger fire. If the fire is burning over too large an area to be extinguished easily, evacuate the area and activate the fire alarm. If your clothing should catch on fire, use the safety shower or a fire blanket to extinguish the flames or stop, drop, and roll.

The Bunsen burner is composed of a base, containing a gas inlet and a needle valve for controlling the flow of gas, to which is attached a chimney, possessing air ports at the bottom for the control of airflow (see Figure 13). The gas and air are mixed in the Bunsen burner prior to being fed into the flame. The relative amounts of gas and air mixed in the Bunsen burner control the size and temperature of the flame.

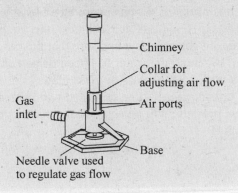

Figure 13
Bunsen burner.

Start with the needle valve and air ports of the chimney closed. Do not force the needle valve; it damages easily. Connect the gas inlet of the Bunsen burner to the gas tap with rubber tubing. Fully turn on the gas at the tap; the gas flow is controlled at the Bunsen burner using the needle valve, not at the gas tap. Light a match, place it at the top of the chimney, and open the needle valve until the gas is lit. The flame will be yellow and unsteady. Opening the needle valve further will increase the size of the flame. Twist the chimney counterclockwise to open the air ports. As more air mixes with the gas, the flame will become bluer and steadier. This adjustment needs to be completed soon after the burner is lit because these parts heat up. Stop adjusting the air ports when two distinct zones are visible in the flame, the inner, smaller of the two being a blue cone. The hottest part of the flame is just above the cone. Position the Bunsen burner as desired for heating.

GAS COLLECTION BY WATER DISPLACEMENT

Gases can be collected in the laboratory by water displacement, provided that they are relatively insoluble in water.

Fill a pneumatic trough (see Figure 14) with enough water to cover the interior shelf. Position the pneumatic trough so that water will drip from the overflow spout into the sink. Fill an Erlenmeyer flask or gas collecting bottle completely with water, cover it with a small watchglass or glass plate so that no air bubbles enter the flask, and place it in an inverted position in the water in the pan. Remove the watchglass, and position the top of the flask above the hole in the shelf. Tubing from the vessel where the gas will be generated should be placed so that it extends under the shelf and into the Erlenmeyer flask as far as possible.

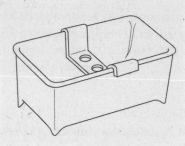

Figure 14
Pneumatic trough.

The reaction of interest then is initiated. As gas is produced by the reaction, it will flow through the tubing into the Erlenmeyer flask. Being less dense than water, the gas will rise to the top of the Erlenmeyer flask. The accumulating gas creates a pressure that pushes the water down and out of the flask into the pneumatic trough. When all the water has been displaced from the flask, remove it from the pan, and immediately seal it with a watchglass or glass plate.

Note: The gas collected in this manner will be saturated with water vapor. According to Dalton's law of partial pressures, the total pressure will be equal to the partial pressure of the desired gas plus the partial pressure of water vapor:

$$P_{total} = P_{H_2O} + P_{gas} \tag{1}$$

The partial pressure of water vapor as a function of temperature is listed in Appendix B.

If the reaction creating the gas involves heating, a trap often is inserted between the reaction and the gas collection flask. This can be any flask that can be sealed with two tubes, one attached to the gas-generating flask and one to the gas-collecting flask. This prevents the suction of water back into the reaction flask as it cools.

pH MEASUREMENTS

The pH is defined as the negative base-10 logarithm of the hydronium ion concentration:

$$pH = -\log[H_3O^+] \tag{2}$$

It is a useful means of quantifying the acidity or basicity of a solution. The pH is conveniently measured in the laboratory using a pH meter (see Figure 15).

To prepare the pH meter for use, it must be standardized against buffer solutions of known pH. Obtain about 25 mL each of pH 7 and pH 4 buffer in two 30 mL beakers. Raise the electrode out of the soaking solution. Rinse the electrode tip over a beaker with deionized water from a squirt bottle. Turn the pH meter on. Set the slope control to 100%. Immerse the electrode about 1 inch into the pH

7 buffer solution. Do not let the electrode hit the bottom of the beaker, contact that could damage the delicate electrode. Swirl the solution gently to wet the electrode. Measure the temperature of the solution, and adjust the temperature control to match. Adjust the standardize control until the pH meter reads 7.00. Remove the electrode from the pH 7 buffer, rinse it over a waste beaker with deionized water, and immerse it in the pH 4 buffer solution. Adjust the slope control until the meter reads 4.00. Remove the electrode from the pH 4 buffer solution, and rinse with deionized water over a waste beaker. Return the electrode to the pH 7 buffer solution. Does the pH meter read 7? If not, repeat the calibration process. Once the calibration process is complete, the pH meter is ready for use.

To measure the pH of a solution, submerge the tip of the electrode into the solution. Swirl the solution. Allow the pH meter to stabilize, and read the displayed pH value.

When the pH meter is not in use, immerse the electrode in the soaking solution. Turn the pH meter off when finished.

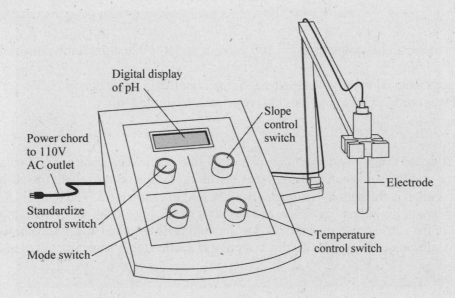

Figure 15
pH meter.

SPECTROMETERS

Spectrometers are devices used to measure the diminution in intensity of electromagnetic radiation as it passes through a sample. A schematic of a spectrometer is shown below:

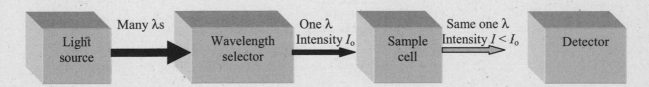

Electromagnetic radiation of many wavelengths is produced by the light source. This light of many wavelengths is passed onto a wavelength selector, more formally known as a *monochromator*. The monochromator allows only the chosen wavelength to pass onto the sample. The selected wavelength passes through the sample. If the sample absorbs light of this wavelength, the intensity of the light transmitted by the sample (I) will be diminished relative to that of the light incident on the sample.

The detector compares the intensities of the incident and transmitted light. The result of this intensity comparison then is displayed on a meter readout.

Data from a spectrometer usually are reported in terms of absorbance or percent transmittance of light at a specific wavelength. A collection of such data at various wavelengths comprises a spectrum. The percent transmittance (T) is defined by the expression

$$T = \left(\frac{I}{I_0}\right)100 \tag{3}$$

Absorbance (A) is defined by the equation

$$A = \log\left(\frac{I_0}{I}\right) \tag{4}$$

A completely transparent sample has $T = 100$ percent and $A = 0$; a completely absorbing sample has $T = 0$ and $A = \infty$.

Absorbance is related to the path length (d) of the sample and to the concentration (c) of the absorbing molecules. For dilute solutions, this relationship has the form

$$A = \varepsilon dc \tag{5}$$

Equation (5) is called the *Beer-Lambert law*, or simply *Beer's law*. The absorbance A is unitless. The concentration c usually is given in units of moles per liter (M). The distance light travels through the sample, the so-called pathlength d, is commonly expressed in centimeters. The quantity ε is called the *molar absorptivity* or *extinction coefficient* and has units of $M^{-1} \cdot cm^{-1}$; the molar absorptivity is the characteristic of the substance that expresses how much light is absorbed at a particular wavelength by a particular substance. To emphasize this last point, Equation (5) is sometimes expressed as

$$A_\lambda = \varepsilon_\lambda dc \tag{6}$$

because the values of A and ε depend on the wavelength of light; if the wavelength is changed, both A and ε change. Indeed, an absorbance spectrum is nothing more than a plot illustrating how A (or ε) varies with wavelength.

Spectrometers capable of analyzing each region of the electromagnetic spectrum have been made and used by scientists. In the general chemistry laboratory, investigations with light typically use the visible spectrum. The visible spectrometer commonly used in general chemistry laboratories is the Spec 20 (see Figure 16).

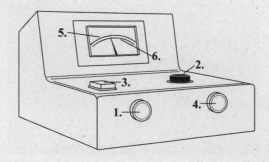

Figure 16
The Spec 20.

Switch the Spec 20 on (#1 in Figure 16), and allow it to warm up for 10 minutes prior to use. Adjust the wavelength control (#2 in Figure 16) to the desired value in the visible spectrum.

It is necessary to standardize the spectrometer prior to use. All species in the solution and the glass of the cuvet itself absorb light. The standardization accounts for this absorption. To standardize the Spec 20, use a blank, which is a cuvet filled three-quarters full of the solvent being used in the solution. Holding the cuvet by the lip, wipe the outside with a clean Kimwipe to remove any fingerprints or stains that might irreproducibly reduce the intensity of light transmitted to the detector. Place the cuvet in the sample compartment (#3 in Figure 16), pushing it fully down. Align the mark on the cuvet with that on the sample holder, and close the sample compartment. Adjust the calibration dial (#4 in Figure 16) so that the spectrometer reads zero absorbance.

Remove the cuvet from the spectrometer, and empty it into a waste beaker. Rinse the cuvet several times with the solution to be analyzed. Fill the cuvet three-quarters full with the solution. Holding the cuvet by the lip, wipe the outside of the cuvet clean with a Kimwipe, place the cuvet in the sample compartment, align the mark on the cuvet with that on the sample holder, and close the sample compartment. Read and record the percent transmittance (#5 in Figure 16) and the absorbance (#6 in Figure 16). Note that the meter has a mirrored back. To the observer's eye, the needle should obscure its reflection in the mirror, or a parallax error will result.

Note: The spectrometer must be restandardized each time the wavelength selector is adjusted.

DIGITAL MULTIMETERS

Digital multimeters (see Figure 17) are devices used to measure voltages, currents, and resistances. In the general chemistry laboratory, they typically are used to measure voltages.

Obtain two electrical leads. Each electrical lead should have an alligator clip at one end and a banana clip at the other. It is convenient if they have different colors but not essential. Insert the banana clip of each electrical lead into the slots on the digital multimeter used for measuring voltage. These slots typically are typically labeled with the symbol "V" for voltage and "COM" for common. Attach the alligator clips to the portion of the electric circuit across which the measured voltage is desired, such as the electrodes in an electrochemical cell.

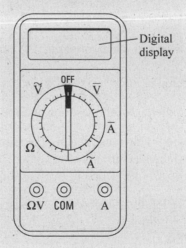

Figure 17
Digital multimeter.

Switch the multimeter on by adjusting the central switch to one of the voltage settings. Be sure to set the switch to one of the direct voltage settings, indicated by $\overline{V}$. (The $\tilde{V}$ indicates alternating voltage, A indicates current, and Ω indicates resistance.) Adjust the central switch to display the voltage on the appropriate scale; for example, a voltage on the order of 1 V should be measured on the 2 V setting, whereas a voltage on the order of 0.01 V should be measured on the 200 mV setting. If the displayed voltage is negative but should be positive, simply reverse the electrical leads.

Remember to turn the digital multimeter off when finished so that its batteries are not wasted.

CENTRIFUGATION

A centrifuge (see Figure 18) greatly accelerates the rate that solid in a solid–liquid mixture sinks to the bottom of a testtube. It does this by rotating the sample at high velocity. The resulting centrifugal force quickly pushes the solid to the bottom of the testtube, compacting it. The supernatant (liquid above the solid) then is decanted.

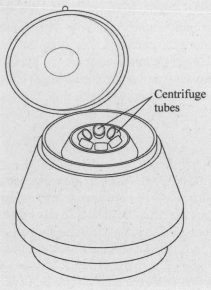

Centrifuge tubes

Figure 18
Centrifuge.

The following rules should be observed when operating the centrifuge:

- Each testtube should be labeled.

- Leave at least 1 cm between the top of the testtube and the liquid.

- Each testtube must be balanced with a testtube on the opposite side containing an amount of water equal to the amount of the mixture in the testtube. Be sure that the holder of the testtubes has no broken glass or other debris in the bottom, or the testtube can crack.

- The centrifuge must never be left unattended. Even well-balanced centrifuges have a tendency to move owing to vibration.

- Do not touch the centrifuge while it is turned on.

- After the centrifuge has been turned off, pressing your hands against the upper portion of the cone can gently slow the rotation. The braking must be very gentle. You must remove your hands before the spinning stops. Otherwise, the sediment will be stirred up. Do not place your hands or anything into the path of the testtubes tops; they will break, and you can get cut.

Laboratory Safety

SAFETY REGULATIONS

The experiments in this Lab Manual have been designed to pose minimal risks. However, the nature of chemical experimentation makes it impossible to eliminate all hazards. For your own safety and that of others, you are required to adhere to the following safety regulations:

 Safety Precautions:

- Safety goggles must be worn at all times in the laboratory.

- Clothing that provides maximum protection must be worn in the laboratory. Shorts, miniskirts, and shirts that leave the midriff exposed are prohibited.

- Long sleeves should be buttoned or rolled back. Scarves and neckties should be removed or secured against the body. Such potentially loose items of clothing can accidentally drape into a Bunsen burner flame and ignite. Long hair should be tied back for this same reason.

- Open-toed shoes, sandals, and thin canvas sneakers provide little to no protection for your feet and are not allowed in the laboratory.

- Wearing contact lenses while working in the laboratory is strongly discouraged because they greatly increase the chance of eye injury in the event of a chemical splash.

- Do not eat, drink, chew gum, or smoke in the lab lest a harmful chemical substance be accidentally ingested.

- Maintain an uncluttered work area. Store all coats, bags, etc. in the designated area.

- Never heat a closed container. The large resulting pressure could lead to the container exploding.

- Do not leave a lit Bunsen burner unattended.

- Extinguish all flames when using flammable, volatile chemicals to minimize the risk of fire.

- Always add 2 to 3 boiling stones or stir with a magnetic stirbar when heating liquids and solutions to prevent "bumping."

- Do not allow liquids or solutions to heat to dryness unless required by the experimental procedure, in which case an evaporating dish should be used.

- Never point a testtube being heated at yourself or others lest the contents bubble out.

- Never fully inhale chemical vapors. If odor detection is necessary, waft a tiny amount of vapor toward your nose.

- Never remove a chemical substance from the laboratory unless explicitly given permission to do so by your instructor.

Safety Precautions: (Continued)

- Always add a concentrated reagent to water, while stirring, when diluting. Never add water to a concentrated reagent. If water is added to a concentrated reagent, localized heating could cause a dangerous chemical splash.

- Never work in the lab without instructor supervision.

- Dispose of all chemical waste as directed by your instructor. When unsure of how to dispose of a chemical, ask.

- When disposing of glass items, use the special container provided to ensure the safety of the custodial staff.

- Read all labels on a container before using, and label all chemical containers properly. Never store a chemical in an unlabeled container.

- Always use a pipet bulb when pipeting. Never pipet by mouth.

- Immediately report all injuries, no matter how small, to your instructor.

- Report and clean up all spills immediately.

- Excessive noise, disruptive behavior, and pranks have no place in the lab.

- Clean your workspace and wash your hands at the end of each laboratory session.

SAFETY EQUIPMENT

Goggles and Aprons

The two pieces of safety equipment that you will use in every laboratory session are safety goggles and lab aprons. Government regulations require that students use protective eyewear while working in the chemistry laboratory, and for good reason: Your eyes are especially sensitive and can be damaged easily. *Protective eyewear approved by your institution must be worn at all times while you are in the laboratory*, even if you are not working with chemicals.

A lab apron will help to protect you from chemical spills and your clothing from chemical staining and damage.

Eyewash Fountain

If a chemical splashes in or near your eyes, immediately flush your eyes with water for at least 15 minutes. An eyewash fountain is provided for this purpose. The eyewash is activated by pushing the attached lever forward, causing gentle streams of water to be dispensed from the two fountain outlets. If the chemical splash affects the eyes, lower your head to place your eyes in the streams of water. Pull back your eyelids and roll your eyes back and forth to ensure a complete flushing of the affected area. If the splash occurs while you are wearing your goggles and the eyes are unaffected, first rinse your face with the goggles on because your eyes can become contaminated if the goggles are not rinsed before being removed. After the goggles are rinsed, flush the eyes for 15 minutes.

If a chemical spill on or near your eyes should occur, immediately attract the attention of your instructor.

Fire Extinguisher

The laboratory is equipped with a fire extinguisher to be used in the event of a large fire. Before activating the fire extinguisher, it is necessary to remove the safety pin. Squeeze the handle to activate the fire extinguisher. The discharge is directed at the base of the flame with a side-to-side motion.

If the fire is burning over too large an area to be extinguished easily, evacuate the area and activate the fire alarm. If you are working in the laboratory when the fire alarm sounds, turn off the gas, shut down your experiment, and leave the building by the nearest safe exit

A fire in a small container can be smothered by covering the vessel with a watchglass or a larger beaker.

Regardless of the size of the fire, immediately inform your instructor of the situation.

Fire Blanket

If a clothing fire occurs, the fire blanket can be used to extinguish the flames. Wrapping the fire blanket tightly around you will smother the flames.

If you happen to be on the far side of the room from the fire blanket and your clothing catches fire, use the "Stop-Drop-Roll" technique for extinguishing the flames. You must stop (do not run!), drop to the floor, and roll to smother the flames.

Regardless of the technique employed to extinguish the fire, immediately notify your instructor of the danger.

Safety Shower

Use the safety shower in the event of a chemical spill affecting large portions of your body. Engage the safety shower by pulling downward on the attached ring and chain. Stand under the resulting flow of water. Remove any clothing affected by the chemical spill. The threads in cloth can act as capillary tubes, quickly spreading the chemical over an even larger portion of your body. Moreover, your clothes will slow the rate at which chemicals can be washed from your body. Don't be unduly modest. Chemical burns are permanent, embarrassment is temporary.

For small chemical spills, rinse with water in the sink for at least 15 minutes. Remove any affected clothing.

Always notify your instructor immediately if you suffer a chemical spill on your skin.

Fume Hood

The vapors of many volatile compounds are toxic. To avoid unnecessary exposure to toxic vapors, lab work with such compounds will be restricted to one of the fume hoods located in the lab. The fume hood is equipped with fans that pull gases up into exhaust ports, away from the user. For maximum protection, the glass sash fronting the fume hood should be lowered to the proper level (indicated on the fume hood with a marking line), and all chemicals should be kept at least 6 inches behind the sash.

If an accidental inhalation of harmful vapors occurs, notify your instructor and leave the lab to a location of fresh air.

First Aid Kit

A first aid kit is present in the lab to treat small cuts and burns, the most common injuries suffered by students in the laboratory. Small cuts should be rinsed thoroughly using water from the sink and examined for traces of foreign matter, glass, chemicals, etc. Remove any foreign matter, and apply an adhesive bandage. Severe cuts should be treated by applying pressure to the wound to control the rate of bleeding. Meanwhile, the instructor or a student designated by the instructor should seek out medical assistance. Small burns should be rinsed with cool water for at least 15 minutes.

Even if the burn or cut is minor, it is still necessary to notify your instructor immediately.

MATERIAL SAFETY DATA SHEETS (MSDS)

Material Safety Data Sheets (MSDS) contain information designed to help you by making you aware of the dangers posed by a chemical, by informing you how to work safely with each compound, and by explaining what to do in the event of an accident. The MSDS for the chemicals used in each experiment will be available for you to read if you so desire. It is also possible to access MSDS online at **http://msds.ehs.cornell.edu/msdssrch.asp.**

All MSDS contain the same basic information, but the manner in which it is organized is left to the chemical supplier. This causes MSDS from different suppliers to have different formats. Despite these various formats, all MSDS contain information on the following topics:

- Product identification
- Ingredients
- Physical data
- Fire and explosion hazards
- Reactivity data
- Health hazards
- Spill, leak, and disposal procedures
- Personal protection information
- Special precautions or other comments

Think Safety! Read the safety guidelines in the lab description and the safety warnings on labels, and anticipate the possible hazards and proper remedies to accidents!

SAFETY QUIZ

1. The use of eye protection in the laboratory is (circle one) optional / required.

2. Are pranks, disruptive behavior, or other acts of mischief within the lab allowed?
   Yes or no (circle one)?

3. True or False (circle one): It is permissible to work in the lab without the supervision of an instructor.

4. True or False (circle one): It is safe to use a Bunsen burner in the presence of flammable, volatile chemicals.

5. True or False (circle one): It is permissible to eat, drink, or smoke within the lab.

6. True or False (circle one): It is safe to work in the lab while wearing sandals.

7. True or False (circle one): Loose sleeves should be rolled up and long hair tied back while working in the lab.

8. True or False (circle one): It is unsafe to allow the lab bench and aisle near you to become cluttered.

9. True or False (circle one): Keeping your eyes closed while rinsing them to remove an accidentally introduced chemical will render the treatment ineffective.

10. How long should the eye wash be used to rinse eyes contaminated with a chemical reagent? 1, 5, 10, 15, or 20 minutes (circle one)

11. For a small chemical spill on the hand or arm, a student should rinse the affected area with water in the sink or the safety shower (circle one)?

12. True or False (circle one): For a large chemical spill, the safety shower should be used while remaining dressed.

13. For chemical spills, the rinsing with water should last at least 1, 5, 10, or 15 minutes (circle one).

14. True or False (circle one): Small fires are to be extinguished by placing a nonflammable object over the mouth of the container.

15. True or False (circle one): Larger fires are to be extinguished with the fire extinguisher.

16. True or False (circle one): If a fire cannot be put out by smothering or with a fire extinguisher, the area should be evacuated and the fire alarm activated.

17. True or False (circle one): At the sound of the fire alarm, turn off the gas, shut down your experiment, and leave the building by the nearest safe exit.

18. True or False (circle one): It is okay to not report a small cut or chemical spill to your instructor.

19. True or False (circle one): It is unsafe to heat a closed container.

20. True or False (circle one): It is okay to take chemicals home with you.

21. True or False (circle one): It is always okay to pour chemical waste down the sink.

22. True or False (circle one): Concentrated acids should be added to water with stirring.

23. True or False (circle one): Shorts and miniskirts are acceptable attire in the lab.

24. True or False (circle one): The fire blanket always should be used when a student's clothing catches fire.

25. True or False (circle one): When performing an experiment in a fume hood, keep the glass sash above head level.

26. Sketch a diagram of your laboratory in the space below. On your drawing clearly indicate the locations of each of the following: exits, eyewash fountain, fire extinguisher, fire blanket, safety shower, fume hood, and first aid kit.

SAFETY CONTRACT

Safety Contract

I have read the Safety Regulations and the section on Safety Equipment. I have completed the Safety Quiz. I agree to conform my conduct to the dictates of these documents. I realize that failure to do so may result in my dismissal from lab with no opportunity for makeup of missed work.

Name (print): _____

Signature: _____ Date: _____

EXPERIMENT 1

Mathematics Review

OBJECTIVE

The basic mathematics necessary for general chemistry will be reviewed and practiced. These include the SI unit system, scientific notation, measurement, accuracy, precision, significant figures, and dimensional analysis.

INTRODUCTION

Throughout your study of chemistry you will make a great many quantitative measurements. Mass, volume, absorbance, time, pressure, temperature: these are just some of the quantities you might measure in a chemistry experiment. You then could use these measured values to calculate other masses, volumes, times, pressures, or temperatures or perhaps densities, concentrations, or rates of reaction. The results of measurements and calculations such as these need to be presented in a way that is readily understandable to others, understandable in terms of both *what* property was measured (i.e., what value was obtained) and *how well* it was measured. This exercise will detail, through both example and application, the manner in which chemists accomplish these two goals.

Part A: The SI Unit System

To make clear what was measured, scientists the world over have agreed to use the International System of Units for measurement, abbreviated SI for the French *Système Internationale d'Unités*. Based on the metric system, the SI system has seven fundamental units (see Table 1.1). These seven fundamental units, and those derived from them, are sufficient for reporting all scientific measurements.

Table 1.1: Fundamental SI Units of Measure

Physical Quantity	Name of Unit	Abbreviation
Mass	kilogram	kg
Length	meter	m
Temperature	kelvin	K
Amount of substance	mole	mol
Time	second	s
Electric current	ampere	A
Luminous intensity	candela	cd

Despite being comprehensive, the sizes of these units are often inconvenient. A meter, for example, is a convenient unit for expressing the height of an adult person (6.00 ft = 1.83 m), but it is cumbersomely large when describing the diameter of a sodium atom (0.000000000372 m) and awkwardly small when describing the diameter of the Earth (12,740,000 m). For this reason, prefixes corresponding to multiplying factors are used to change the sizes of the SI units (see Table 1.2). So the diameter of a sodium atom could be expressed in a much tidier manner as 0.372 nm and that of the Earth as 12.74 Mm.

Table 1.2: Common Prefixes for SI Units

Multiplying Factor	Prefix	Symbol	Example
$1,000,000,000 = 10^9$	giga	G	$1\ Gm = 10^9\ m$
$1,000,000 = 10^6$	mega	M	$1\ MA = 10^6\ A$
$1,000 = 10^3$	kilo	k	$1\ km = 10^3\ m$
$0.1 = 10^{-1}$	deci	d	$1\ ds = 10^{-1}\ s$
$0.01 = 10^{-2}$	centi	c	$1\ cm = 10^{-2}\ m$
$0.001 = 10^{-3}$	milli	m	$1\ mmol = 10^{-3}\ mol$
$0.000001 = 10^{-6}$	micro	μ	$1\ \mu cd = 10^{-6}\ cd$
$0.000000001 = 10^{-9}$	nano	n	$1\ nK = 10^{-9}\ K$
$0.000000000001 = 10^{-12}$	pico	p	$1\ ps = 10^{-12}\ s$
$0.000000000000001 = 10^{-15}$	femto	f	$1\ fA = 10^{-15}\ A$

Part B: Scientific Notation

An alternate notation of convenience for expressing numerical results is the format called *scientific notation*. Scientific notation is an exponential notation that expresses numbers using powers of 10. For example, 358,000,000 m is expressed in scientific notation as 3.58×10^8 m, 3.58 m multiplied by 10 eight times:

$$3.58\ m \times 10 \times 10 \times 10 \times 10 \times 10 \times 10 \times 10 \times 10 = 3.58 \times 10^8\ m \tag{1}$$

Each multiplication moves the decimal point one place to the right,

$$
\begin{aligned}
3.58\ m \times 10 &= 35.8\ m \\
35.8\ m \times 10 &= 358.\ m \\
358.\ m \times 10 &= 3580.\ m
\end{aligned}
\tag{2}
$$

so the exponent can be interpreted as the number of places the decimal point has to be moved; i.e., in converting 3.58×10^8 m from scientific to standard notation the decimal point is moved eight places to the right.

Numbers less than 1 are converted to scientific notation through division by multiples of 10. For example, 0.00000453 mol is expressed as 4.53×10^{-6} mol, 4.53 mol divided by 10 six times:

$$\frac{4.53\ mol}{10 \times 10 \times 10 \times 10 \times 10 \times 10} = 4.53 \times 10^{-6}\ mol \tag{3}$$

Again, the exponent can be interpreted as the number of places the decimal point has to be moved, only now the decimal point is moved in the opposite direction; i.e., to convert 4.53×10^{-6} mol from scientific to standard notation the decimal point is moved six places to the left.

Part C: Measurement, Accuracy, and Precision

How well a quantity was measured is usually expressed in terms of precision and accuracy. Precision is an indication of the reproducibility of a result. Accuracy refers to how close a measured value is to the accepted or "true" value. Precision relates to the quality of an operation by which a result is obtained. Accuracy relates to the quality of the result itself. To illustrate the difference between precision and accuracy consider the analogy of a marksman for which the "truth" is represented by a bullseye. In Figure 1.1 the marksman achieved reproducibility, but the results differ from the "true" value. Thus the results can be described as precise but inaccurate. In Figure 1.2 the results are not reproducible and must be considered imprecise. However, the results are clustered around the "true" value, making them accurate. The results in Figure 1.3 are both precise and accurate. Those in Figure 1.4 are neither precise nor accurate.

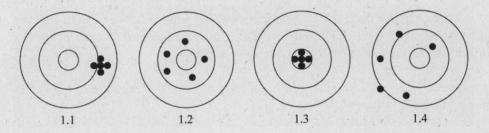

1.1   1.2   1.3   1.4

Figures 1.1–1.4
Targets illustrating the results of four marksmen provide examples of the difference between precision and accuracy.

In cases where the "true" or accepted value is known, the accuracy of a measured value can be quantified by calculating the percent error:

$$\text{Percent Error} = \left| \frac{\text{accepted value} - \text{measured value}}{\text{accepted value}} \right| \times 100 \qquad (4)$$

The two vertical lines indicate an absolute value. The smaller the percent error, the greater the accuracy.

The precision of a measured value is expressed by recording all the digits that are certain plus one additional digit that is an estimate. In this way, only the uncertain digit (the last one) should vary if the measurement is repeated using the same instrument. For example, using the "ruler" below, the black bar is measured by a student to have a length of 6.3 cm. The first digit is certain. The second digit is an estimate.

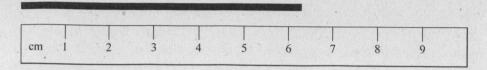

Figure 1.5
A "ruler" and a black bar. The uncertainty of the ruler lies in the tenths place.
Using this ruler, the black bar has a length of 6.3 cm.

Part D: Significant Figures

The total number of digits in the measurement is called the *number of significant figures*. The length of the bar (6.3 cm) has two significant figures. The diameter of the Earth (12.74 Mm) has four. All the digits but the last are certain. The last digit is an estimate; it is generally assumed to have an error of plus or minus 1; i.e., the length of the bar is 6.2–6.4 cm.

In most cases, recording a measured value to convey the proper number of significant figures is easy. Similarly, it is typically clear when reading a reported value how many significant figures are present. Confusion can arise, however, when a value begins or ends in one or more zeros, e.g., 0.000345 g. Clearly, the last three digits are significant, but what about the leading zeros? Are all, none, or some significant? What about 8400 m? Does this value contain two, three, or four significant figures? The following rules answer these questions:

1. *Zeros in the middle of a number are always significant.* For example, the value 704 s has three significant figures.

2. *Leading zeros are not significant.* Thus 0.000345 g has only three significant figures. The leading zeros serve only to fix the location of the decimal point.

3. *Zeros at the end of a number are only significant if the number contains a decimal point.* So the value 8400 m has only two significant figures, the value 82.100 kg has five significant figures, the value 300. K has three significant figures, and the value 2.50×10^{-3} L has three significant digits. This third rule is the one most often violated when reporting measured values. For example, a student might record a temperature reading as 20°C, which has only one significant figure, implying that the temperature is somewhere between 10 and 30°C when he means 20.°C, that the temperature is between 19 and 21°C.

4. *Exact numbers effectively have an infinite number of significant figures.* One situation in which exact numbers arise is counting: 2 beakers, 4 Erlenmeyer flasks, etc. There is no uncertainty associated with these values. Another situation in which exact numbers appear are definitions. By definition, each minute contains 60 seconds. These exact numbers will not limit the number of significant figures when used in a calculation.

There are also rules for determining the number of significant figures when two or more measured values are combined arithmetically. These rules ensure that precision isn't gained falsely or lost simply by combining measured values arithmetically.

1. *For multiplication or division the result can't have more significant figures than either of the original numbers.* For example,

$$2.41\,\text{m}\times3.2\,\text{m}=7.712\,\text{m}^2 \xrightarrow{\text{corrected to the proper number of significant figs}} 7.72\,\text{m}^2 \tag{5}$$

2. *For addition or subtraction the result can't have more decimal places than either of the original numbers.* For example,

$$8.92\,\text{mL}+94.4\,\text{mL}=103.32\,\text{mL} \xrightarrow{\text{corrected to the proper number of significant figs}} 103.3\,\text{mL} \tag{6}$$

Notice that in each of these cases it was necessary to round off in order to achieve the correct number of significant figures. The rules for rounding:

1. *In a series of calculations carry the extra digits through to the final result and then round.*

2. *If the digit to be removed*
   (a) is less than 5, round down by dropping it and all the following digits;
   (b) is equal to or greater than 5, round up by adding 1 to the digit on the left.
   For example,

$$\frac{(9.2\,\text{mL})(18.37\,\text{mol/L})}{25.036\,\text{mL}}=6.7504394\,\text{mol/L}\longrightarrow 6.8\,\text{mol/L} \tag{7}$$

Rounding was postponed until the end of the calculation. The final value was rounded to two significant figures because that is how many the term with the smallest number of significant figures (9.2 mL) contained. The value was rounded up to 6.8 mol/L because the digit to be removed was a 5.

Part E: Dimensional Analysis

It is often necessary to convert a value from one system of units to another. The standard method of accomplishing this is known as *dimensional analysis*. In dimensional analysis the unit conversion is accomplished through multiplying by a conversion factor. A conversion factor is based on an equality between two different sets of units. For example, the equality

$$1\,\text{in}=2.54\,\text{cm} \tag{8}$$

generates a conversion factor when both sides of the equation are divided by one of the two terms:

$$1=\frac{2.54\,\text{cm}}{1\,\text{in}} \quad\text{or}\quad \frac{1\,\text{in}}{2.54\,\text{cm}}=1 \tag{9}$$

Note that this conversion factor, as is true of all conversion factors, is numerically equal to 1. This ensures that multiplication by a conversion factor only changes the units the value is expressed in, not the value itself.

When using a conversion factor treat the units as algebraic variables. Choose the version of the conversion factor (2.54 cm/1 in or 1 in/2.54 cm) that causes the original units to be divided out but the desired units to remain. For example, the conversion of 45.8 cm to inches:

$$45.8 \text{ cm} \left(\frac{1 \text{ in}}{2.54 \text{ cm}} \right) = 18.0 \text{ in} \tag{10}$$

(*Note:* The equality 1 in = 2.54 cm is a definition; hence it will not limit the number of significant figures in the result.) Of course, it is possible to perform a unit conversion using more than one conversion factor. For example, the conversion of 5.8 ft to centimeters:

$$5.8 \text{ ft} \left(\frac{12 \text{ in}}{1 \text{ ft}} \right) \left(\frac{2.54 \text{ cm}}{1 \text{ in}} \right) = 180 \text{ cm} \tag{11}$$

Completing the worksheet on the following pages will increase your understanding of these concepts.

Name: _____   Date: _____

Lab Instructor: _____   Lab Section: _____

EXPERIMENT 1

Mathematics Review

RESULTS/OBSERVATIONS

Part A: The SI Unit System

1. What SI prefix symbols correspond to the following multiplying factors:

    (a) 10^{-15} _____   (b) 10^6 _____

    (c) 10^{-3} _____   (d) 10^{-9} _____

2. What multiplying factors correspond to each of the following SI prefixes?

    (a) μ _____   (b) k _____

    (c) p _____   (d) c _____

Part B: Scientific Notation

3. Express the following values in scientific notation:

    (a) 252.37 mL _____   (b) 493.700 m _____

    (c) 0.000000082 g _____

4. Convert the following values to standard notation:

    (a) 5.35×10^5 kg _____   (b) 1.94×10^{-3} cm _____

    (c) 4.77×10^{-9} m _____

Name: _____   Date: _____

Lab Instructor: _____   Lab Section: _____

Part C: Measurement, Accuracy, and Precision

5. The concentration of a solution was determined by chemist three different times. The data are below. Are these results precise? Explain.

Trial #	Concentration (mol/L)
1	0.0147
2	0.0146
3	0.0148

6. Experimentally a chemist determined the molecular weight of the compound aniline to be 94.45 g/mol. Calculate the percent error in this determination. The accepted value of the molecular weight of aniline is 93.13 g/mol. Is the result accurate? Explain.

7. Using the ruler below, what is the length of the black bar? _____

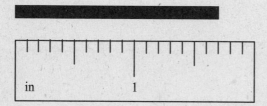

Part D: Significant Figures

8. How many significant figures does each of the following values contain?

   (a)   476 ps   _____   (b)   5,017 km .   _____

   (c)   0.0012 m   _____   (d)   40 g   _____

   (e)   350. K   _____   (f)   6.0200 A   _____

Name: _____   Date: _____

Lab Instructor: _____   Lab Section: _____

9. Round the following quantities to the number of significant figures indicated:

   (a) 39471 km (3 sig figs) _____

   (b) 5.614578 mL (2 sig figs) _____

   (c) 0.00031445 g (4 sig figs) _____

   (d) 8.94529×10^8 kg (3 sig figs) _____

10. Express the results of the following calculations to the correct number of significant figures:

   (a) 5.39 cm $\times$ 18.22 cm = _____

   (b) 5.33 m $\times$ 9.9629 $\times 10^{-3}$ m $\times$ 18.943 m = _____

   (c) 543.7 g $\div$ 568.45627 mL = _____

   (d) 14.37 mL + 251.5 mL = _____

   (e) 5984.36 mg – 7.76 mg = _____

   (f) (12.1 mL)(1.0076 g/mL) – 1.05 g = _____

Part E: Dimensional Analysis

11. Perform the indicated conversion in each of the following:

   (a) Convert the mass of an electron, 9.109×10^{-31} kg, to lbs.

Name: _____   Date: _____

Lab Instructor: _____   Lab Section: _____

(b) Convert 1.0 gallon to milliliters.

(c) Express the speed of light, 3.00×10^8 m/s, in miles per hour.

(d) Convert 359 m^3 to ft^3.

EXPERIMENT 2

Temperature Scales

OBJECTIVE

To determine a mathematical relationship between the Celsius and Fahrenheit temperature scales both graphically and algebraically.

INTRODUCTION

People in the United States normally refer to temperature using the Fahrenheit scale, but we're almost the only ones; most of the rest of the world uses the Celsius temperature scale. Scientific measurements are reported routinely in degrees Celsius, even in this country. Of course, both units of temperature measure the same thing. Water may boil at 212 degrees on the Fahrenheit scale and 100 degrees on the Celsius, but in both cases it is still the same phenomenon, only the scale on which it is measured and units in which it is reported differ. Accordingly, there must be some mathematical relationship between these two temperature scales, some means by which a temperature measured in one set of units can be expressed in terms of the other. In this experiment you will determine a mathematical relationship between these two temperature scales, first graphically and then algebraically

ADDITIONAL READING

Read the section in the "Laboratory Techniques" chapter at the beginning of this Lab Manual on heating liquids and solutions prior to performing this experiment.

Safety Precautions:

- Protective eyewear approved by your institution must be worn at all times while you are in the laboratory.

- The heating surface of the hotplate displays no visible evidence of being hot. To avoid burns do not touch it. If a burn does occur, run cold water over the affected area. Have a classmate inform your instructor.

PROCEDURE

Fill a 400 mL beaker about two-thirds full of cold water, place it on a stirrer-hotplate, and add a magnetic stirbar. Turn the stirrer on. Move the beaker around until the stirbar rotates freely without hitting the sides of the beaker. Obtain a digital thermometer and probe. Clamp the thermometer probe to a ring stand. Immerse the thermometer probe in the water in the beaker (see Figure 2.1). Keep the tip of the probe at least 1 in above the bottom of the beaker. The thermometer probe is very delicate. Forceful contact between it and the stirbar is to be avoided.

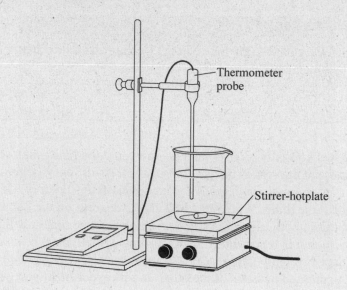

Figure 2.1
Experimental setup for measuring the water temperature in both degrees Celsius and Fahrenheit over a range of values.

The digital thermometer can measure temperature in both degrees Celsius and Fahrenheit. To measure the temperature in degrees Celsius simply turn on the digital thermometer by pressing the on/off switch; the initial reading is in degrees Celsius. To change to degrees Fahrenheit press the F/C button. (To return to Celsius press the button again.) Take an initial reading of the temperature of the cold water in both degrees Celsius and degrees Fahrenheit and record the measurements.

Check that the plastic cord of the thermometer probe is not in contact with the hotplate; otherwise, it will melt when the heat is turned on. Turn on the hotplate, and heat the water gently until it has reached approximately 30°C. Record the exact thermometer reading of the temperature in both °C and °F. Continue heating and recording Celsius and Fahrenheit temperatures at 10° (Celsius) intervals until the water has reached 90°C. Be sure to record the exact temperatures you read from the thermometer, not the approximate temperatures. Best results in this exercise are achieved when the water is heated slowly.

Turn off both the magnetic stirrer control and the hotplate control. Allow the beaker to cool before removing it from the stirrer-hotplate.

Name: _____   Date: _____

Lab Instructor: _____   Lab Section: _____

EXPERIMENT 2

Temperature Scales

RESULTS/OBSERVATIONS

1. Temperature data:

Approximate Temperature (°C)	Actual Temperature (°C)	Actual Temperature (°F)
Initial temperature		
30		
40		
50		
60		
70		
80		
90		

2. Use the temperature data to prepare a graph of Fahrenheit temperature (y-axis) versus Celsius temperature (x-axis). Read Appendix A: Handling Data (at the end of this Lab Manual) for instructions on properly preparing and interpreting graphs. Graph paper is provided on the following page.

3. Based on the graph, what is the specific form of the algebraic equation that states the relation between temperatures in the Fahrenheit and Celsius scales; i.e., what are the values of m and b in the equation $°F = m(°C) + b$?

4. Using the formula you obtained in question 2, find the Celsius temperature equivalent of 74.0°F. Also read the temperature off your graph. Are the two values the same?

5. Using the accepted formula for the relationship between temperature in the Fahrenheit and Celsius scales, $°F = (9/5)(°C) + 32$, what is the Celsius equivalent of 74.0°F?

Name: _____ Date: _____

Lab Instructor: _____ Lab Section: _____

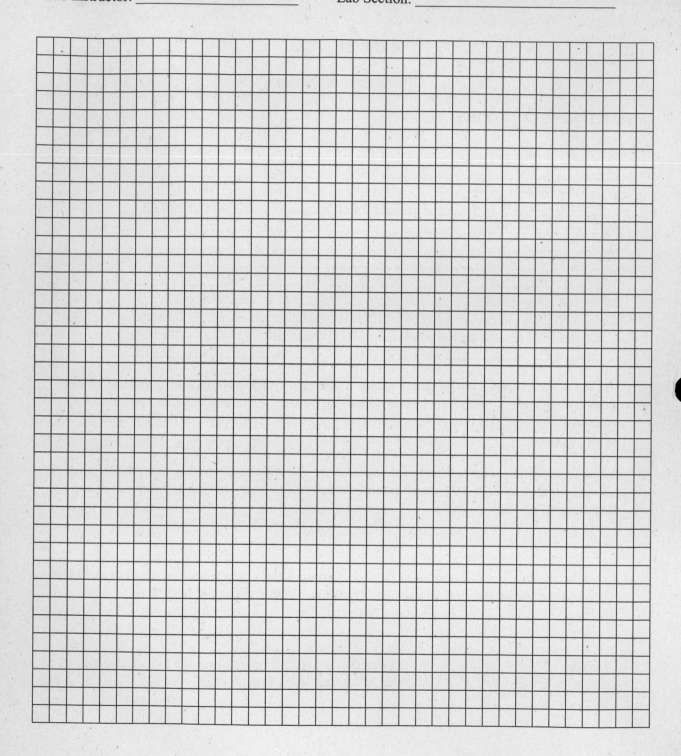

Experiment 3

Glassware Calibration

OBJECTIVE

Three commonly used pieces of glassware, a beaker, a graduated cylinder, and a pipet, will be calibrated. The relative precisions of these three pieces of glassware will be determined.

INTRODUCTION

Chemistry is an experimental science. It requires careful observation and the use of good laboratory techniques. But even with the best laboratory technique, no measurement can be made exactly. Every measurement is really an approximation of the exact answer. Improvements in technique can lessen the amount of uncertainty in a measurement but never eliminate it entirely. As an experimenter, it is necessary to know the degree of uncertainty associated with each measurement and the relationship between these uncertainties and the reliability of the final result.

Precision and *accuracy* are the terms used commonly to describe the degree of uncertainty in a measurement. *Precision* refers to the degree of refinement in the performance of an operation or the degree of perfection in the instruments and methods used to obtain a result; it is an indication of the reproducibility of a result. *Accuracy* refers to how close a measured value is to the accepted or "true" value. Precision relates to the quality of an operation by which a result is obtained. Accuracy relates to the quality of a result. To illustrate the difference between precision and accuracy, consider the analogy of a marksman for which the "truth" is represented by a bullseye. In Figure 3.1 the marksman achieved reproducibility, but the results differ from the "true" value. Thus the results can be described as precise but inaccurate. In Figure 3.2 the results are not reproducible and must be considered imprecise. However, the results are clustered around the "true" value, making them accurate. The results in Figure 3.3 are both precise and accurate. Those in Figure 3.4 are neither precise nor accurate.

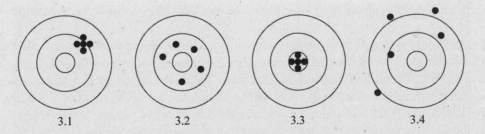

3.1   3.2   3.3   3.4

Figures 3.1–3.4
Targets illustrating the results of four marksmen provide examples of the difference between precision and accuracy.

In the preceding example the data were presented pictorially. However, a pictorial representation of data isn't always possible. And even when it is possible, it is often an inconvenient way of presenting data. When repetitive measurements are made in an experiment, it is typical to express the data as a single representative number. This number is called the *mean* $(\bar{x})$ or *average*. The mean is defined by the equation

$$\bar{x} = \frac{\sum_i x_i}{N} \qquad (1)$$

where each x_i is the result of an individual measurement. The symbol Σ denotes summation:

$$\sum_i x_i = x_1 + x_2 + x_3 + \cdots + x_N \qquad (2)$$

The mean is the sum of the measured values divided by the total number of values (N).

Precision is frequently expressed quantitatively in terms of deviation from the mean $(\bar{x} - x_i)$. This is the difference between an individual measured value and the mean value. The average deviation $(\Delta\bar{x})$,

$$\Delta\bar{x} = \frac{\sum_i |\bar{x} - x_i|}{N} \qquad (3)$$

is used to express precision when the data set is small (less than five repetitive measurements). A large average deviation indicates imprecision. An average deviation that is small relative to the mean indicates a high degree of precision.

In cases where the "true" or accepted value is known, the accuracy of a result can be quantified by calculating the percent error:

$$\text{Percent error} = \left| \frac{\text{accepted value} - \text{mean value}}{\text{accepted value}} \right| \times 100 \qquad (4)$$

The two vertical lines indicate an absolute value. The smaller the percent error, the greater the accuracy.

In this experiment you will examine the issue of uncertainty in measurement by calibrating three pieces of glassware commonly used in the laboratory to measure volumes: a beaker, a graduated cylinder, and a pipet. To calibrate a piece of glassware, measure and dispense a given volume of water. The actual volume of water dispensed will be calculated using the mass of water and its density (see Table 3.1). This will repeated in triplicate. The average volume and average deviation will be determined. These values then will be used to compare the precisions of the three pieces of glassware.

Table 3.1: Density of Water at Various Temperatures

Temperature (°C)	Density (g/mL)
15	0.999099
16	0.998943
17	0.998774
18	0.998595
19	0.998405
20	0.998203
21	0.997992
22	0.997770
23	0.997538
24	0.997296
25	0.997044
26	0.996783
27	0.996512
28	0.996232

ADDITIONAL READING

Read the section in the "Laboratory Techniques" chapter at the beginning of this Lab Manual on pipets prior to performing this experiment.

 Safety Precautions:

- Protective eyewear approved by your institution must be worn at all times while you are in the laboratory.

PROCEDURE

Part A: Beaker Calibration

Weigh a 200 mL beaker. Record the mass. Use a 50 mL beaker to measure 10 mL of distilled water. Add the 10 mL of water to the 200 mL beaker. Determine and record the mass of the filled beaker. Calculate the mass of water delivered to the 200 mL beaker by difference. Again, use the 50 mL beaker to measure 10 mL of distilled water. Add this water to that already in the 200 mL beaker. Again, determine and record the mass of the filled 200 mL beaker, and then calculate the mass of water delivered by difference. Repeat a third time. Measure and record the temperature of the water in the 200 mL beaker.

Use the water density values in Table 3.1 and the mass of water to calculate the volume of water actually delivered in each trial. Determine the average volume of water delivered and the average deviation.

Part B: Graduated Cylinder Calibration

Empty the 200 mL beaker used in Part A. Repeat the procedure of Part A, only use a graduated cylinder to measure 10 mL of distilled water.

   Use the water density values in Table 3.1 and the mass of water to calculate the volume of water actually delivered in each trial. Determine the average volume of water delivered and the average deviation.

Part C: Pipet Calibration

Repeat the procedure of Part A, only use a pipet to measure the 10 mL of distilled water.

   Use the water density values in Table 3.1 and the mass of water to calculate the volume of water actually delivered in each trial. Determine the average volume of water delivered and the average deviation.

   Use the results to rank the three measuring devices according to increasing precision.

Name: _____   Date: _____

Lab Instructor: _____   Lab Section: _____

EXPERIMENT 3

Glassware Calibration

PRELABORATORY QUESTIONS

1. Explain the difference between precision and accuracy.

2. A chemist performed an experiment to determine the melting point of sodium chloride with the following results:

Trial Number	Melting Point ($^{\circ}$C)
1	702
2	701
3	703
4	702
5	704

  (a) Determine the mean and average deviation of these determinations. Are these determinations precise?

Name: _____   Date: _____

Lab Instructor: _____   Lab Section: _____

(b) Determine the percent error in these results. The accepted value for the melting point of sodium chloride is 801°C. Are these results accurate?

3. Why should repetitive measurements always be made, if possible, when performing an experiment?

4. Use Table 3.1 to determine the volume occupied by 25.139 g of water at 27.0°C.

Name: _____   Date: _____

Lab Instructor: _____   Lab Section: _____

EXPERIMENT 3

Glassware Calibration

RESULTS/OBSERVATIONS

Part A: Beaker Calibration

1. (*Note:* The initial mass of the beaker for trial #2 will be the same as the beaker mass after adding 10 mL of water in trial #1. Similarly, the initial beaker mass in trial #3 will be the same as the beaker mass after adding 10 mL of water in trial #2.)

Trial #	Initial Beaker Mass (g)	Beaker Mass after Adding 10 mL of Water (g)	Mass of Water (g)	Calculated Volume of Water (mL)
1				
2				
3				

2. Temperature of the water: _____

3. Use the water density values in Table 3.1 and the mass of water to complete the column "Calculated Volume of Water (mL)" in the data table in above. Show a sample calculation for the trial #1 data.

Part B: Graduated Cylinder Calibration

4.

Trial #	Initial Beaker Mass (g)	Beaker Mass after Adding 10 mL of Water (g)	Mass of Water (g)	Calculated Volume of Water (mL)
1				
2				
3				

Name: _____   Date: _____

Lab Instructor: _____   Lab Section: _____

5. Temperature of the water: _____

6. Use the water density values in Table 3.1 and the mass of water to complete the column "Calculated Volume of Water (mL)" in the data table in above. Show a sample calculation for the trial #1 data.

Part C: Pipet Calibration

7.

Trial #	Initial Beaker Mass (g)	Beaker Mass after Adding 10 mL of Water (g)	Mass of Water (g)	Calculated Volume of Water (mL)
1				
2				
3				

8. Temperature of the water: _____

9. Use the water density values in Table 3.1 and the mass of water to complete the column "Calculated Volume of Water (mL)" in the data table in above. Show a sample calculation for the trial #1 data.

Name: _____   Date: _____

Lab Instructor: _____   Lab Section: _____

10. Complete the table below. Show sample calculations for the data from Part A.

	Average Calculated Volume of Water (mL)	Average Deviation in Calculated Volume of Water (mL)
Part A: Beaker		
Part B: Graduated Cylinder		
Part C: Pipet		

11. Rank the three measuring devices according to increasing precision.

_____ < _____ < _____

Name: _____   Date: _____

Lab Instructor: _____   Lab Section: _____

EXPERIMENT 3

Glassware Calibration

POSTLABORATORY QUESTIONS

1. A calibration weight with a known mass of 25.000 g was weighed on two different balances with the following results:

Trial #	Balance 1 (g)	Balance 2 (g)
1	25.007	25.06
2	25.005	25.07
3	25.008	25.04

(a) Calculate the mean and average deviation for each set of measurements on each balance.

(b) Which balance is more precise?

(c) Calculate the percent error for each set of measurements.

(d) Which set of measurements is more accurate?

Name: _____   Date: _____

Lab Instructor: _____   Lab Section: _____

2. It isn't always best to use the instrument with the highest precision. For example, instruments of high precision almost always cost more than similar instruments of low precision. You wouldn't use a high-precision instrument and run the risk of breaking it when a low-precision instrument would suffice. In the following situations decide whether a high-precision volume-measuring instrument is necessary or whether a low-precision instrument will suffice.

(a) A volume of solution needs to be obtained for rinsing a buret.

(b) 15 mL of ethanol to wash a solid precipitate.

(c) 25 mL of sodium hydroxide solution to be used in a titration experiment to determine the concentration of an acid solution.

(d) 50 mL of cyclohexanol that is to be weighed on a high-precision balance to determine the amount present.

EXPERIMENT 4

Densities of
Liquids and Solids

OBJECTIVE

To design and perform an experiment to determine the densities of an unknown liquid and an unknown solid. The unknown solid and liquid densities must be determined in the most accurate and precise manner possible.

INTRODUCTION

There is a degree of uncertainty in every measurement: nothing—be it the mass of a solid sample, the volume of an aqueous solution, or any other physical observable—can be measured exactly. As an experimenter, it is necessary to know the degree of uncertainty associated with each measurement and the relationship between these uncertainties and the reliability of the final result.

 Precision and *accuracy* are the terms used commonly to describe the degree of uncertainty in a measurement. *Precision* refers to the degree of refinement in the performance of an operation or the degree of perfection in the instruments and methods used to obtain a result; it is an indication of the reproducibility of a result. *Accuracy* refers to how close a measured value is to the accepted or "true" value. Precision relates to the quality of an operation by which a result is obtained. Accuracy relates to the quality of a result. To illustrate the difference between precision and accuracy, consider the analogy of a marksman for which the "truth" is represented by a bullseye. In Figure 4.1 the marksman achieved reproducibility, but the results differ from the "true" value. Thus the results can be described as precise but inaccurate. In Figure 4.2 the results are not reproducible and must be considered imprecise. However, the results are clustered around the "true" value, making them accurate. The results in Figure 4.3 are both precise and accurate. Those in Figure 4.4 are neither precise nor accurate.

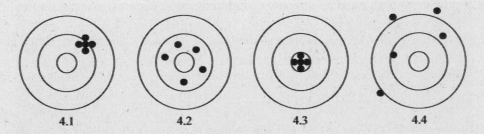

4.1   4.2   4.3   4.4

Figures 4.1–4.4
Targets illustrating the results of four marksmen provide examples of the difference between precision and accuracy.

 In the preceding example, the data were presented pictorially. However, a pictorial representation of data isn't always possible. And even when it is possible, it is often an inconvenient way of presenting data. When repetitive measurements are made in an experiment, it is typical to express the data as a single representative number. This number is called the *mean* $(\bar{x})$ or *average*. The mean is defined by the equation

$$\bar{x} = \frac{\sum_i x_i}{N} \tag{1}$$

where each x_i is the result of an individual measurement. The symbol Σ denotes summation:

$$\sum_i x_i = x_1 + x_2 + x_3 + \cdots + x_N \tag{2}$$

The mean is the sum of the measured values divided by the total number of values (N). Precision is frequently expressed quantitatively in terms of deviation from the mean $(\bar{x} - x_i)$. This is the difference between an individual measured value and the mean value. The average deviation $(\Delta \bar{x})$,

$$\Delta \bar{x} = \frac{\sum_i |\bar{x} - x_i|}{N} \tag{3}$$

is used to express precision when the data set is small (less than five repetitive measurements). A large average deviation indicates imprecision. An average deviation that is small relative to the mean indicates a high degree of precision.

In cases where the "true" or accepted value is known, the accuracy of a result can be quantified by calculating the percent error:

$$\text{Percent error} = \left| \frac{\text{accepted value} - \text{mean value}}{\text{accepted value}} \right| \times 100 \tag{4}$$

The two vertical lines indicate an absolute value. The smaller the percent error, the greater is the accuracy.

In this experiment you will examine the issue of uncertainty in measurement by determining the densities of an unknown liquid and an unknown solid in the most accurate and precise manners possible.

The density (D) of a substance is defined as the ratio of its mass to the volume it occupies:

$$D = \frac{m}{V} \tag{5}$$

where m is the mass of the substance and V is the volume is occupies. In chemistry, the density of a substance typically is reported in units of grams per milliliter.

Density is an example of an intensive property, a property independent of the amount of substance present: 5 g of lead has the same density as 0.0003 g of lead. Being an intensive property makes density a useful attribute for characterizing substances. In this experiment you will design and implement a procedure to determine the densities of an unknown liquid and an unknown solid in the most accurate and precise manners possible. Using these densities, you will identify the unknown solid and unknown liquid from the list of possibilities in Table 4.1.

Table 4.1: Solid and Liquid Densities

Solid	Density (g/mL)
Rubidium	1.5
Magnesium	1.8
Aluminum	2.7
Scandium	3.0
Zinc	7.1
Tin	7.3
Steel	7.8
Monel metal	8.9
Wood's metal	9.7
Lead	11.4

Liquid	Density (g/mL)
Ethanol	0.7893
Hexane	0.6603
Cyclohexanol	0.9624
Ethyl acetate	0.9003
Methyl benzoate	1.0888

ADDITIONAL READING

Read the sections in the "Laboratory Techniques" chapter at the beginning of this Lab Manual on cleaning glassware, handling chemicals, weighing, and pipets prior to performing this experiment.

 Safety Precautions:

- Protective eyewear approved by your institution must be worn at all times while you are in the laboratory.

- Many of the liquid unknowns used in this experiment are flammable. No open flames should be used during this experiment. In the event of a fire, notify your instructor. A fire in a small container can be smothered by covering the vessel with a watchglass. Use a fire extinguisher to eliminate a larger fire. If the fire is burning over too large an area to be extinguished easily, evacuate the area and activate the fire alarm. If your clothing should catch on fire, use the safety shower or a fire blanket to extinguish the flames.

EXPERIMENT

You are to design and implement a procedure that allows the densities of an unknown liquid and an unknown solid to be determined in the most accurate and precise way possible. Use these densities to identify the unknown solid and unknown liquid from the list of possibilities in Table 4.1.

In order to be confident of the efficacy of your procedure for determination of the liquid density, it is mandatory that you first test your procedure on a liquid of known density, water in this case. The density of water is 1.00 g/mL (Table 3.1 lists the actual temperature dependence of the density of water).

EQUIPMENT AND REAGENTS

To perform this experiment, you will have access to all the equipment in your lab drawer, a graduated cylinder, a 10 mL pipet, and the electronic balances.

Waste Disposal. Dispose of all chemical waste as directed by your instructor.

Name: _____   Date: _____

Lab Instructor: _____   Lab Section: _____

EXPERIMENT 4

Densities of Liquids and Solids

PRELABORATORY QUESTIONS

1. A student performed an experiment to determine the density of neopentyl alcohol with the following results:

Trial #	Density (g/mL)
1	0.815
2	0.982
3	0.641

(a) Determine the mean and average deviation of these determinations. Are these determinations precise?

(b) Determine the percent error in these results. The accepted value of the density of neopentyl alcohol is 0.812 g/mL. Are these results accurate?

Name: _____   Date: _____

Lab Instructor: _____   Lab Section: _____

2. A student determined the mass of a liquid sample using two different electronic balances. Using balance #1 the liquid sample was determined to have a mass of 15.1 g. Using balance #2 the mass of the liquid sample was determined to be 15.072 g. Are these two mass measurements equally precise? If not, which is more precise?

3. If possible, why should repetitive measurements always be made when performing an experiment?

4. In this experiment, how will you gauge the precision of your result?

5. In this experiment, how will you determine the accuracy of your result?

6. Choose one of the five liquids listed in Table 4.2. As listed in the Material Safety Data Sheets (MSDS) (**http://msds.ehs.cornell.edu/msdssrch.asp**), what are the dangers of your chosen liquid?

Name: _____   Date: _____

Lab Instructor: _____   Lab Section: _____

EXPERIMENT 4

Densities of Liquids and Solids

RESULTS/OBSERVATIONS

1. Unknown liquid identification number: _____

2. Unknown solid identification number: _____

3. Write any pertinent data in the space below. Clearly indicate both the property and the amount. If reasonable, use the table below.

4. Perform any necessary calculations in the space below. Clearly indicate the quantity being calculated.

5. Identity of unknown solid: _____

6. Identity of unknown liquid: _____

Name: _____   Date: _____

Lab Instructor: _____   Lab Section: _____

EXPERIMENT 4

Densities of Liquids and Solids

POSTLABORATORY QUESTIONS

1. Two students performed this experiment to determine the density of the same unknown liquid with the following results:

Trial #	Density (g/mL), Student A	Density (g/mL), Student B
1	0.6572	0.674
2	0.6613	0.662
3	0.6594	0.653

   (a) Which student determined the density of the unknown liquid more precisely? Justify your answer.

   (b) What is the identity of the unknown liquid?

   (c) Which student's density determination was more accurate? Justify your answer.

EXPERIMENT 5

Synthesis of Potassium Ferric Oxalate Trihydrate

OBJECTIVE

The compound potassium ferric oxalate trihydrate [$K_3Fe(C_2O_4)_3 \cdot 3H_2O$] will be synthesized in a two-step process. The actual, theoretical, and percent yields of $K_3Fe(C_2O_4)_3 \cdot 3H_2O$ will be determined.

INTRODUCTION

A chemical synthesis is the use of one or more chemical reactions to bring about the construction of a desired chemical product or products. The reasons for carrying out a chemical synthesis are varied. It might be the production of a useful substance not found in nature, plastics, for example, or the production of a substance found in nature but difficult or expensive to isolate in large quantities, such as ammonia. Then again, it might be the desire to create a substance with new, useful properties, say, a new medication. Whatever the motivation for the synthesis, maximizing the conversion of reactants to products is always a priority.

In this experiment you will synthesize the compound potassium ferric oxalate trihydrate [$K_3Fe(C_2O_4)_3 \cdot 3H_2O$] in a two-step process. The first step involves the reaction of ferrous ammonium sulfate hexahydrate [$Fe(NH_4)_2(SO_4)_2 \cdot 6H_2O$] with oxalic acid ($H_2C_2O_4$):

$$Fe(NH_4)_2(SO_4)_2 \cdot 6H_2O(s) + H_2C_2O_4(aq) \rightarrow$$
(Pale green)

$$FeC_2O_4 \cdot 2H_2O(s) + (NH_4)_2SO_4(aq) + H_2SO_4(aq) + 4H_2O(l) \qquad (1)$$
(Yellow)

Where applicable, the colors of reactants and products are indicated in parentheses below their molecular formula.

The ferrous oxalate dihydrate ($FeC_2O_4 \cdot 2H_2O$) produced in the Reaction (1) will be separated from the other products by decantation. In the second step, the ferrous oxalate dihydrate will be converted to potassium ferric oxalate trihydrate [$K_3Fe(C_2O_4)_3 \cdot 3H_2O$] through reaction with oxalic acid, hydrogen peroxide (H_2O_2), and potassium oxalate ($K_2C_2O_4$):

$$2\,FeC_2O_4 \cdot 2H_2O(s) + H_2C_2O_4(aq) + H_2O_2(aq) + 3\,K_2C_2O_4(aq) \rightarrow$$

$$2\,K_3Fe(C_2O_4)_3 \cdot 3H_2O(s) \qquad (2)$$
(Green)

The potassium ferric oxalate trihydrate crystals then are separated from solution by vacuum filtration.

Your objective is to prepare as great a yield of potassium ferric oxalate trihydrate as possible. To measure the effectiveness with which this objective is met, you will calculate the percent yield:

$$\text{Percent yield} = \frac{\text{actual yield}}{\text{theoretical yield}} \times 100 \qquad (3)$$

The theoretical yield is the calculated maximum amount of product that might be obtained under ideal conditions from the reactants. In an experiment, the theoretical yield is seldom, if ever, reached. In this experiment, the theoretical yield is the maximum number of grams of product [$K_3Fe(C_2O_4)_3 \cdot 3H_2O$] that might be obtained from the specified amounts of reactants. To determine the percent yield it is necessary to calculate the moles of each reactant and then find the limiting reagent. The limiting reagent determines the theoretical yield of potassium ferric oxalate trihydrate.

ADDITIONAL READING

Read the sections in the "Laboratory Techniques" chapter at the beginning of this Lab Manual on cleaning glassware, handling chemicals, weighing, heating liquids and solutions, decantation, and gravity and vacuum filtration prior to performing this experiment.

 Safety Precautions:

- Protective eyewear approved by your institution must be worn at all times while you are in the laboratory.

- Sulfuric acid (H_2SO_4) is a product of the first step of the synthesis. Chemical burns can result if H_2SO_4 (a component of the hot supernatant liquid) comes in contact with your skin. If the hot supernatant liquid spills on your skin, immediately wash the affected area with water. Continue washing with water for 15 minutes. Have a classmate notify your instructor.

- The acetone and ethanol used in week 2 of this experiment are flammable liquids. No open flames should be used during the second week of this experiment. In the event of a fire, notify your instructor. A fire in a small container can be smothered by covering the vessel with a watchglass. Use a fire extinguisher to eliminate a larger fire. If the fire is burning over too large an area to be extinguished easily, evacuate the area and activate the fire alarm. If your clothing should catch on fire, use the safety shower or a fire blanket to extinguish the flames.

PROCEDURE

Week 1

Measure and record the mass of a 200 mL beaker. Add approximately 5 g of ferrous ammonium sulfate hexahydrate [$Fe(NH_4)_2(SO_4)_2 \cdot 6H_2O$] to the beaker. Measure and record this combined mass. Determine the mass of $Fe(NH_4)_2(SO_4)_2 \cdot 6H_2O$ in the beaker by subtraction. Next, dissolve this solid sample in 15 mL of distilled water, measured by graduated cylinder, to which 5 drops of 3 M H_2SO_4 have been added (to prevent premature oxidation of Fe^{2+} to Fe^{3+} by O_2 in the air). To this solution add 25 mL of 1 M $H_2C_2O_4$ (oxalic acid) solution, measured by graduated cylinder.

Heat the mixture to boiling on a hotplate. Stir the mixture continuously with a glass stirring rod to prevent spattering. This solution is very susceptible to boiling over; do not leave it unattended. If it should start to boil over, immediately reduce the heat and remove the beaker from the heat source using large beaker tongs.

After the mixture has come to a boil, turn the heat off and allow the yellow ferrous oxalate dihydrate ($FeC_2O_4 \cdot 2H_2O$) precipitate to settle.

CAUTION: The hot supernatant liquid contains H_2SO_4. Chemical burns can result if H_2SO_4 comes in contact with your skin. If the hot supernatant liquid spills on your skin, immediately wash the affected area with water. Continue washing with water for 15 minutes. Have a classmate notify your instructor.

Using large beaker tongs, carefully decant the hot supernatant liquid (the liquid remaining above the solid) into a waste beaker, retaining the solid in the 200 mL beaker. Add 20 mL of distilled water, measured by graduated cylinder, to wash the precipitate, warm, stir, allow the solid to resettle, and again decant the liquid into the waste beaker, retaining the solid.

To the solid $FeC_2O_4 \cdot 2H_2O$ in the beaker add 20 mL of saturated potassium oxalate solution (300 g $K_2C_2O_4 \cdot H_2O$ per liter), measured by graduated cylinder. Support a thermometer using a ring stand, a split one-holed stopper, and a clamp (see Figure 5.1).

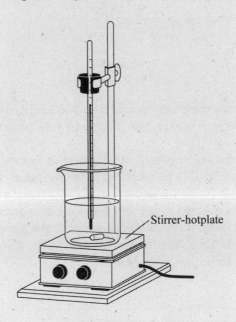

Stirrer-hotplate

Figure 5.1
A thermometer is used to monitor the solution temperature while heating. Support the thermometer using a ringstand, clamp, and split one-holed stopper.

Place the thermometer in the solution. Carefully heat the solution to 40°C. Obtain 20 mL of 3% H_2O_2 (hydrogen peroxide) in a 25 mL graduated cylinder. Add the 20 mL of H_2O_2 very slowly, a few drops at a time, stirring continuously and keeping the temperature near 40°C. A small amount of a slimy red precipitate [iron(III) hydroxide, $Fe(OH)_3$] might appear briefly. After adding all the hydrogen peroxide, heat to boiling. To the boiling solution add 8 mL of 1 M $H_2C_2O_4$—the first 5 mL all at once and the last 3 mL very slowly—keeping the solution boiling. Set up a funnel with filter paper on a ring stand and gravity filter the boiling solution into a clean 100 mL beaker (see Figure 5.2)

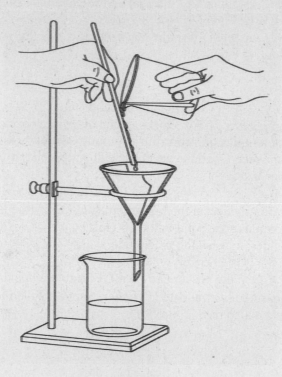

Figure 5.2
Gravity filtration.

Allow the solution to cool to room temperature, and then add 10 mL of ethanol to the beaker. Cover your sample beaker with your largest beaker, and place it in your laboratory desk until the next laboratory period.

The reason that the beaker should be placed in your desk is that the product, potassium ferric oxalate trihydrate [$K_3Fe(C_2O_4)_3 \cdot 3H_2O$], decomposes in the presence of light. The products of the decomposition are carbon dioxide [from the oxidation of $C_2O_4^{2-}$] and a complex of Fe^{2+} (from the reduction of Fe^{3+}). The beaker is covered to prevent complete evaporation of the water. Complete evaporation of the water would make it impossible to separate the pure $K_3Fe(C_2O_4)_3 \cdot 3H_2O$ crystals from the other salts, which would have remained in solution. Also, slower solvent evaporation encourages the development of larger, purer crystals.

Week 2

Assemble a vacuum filtration setup (see Figure 5.3). Using a stirring rod equipped with a rubber policeman, transfer the crystals of $K_3Fe(C_2O_4)_3 \cdot 3H_2O$ from the beaker to the filter paper in the funnel. Wash the crystals with 10 mL of a 50:50 ethanol–water mixture, followed by 10 mL of acetone. Allow the crystals to air-dry.

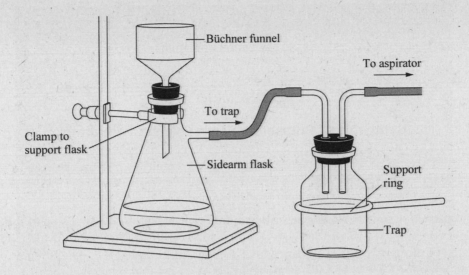

Figure 5.3
Vacuum filtration setup.

Determine the mass of a stoppered glass vial. Transfer the dry crystals from the filter paper into the preweighed glass vial. Obtain the combined mass of the glass vial and crystals. Determine the mass of $K_3Fe(C_2O_4)_3 \cdot 3H_2O$ by subtraction. Affix a label with your name, the date, your lab section, and your lab instructor's name to the container. Submit your labeled container of crystals to your instructor.

Waste Disposal. Dispose of all chemical waste as directed by your instructor.

Name: _____   Date: _____

Lab Instructor: _____   Lab Section: _____

EXPERIMENT 5

Synthesis of Potassium Ferric Oxalate Trihydrate

PRELABORATORY QUESTIONS

1. A student carried out the synthesis described in this experiment. In the first step of the synthesis the student combined 4.918 g of $Fe(NH_4)_2(SO_4)_2 \cdot 6H_2O$ with 25.0 mL of 1.0 M $H_2C_2O_4$. In the second step of the synthesis the student added 20.0 mL of saturated $K_2C_2O_4$, 20.0 mL of 3% H_2O_2, and 8.0 mL of 1.0 M $H_2C_2O_4$ to the solid $FeC_2O_4 \cdot 2H_2O$ produced in the first step.

   (a) Calculate the number of moles of $Fe(NH_4)_2(SO_4)_2 \cdot 6H_2O$ in 4.918 g of $Fe(NH_4)_2(SO_4) \cdot 6H_2O$.

   (b) Calculate the number of moles $H_2C_2O_4$ in 25.0 mL of 1.0 M $H_2C_2O_4$.

   (c) Calculate the number of moles of $K_2C_2O_4$ in 20.0 mL of a saturated $K_2C_2O_4$ (300 g $K_2C_2O_4 \cdot H_2O$ per liter) solution.

   (d) Calculate the number of moles of H_2O_2 in 20.0 mL of a solution that is 3% H_2O_2 by weight. Assume that the density of the solution is 1.01 g/mL.

   (e) Calculate the number of moles $H_2C_2O_4$ in 8.0 mL of 1.0 M $H_2C_2O_4$.

Name: _____   Date: _____

Lab Instructor: _____   Lab Section: _____

(f) What is the limiting reactant of this two-step reaction?

(g) What is the theoretical yield, in grams of $K_3Fe(C_2O_4)_3 \cdot 3H_2O$?

(h) If the actual yield of $K_3Fe(C_2O_4) \cdot 3H_2O$ was 4.827 g, what is the percent yield?

Name: _____   Date: _____

Lab Instructor: _____   Lab Section: _____

EXPERIMENT 5

Synthesis of Potassium Ferric Oxalate Trihydrate

RESULTS/OBSERVATIONS

1. Mass of 200 mL beaker: _____

2. Combined mass of 200 mL beaker and $Fe(NH_4)_2(SO_4)_2 \cdot 6H_2O$: _____

3. Mass of $Fe(NH_4)_2(SO_4)_2 \cdot 6H_2O$ (calculated by difference): _____

4. Mass of stoppered glass vial: _____

5. Combined mass of stoppered glass vial and the $K_3Fe(C_2O_4)_3 \cdot 3H_2O$: _____

6. Mass of $K_3Fe(C_2O_4)_3$ 3 H_2O (calculated by difference): _____

7. Calculation of theoretical yield of $K_3Fe(C_2O_4)_3 \cdot 3 H_2O$ assume $Fe(NH_4)_2(SO_4)_2 \cdot 6H_2O$ is the limiting reactant,

8. Percent yield of $K_3Fe(C_2O_4)_3 \cdot 3 H_2O$: _____

Name: _____   Date: _____

Lab Instructor: _____   Lab Section: _____

EXPERIMENT 5

Synthesis of Potassium Ferric Oxalate Trihydrate

POSTLABORATORY QUESTIONS

1. Speculate as to why your percent yield was less than 100%.

2. What advantage is there to calculating the percent yield for each step of a multistep synthesis rather than determining the percent yield of the final product only?

3. The $K_3Fe(C_2O_4)_3 \cdot 3H_2O$ synthesized in this experiment can be analyzed by oxidation of the oxalate ion $(C_2O_4^{2-})$ by the permanganate ion (MnO_4^-) according to the following reaction:

$$16\,H^+ + 5\,C_2O_4^{2-} + 2\,MnO_4^- \rightarrow 10\,CO_2 + 2\,Mn^{2+} + 8\,H_2O$$

   (a) Calculate the percent by weight of oxalate in $K_3Fe(C_2O_4)_3 \cdot 3H_2O$.

Name: _____   Date: _____

Lab Instructor: _____   Lab Section: _____

(b) A student synthesized 0.832 g of $K_3Fe(C_2O_4)_3 \cdot 3H_2O$ by following the same instructions as given in this experiment. This 0.832 g sample reacted with exactly 14.01 mL of 0.138 M $KMnO_4$ according to the balanced equation above. Calculate the percent by weight of oxalate in the sample and compare it with the answer to part (a).

4. The experimental procedure directed you to cover the stored beaker to prevent complete evaporation of the water. Evaporation of all the water would have made it impossible to separate the desired product from "other salts, which would have remained in solution." Assuming that $Fe(NH_4)_2(SO_4)_2 \cdot 6H_2O$ is the limiting reactant, what other oxalate salt most likely would be in solution?

EXPERIMENT 6

Analysis of Potassium Ferric Oxalate Trihydrate

OBJECTIVE

The molarity of a $KMnO_4$ solution will be determined by titration with sodium oxalate ($Na_2C_2O_4$). The $K_3Fe(C_2O_4)_3 \cdot 3H_2O$ sample synthesized in Experiment 5 will be analyzed by titration with the standardized $KMnO_4$ solution. The results of this titration will be used to calculate the percent by weight of oxalate in the $K_3Fe(C_2O_4)_3 \cdot 3H_2O$ sample. Comparison of this value with the theoretical value will measure the purity of the $K_3Fe(C_2O_4)_3 \cdot 3H_2O$ sample.

INTRODUCTION

At the end of Experiment 5: Synthesis of Potassium Ferric Oxalate Trihydrate the percent yield was calculated. This calculated percent yield was based on the assumption that all the solid recovered and isolated was potassium ferric oxalate trihydrate [$K_3Fe(C_2O_4)_3 \cdot 3H_2O$]. Was this assumption justified? Can a chemist assume the purity of a product? In general, the answer to these questions is, "No, it is not reasonable to make assumptions regarding the purity of a product." The purity must be analyzed in some manner. In this experiment the purity of the $K_3Fe(C_2O_4)_3 \cdot 3H_2O$ sample synthesized in Experiment 5 will be analyzed by titration with permanganate (MnO_4^-).

For a titration analysis involving MnO_4^- to be useful it is necessary to know the concentration of the MnO_4^- solution. However, freshly prepared MnO_4^- solutions are unstable; dust and organic matter dissolved in the water react with MnO_4^- to produce such species as solid MnO_2, which acts to catalyze further decomposition. Fortunately, it is rather simple to remove the solid MnO_2 by carefully siphoning off the $KMnO_4$ solution after it has been allowed to react for a day or two. The resulting $KMnO_4$ solution is stable for weeks. The concentration of the stable $KMnO_4$ solution then must be determined by standardization.

In this experiment you will standardize an MnO_4^- solution by titration with sodium oxalate ($Na_2C_2O_4$):

$$5\ C_2O_4^{2-}(aq) + 2\ MnO_4^-(aq) + 16\ H^+(aq) \rightarrow 10\ CO_2(g) + 2\ Mn^{2+}(aq) + 8\ H_2O(l) \tag{1}$$
$$\text{(purple)}$$

Since permanganate is the only colored species involved in this reaction, it will serve as its own indicator: When all the $C_2O_4^{2-}$ has been consumed in the reaction, a drop of excess MnO_4^- will lead to a persistent pink coloration in the solution (the color at the end-point will be pink instead of purple because the MnO_4^- will be diluted significantly).

The standardized MnO_4^- solution then will be used to titrate the oxalate ($C_2O_4^{2-}$) portion of the $K_3Fe(C_2O_4)_3 \cdot 3H_2O$ sample. The net ionic equation for this reaction is the same as Reaction (1). From the volume and molarity of the $KMnO_4$ solution, the moles of MnO_4^- consumed in the titration can be determined. The moles of MnO_4^- and Reaction (1) allow the moles of $C_2O_4^{2-}$ and the percent by weight of oxalate in the sample to be calculated. This value will be compared with the theoretical percent by weight of oxalate in $K_3Fe(C_2O_4)_3 \cdot 3H_2O$ to estimate the purity of the sample.

This experiment involves two parts. In Part A the permanganate solution will be standardized by titration with sodium oxalate. Concentration values will be collected from each student in the class. This set of values will be averaged in order to obtain a more precise value of the molarity of the $KMnO_4$ solution. In Part B the $K_3Fe(C_2O_4)_3 \cdot 3H_2O$ sample synthesized in Experiment 5 will be analyzed by titration with the standardized $KMnO_4$ solution. The results of this titration will be used to calculate the percent by weight of oxalate in the $K_3Fe(C_2O_4)_3 \cdot 3H_2O$ sample. Comparison of this value with the theoretical value will measure the purity of the $K_3Fe(C_2O_4)_3 \cdot 3H_2O$ sample.

ADDITIONAL READING

Read the sections in the "Laboratory Techniques" chapter of this Lab Manual on cleaning glassware, handling chemicals, weighing, heating liquids and solutions, and burets prior to performing this experiment.

 Safety Precautions:

- Protective eyewear approved by your institution must be worn at all times while you are in the laboratory.

- Chemical burns can result if the 3 M H_2SO_4 used in this experiment comes in contact with your skin. If you spill 3 M H_2SO_4 on your skin, immediately wash the affected area with water. Continue washing with water for 15 minutes. Have a classmate notify your instructor.

- The glass weighing bottle becomes hot after resting in the drying oven. Remove it from the oven with care. If a burn does occur, run cold water over the affected area. Have a classmate inform your instructor.

PROCEDURE

Part A: Standardization of KMnO₄

Clean a weighing bottle, and label it with your name. Dry it thoroughly in an oven, which should be held at a temperature of about 120°C. Weigh the empty bottle on a balance, recording the mass. Add approximately 0.5 g of sodium oxalate ($Na_2C_2O_4$) to the weighing bottle. Place the bottle of $Na_2C_2O_4$, with the cover off, in the oven to dry for at least an hour to remove any loosely bound waters of hydration. (*While waiting for the $Na_2C_2O_4$ to dry perform Part B.*)

Cap the weighing bottle after removing it from the oven.

AUTION: Handle the weighing bottle with care. The glass will be hot.

Weigh the bottle of dried $Na_2C_2O_4$. Record this mass. Determine the mass of $Na_2C_2O_4$ by difference.

Clean a 250 mL Erlenmeyer flask. Weigh 0.12–0.15 g of $Na_2C_2O_4$ from the weighing bottle into the Erlenmeyer flask. Immediately seal the weighing bottle and weigh it on the electronic balance. Record this mass. Determine the exact mass of $Na_2C_2O_4$ transferred to the Erlenmeyer flask by difference.

Add 40–50 mL of distilled water to the Erlenmeyer flask and swirl the contents. Add about 10 mL of 3 M H_2SO_4, measured by graduated cylinder, to the Erlenmeyer flask.

CAUTION: Chemical burns can result if 3 M H_2SO_4 comes in contact with skin.

Add a magnetic stirbar, place the flask on a stirrer-hotplate, and stir. The solid should dissolve.

Fill the buret with the $KMnO_4$ solution, cover the $KMnO_4$ flask with the inverted beaker, and return it to its storage space. Record the initial buret reading to the correct number of decimal places. Position the ringstand so that the buret tip is in the neck of the Erlenmeyer flask. Add 8–10 mL of the $KMnO_4$ solution to the flask. Heat the flask using the stirrer hotplate until the violet color of $KMnO_4$ disappears. Turn off the heat, and titrate the hot solution until the end point is reached, a pink coloration that lasts for 20–30 seconds. Record the final buret reading.

Part B: Titration of $C_2O_4^{2-}$ in the $K_3Fe(C_2O_4)_3 \cdot 3H_2O$ Sample

Thoroughly clean and dry a 400 mL beaker and a 300 mL Florence flask. Use the beaker to obtain about 300 mL of the $KMnO_4$ solution from your instructor. Transfer the $KMnO_4$ solution to the Florence flask, and cover the top with an inverted 50 mL beaker. Label the Florence flask with your name, and store it in a dark place, such as your lab drawer ($KMnO_4$ is light-sensitive).

Clean, dry, and weigh a small beaker. Record the beaker mass. Clean and dry a mortar and pestle. Weigh about 1 g of the $K_3Fe(C_2O_4)_3 \cdot 3H_2O$ sample prepared in Experiment 5 into the mortar. Grind the sample to powder, and then transfer it to the small beaker. Measure the mass of the beaker and sample. Record this mass. Determine the mass of the sample by difference.

Clean a 250 mL Erlenmeyer flask. Place 0.15–0.20 g of the sample in the Erlenmeyer flask. Weigh the beaker and remaining sample on the electronic balance. Record this mass. Determine the exact mass of sample transferred to the Erlenmeyer flask by difference. Add about 30 mL of distilled water to the Erlenmeyer flask, and swirl the contents. Add about 10 mL of 3 M H_2SO_4, measured by graduated cylinder, to the Erlenmeyer flask.

CAUTION: Chemical burns can result if 3 M H_2SO_4 comes in contact with skin.

Add a magnetic stirbar, place the flask on stirrer-hotplate, and stir. The solid should dissolve. If a brownish color persists in the solution, add a bit more 3 M H_2SO_4 to remove it.

Set up a clean buret on a ringstand [see Figure 6.1(a)].

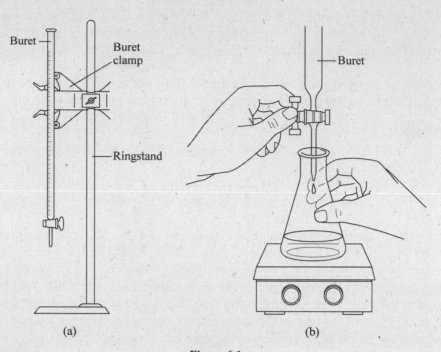

(a)   (b)

Figure 6.1
Using a buret: (a) support a buret with a buret clamp attached to a ringstand; (b) a buret being
used in a titration experiment.

Rinse it with 4–5 mL of the $KMnO_4$ solution. Fill the buret with the $KMnO_4$ solution, cover the
$KMnO_4$ flask with the inverted beaker, and return it to its storage space. Record the initial buret
reading to the correct number of decimal places. (*Note:* Because $KMnO_4$ is opaque, the volume
should be read at the top of the meniscus.) Position the ringstand so that the buret tip is in the neck of
the Erlenmeyer flask [see Figure 6.1(b)]. Add 8–10 mL of the $KMnO_4$ solution to the flask. The
reaction between MnO_4^- and $C_2O_4^{2-}$ is very slow at first, but it is catalyzed by the Mn^{2+} that forms. To
get the reaction started, heat the flask using the stirrer-hotplate until the violet color of $KMnO_4$
disappears. Turn off the heat, and titrate the hot solution until the end-point is reached, a pink
coloration that lasts for 20–30 seconds. Record the final buret reading.

Repeat this titration twice more.

When you have completed these titrations, empty any remaining $KMnO_4$ solution in the buret
into the Florence flask. Cover the flask, and return it to its storage space.

Waste Disposal. Dispose of all chemical waste as directed by your instructor.

Name: _____   Date: _____

Lab Instructor: _____   Lab Section: _____

EXPERIMENT 6

Analysis of Potassium Ferric Oxalate Trihydrate

PRELABORATORY QUESTIONS

1. If 29.74 of $KMnO_4$ solution were required to titrate a 0.137g sample of $Na_2C_2O_4$, what is the concentration of the $KMnO_4$ solution?

2. Why is the sodium oxalate dried in the oven prior to use? What negative effect could result from not drying the sodium oxalate prior to use?

3. What will signal the end-point in the titrations performed in this experiment?

4. Why must the $KMnO_4$ solution be stored in a dark place while not in use?

5. Why must the 400 mL beaker and 300 mL Florence flask used to obtain and store the $KMnO_4$ solution be dry?

6. Why does it not matter that the 250 mL Erlenmeyer flask be dry prior to adding the sample?

Name: _____   Date: _____

Lab Instructor: _____   Lab Section: _____

EXPERIMENT 6

Analysis of Potassium Ferric Oxalate Trihydrate

RESULTS/OBSERVATIONS

Part A: Standardization of $KMnO_4$

1. Mass of the empty weighing bottle: _____

2. Mass of the weighing bottle and $Na_2C_2O_4$: _____

3. Mass of $Na_2C_2O_4$: _____

4. Mass of the weighing bottle and $Na_2C_2O_4$ after transferring some $Na_2C_2O_4$ to the flask: _____

5. Mass of $Na_2C_2O_4$ transferred to the Erlenmeyer flask: _____

6. Initial buret reading: _____

7. Final buret reading: _____

8. Calculation of the $KMnO_4$ solution molarity. (Report this value to your instructor before leaving class.)

9. The class average $KMnO_4$ solution molarity: _____

Name: _____   Date: _____

Lab Instructor: _____   Lab Section: _____

Part B: Titration of $C_2O_4^{2-}$ in the $K_3Fe(C_2O_4)_3 \cdot 3H_2O$ Sample

10. Mass of the small beaker: _____

11. Mass of the small beaker and sample: _____

12. Mass of sample: _____

13. Titration data:

Trial #	Mass of Beaker and Sample after Transferring Some Solid to the Erlenmeyer Flask (g)	Mass of Sample Transferred to the Erlenmeyer Flask (g)	Initial Buret Reading (mL)	Final Buret Reading (mL)
1				
2				
3				

14. Show the calculation of the percent by weight of oxalate in the $K_3Fe(C_2O_4)_3 \cdot 3H_2O$ sample for trial #1.

15. Show the results of your calculations of the percent by weight of oxalate in the $K_3Fe(C_2O_4)_3 \cdot 3H_2O$ sample for all trials in the table below.

Trial #	% by Weight of Oxalate in the $K_3Fe(C_2O_4)_3 \cdot 3H_2O$ Sample
1	
2	
3	

16. Average percent by weight of oxalate in the $K_3Fe(C_2O_4)_3 \cdot 3H_2O$ sample: _____

Name: _____   Date: _____

Lab Instructor: _____   Lab Section: _____

17. Using the theoretical percent by weight of oxalate in $K_3Fe(C_2O_4)_3 \cdot 3H_2O$ (you calculated this in question 3 of the Postlaboratory Questions for Experiment 5) as the accepted value, calculate a percent error.

18. Based on your percent error, comment on the purity of your $K_3Fe(C_2O_4)_3 \cdot 3H_2O$ sample.

Name: _____   Date: _____

Lab Instructor: _____   Lab Section: _____

EXPERIMENT 6

Analysis of Potassium Ferric Oxalate Trihydrate

POSTLABORATORY QUESTIONS

1. Given that in the titration of the sample the reactants are $K_3Fe(C_2O_4)_3$, $KMnO_4$, and H_2SO_4 and the products are gaseous CO_2, $Fe_2(SO_4)_3$, $MnSO_4$, K_2SO_4, and H_2O, write a balanced equation for the reaction, including both reacting species and spectator ions.

2. How would each of the following affect your determined percent by weight of oxalate in the $K_3Fe(C_2O_4)_3 \cdot 3H_2O$ sample?

    (a) The sample was contaminated with $FeC_2O_4 \cdot 2H_2O$.

    (b) Some of the $KMnO_4$ solution decomposed to MnO_2 after titrating the sample three times but before standardization.

Determining the Percent of Sodium Hypochlorite in Bleach

OBJECTIVE

To design and implement a titration procedure that allows the percent by mass of sodium hypochlorite in a bottle of commercially available bleach to be calculated. The reactions used in the titration procedure are those discussed in the introduction to the experiment. The determined percent by mass of sodium hypochlorite will be compared with the manufacturer's stated value.

INTRODUCTION

Nearly all manufacturers test selected product samples to ensure their quality. The nature of these quality control tests depends on the product and the means of manufacture. An aerosol air freshener would be evaluated on aroma, liquid dish detergent on its percent water content, a motor oil on its viscosity. In this experiment you will perform a quality control test on a commercially available bleaching solution, such as Clorox™. The quality control test will involve verifying that the bleaching solution contains the manufacturer's stated amount of active ingredient.

Ordinary bleaching solutions are made by dissolving chlorine gas in aqueous sodium hydroxide:

$$Cl_2(g) + 2 NaOH(aq) \rightarrow NaOCl(aq) + NaCl(aq) + H_2O(l) \tag{1}$$

The active ingredient in such products is sodium hypochlorite (NaOCl), which is dissociated completely into Na^+ and OCl^- ions in aqueous solution. Hypochlorite ion (OCl^-) is an oxidizing agent. In bleaching laundry, clothing stains are oxidized by the hypochlorite ion, and hypochlorite, in turn, is reduced to chloride ion (Cl^-).

In this experiment you will use the oxidizing ability of hypochlorite ion to verify that the manufacturer's stated amount of active ingredient (i.e., percent by mass of sodium hypochlorite) in a bottle of commercially available bleach is accurate. The method you will use involves the following reactions:

Mixing a sample of bleach with a solution containing an excess of iodide ion (I^-), a reducing agent, and acid will lead to the reduction of hypochlorite ion to chloride ion, and iodide ion will become oxidized to iodine (I_2), which has a brown color in aqueous solution. The (unbalanced) net ionic equation for this reaction is

$$OCl^-(aq) + I^-(aq) \rightarrow I_2(aq) + Cl^-(aq) \tag{2}$$
$$\text{(Brown)}$$

The number of moles of iodine produced in Reaction (2) is stoichiometrically related to the number of moles of hypochlorite ion in the initial sample of bleach. Given excesses of iodide and acid, it is possible to determine the mole amount of hypochlorite ion initially present from the amount of iodine produced.

It is possible to measure the amount of iodine produced via Reaction (2) by titrating the iodine with a standardized solution of thiosulfate ion ($S_2O_3^{2-}$). This is an example of an oxidation-reduction, or redox, titration. The (unbalanced) net ionic equation for the reaction used in this titration is

$$I_2(aq) + S_2O_3^{2-}(aq) \rightarrow I^-(aq) + S_4O_6^{2-}(aq) \tag{3}$$
(Brown)

At the equivalence point the number of moles of iodine is stoichiometrically equivalent to the number of moles of thiosulfate. To carry out the titration, place the standardized thiosulfate solution (i.e., a thiosulfate solution of known concentration), which is colorless, in a buret, and add it progressively to a flask containing aqueous iodine produced by Reaction (2), which is brown [see Figure 7.1(a) and (b)]. As the thiosulfate combines with the iodine the brown color of the iodine solution will lessen gradually because the reaction products, iodide and tetrathionate ion ($S_4O_6^{2-}$), are both colorless. As the equivalence point is neared the solution will change from brown to a pale-yellow color. The end point is reached when the solution becomes completely colorless.

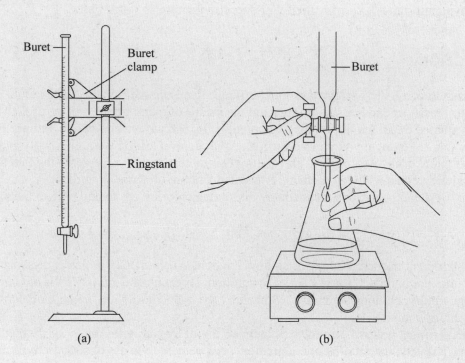

(a)   (b)

Figure 7.1
Using a buret: (a) support a buret with a buret clamp attached to a ringstand; (b) a buret being used in a titration experiment.

Your purpose in this experiment is to design a titration procedure based on Reactions (2) and (3) that will allow for the verification of the manufacturer's stated percent by mass of sodium hypochlorite in a bottle of commercially available bleach.

ADDITIONAL READING

Read the sections in the "Laboratory Techniques" chapter at the beginning of this Lab Manual on cleaning glassware, handling chemicals, pipets, and burets prior to performing this experiment.

Safety Precautions:

■ Protective eyewear approved by your institution must be worn at all times while you are in the laboratory.

■ Chemical burns can result when 6 M HCl comes in contact with skin. If you spill 6 M HCl on your skin, immediately wash the affected area with water. Continue washing with water for 15 minutes. Have a classmate notify your instructor.

EXPERIMENT

You are to design and implement a titration procedure using the reactions discussed in the introduction that will allow you to calculate the percent by mass of sodium hypochlorite in a bottle of commercially available bleach. Repeat your titration procedure at least three times to ensure sufficient experimental precision.

EQUIPMENT AND REAGENTS

To perform this experiment, you will have access to all the equipment in your lab drawer, a 10 mL pipet, a 25 mL pipet, a 50 mL buret, and the electronic balances. Reagents include

10% by weight KI
6 M HCl
standardized thiosulfate solution (~1.0 M, but the actual concentration will be listed on the bottle)
a bottle of bleach (~6.0% by weight NaOCl, but the actual percent by weight is listed on the bottle)

Waste Disposal. Dispose of all chemical waste as directed by your instructor.

Name: _____   Date: _____

Lab Instructor: _____   Lab Section: _____

EXPERIMENT 7

Determining the Percent of Sodium Hypochlorite in Bleach

PRELABORATORY QUESTIONS

1. Balance Reactions (2) and (3) given in the introduction. Assume that these oxidation-reduction reactions occur in acidic solution. What are the oxidizing and reducing agents in Reaction (3)?

2. In order to achieve acceptable precision in this experiment, a pipet rather than a graduated cylinder should be used to measure the amount of bleach solution that will be reacted with excess iodide and acid to produce iodine, which afterward will be titrated with a standardized thiosulfate solution. Both 10 and 25 mL pipets are available to you. Answering the questions in parts (a) through (c) below will enable you to decide which pipet to use in this experiment.

   (a) How many moles of OCl^- are in 10.00 mL of 6.0% by weight bleach solution? How many moles of OCl^- are in 25.00 mL of 6.0% by weight bleach solution? Assume that the bleach solution has a density of 1.1 g/mL.

Name: _____   Date: _____

Lab Instructor: _____   Lab Section: _____

(b) How many milliliters of 1.0 M $Na_2S_2O_3$ are required to titrate 10.00 mL of 6% by weight bleach solution that has been reacted with excess iodide and acid to produce iodine? How many milliliters of 1.0 M $Na_2S_2O_3$ are required to titrate 25.00 mL of 6% by weight bleach solution that has been reacted with excess iodide and acid to produce iodine?

(c) The amount of titrant (the solution in the buret, thiosulfate in this case) to be used in each titration is determined by the desire to maximize experimental precision and not exceed the capacity of the buret. Only a 50 mL buret is available to you, so significantly less than 50.00 mL of thiosulfate should be used. Maximizing precision means using at least 10.00 mL; this will give the volume measurement four rather than three significant digits. Thus, for this experiment, between 15.00 and 40.00 mL of titrant should be used. Which should be used in this experiment to measure the bleach solution, the 10 or 25 mL pipet? Explain.

3. The amounts of KI and HCl used in this experiment are determined by the need to make OCl^- the limiting reagent. The calculations in parts (a) through (c) below will enable you to decide what amounts of KI and HCl to use in this experiment.

(a) How many moles of KI are necessary to completely react with the OCl^-? How many moles of HCl are necessary to completely react with the OCl^-? [*Remember:* The moles of OCl^- depend on your answer to question 2(c), i.e., whether you will be using 10 or 25 mL of 6.0% by weight bleach solution.]

Name: _____   Date: _____

Lab Instructor: _____   Lab Section: _____

(b) In order to be sure that you have a stoichiometric excess of KI, use twice the moles of KI necessary to completely react with OCl⁻. What volume of 10% by weight KI solution contains this number of moles? Assume that the density of the KI solution is 1.1 g/mL.

(c) In order to be sure that you have a stoichiometric excess of HCl, use twice the moles of HCl necessary to completely react with OCl⁻. What volume of 6 M HCl contains this number of moles?

4. Since the KI and HCl are present in stoichiometric excess, is it necessary to measure the amounts of these solutions by pipet, or is a graduated cylinder sufficient? Explain.

5. Look up the Material Safety Data Sheet (MSDS; **http://msds.ehs.cornell.edu/msdssrch.asp**) for sodium hypochlorite. What should you do if some bleach solution spills on your hands?

Name: _____   Date: _____

Lab Instructor: _____   Lab Section: _____

EXPERIMENT 7

Determining the Percent of Sodium Hypochlorite in Bleach

RESULTS/OBSERVATIONS

1. Mass percent of NaOCl in bleach solution as listed on the label: _____

2. Volume of bleach solution: _____ (Trial #1) _____ (Trial #2)

    _____ (Trial #3) _____ (Trial #4)

3. Concentration of $Na_2S_2O_3$ solution: _____

4. Volume of 10% KI solution: _____ (Trial #1) _____ (Trial #2)

    _____ (Trial #3) _____ (Trial #4)

5. Volume of 6 M HCl solution: _____ (Trial #1) _____ (Trial #2)

    _____ (Trial #3) _____ (Trial #4)

6. Volume of $Na_2S_2O_3$ solution: _____ (Trial #1) _____ (Trial #2)

    _____ (Trial #3) _____ (Trial #4)

7. Write any other pertinent data in the space below. Clearly indicate both the property and the amount.

Name: _____   Date: _____

Lab Instructor: _____   Lab Section: _____

8. Calculation of percent NaOCl in bleach using the data collected in trial #_____ (choose a trial).

9. Percent hypochlorite in bleach: _____ (Trial #1) _____ (Trial #2)

   _____ (Trial #3) _____ (Trial #4)

10. Mean percent hypochlorite in bleach: _____ (average of three trials)

11. Is the manufacturer stated percent NaOCl by mass in agreement with your determination? Explain.

Name: _____   Date: _____

Lab Instructor: _____   Lab Section: _____

EXPERIMENT 7

Determining the Percent of Sodium Hypochlorite in Bleach

POSTLABORATORY QUESTIONS

1. Calculate the percent by mass of NaOCl if 18.15 mL of 1.059 M $Na_2S_2O_3$ were required to titrate a 10.00 mL sample of bleach to which excess KI and HCl had been added.

2. Is a 6.0% manufacturer-stated percent by mass of NaOCl valid for the sample titrated in question 1? Explain.

3. How would using too little KI solution affect the determination of the percent by mass NaOCl; i.e., if KI were the limiting reactant, would the determined percent by mass of NaOCl be too small, too large, or unaffected? Explain.

4. How would using too little HCl solution affect the determination of the percent by mass NaOCl; i.e., if HCl were the limiting reactant, would the determined percent by mass of NaOCl be too small, too large, or unaffected? Explain.

Recycling Aluminum

OBJECTIVE

To scale down the given procedure in order to prepare a theoretical yield of 18 g of alum. The synthesis of alum from aluminum requires performing both gravity and vacuum filtrations. After being crystallized and isolated, the percent yield and melting point of the solid alum will be determined.

INTRODUCTION

Aluminum is prized for its low density, nontoxicity, resistance to corrosion, malleability, ductility, and, when alloyed with small amounts of copper, magnesium, silicon, manganese, and other elements, reasonable strength. The primary source of naturally occurring aluminum is bauxite ore ($Al_2O_3 \cdot 2H_2O$). Despite being the third most abundant element, and the most abundant metal, in the Earth's crust, aluminum was a rare and expensive metal until 1886 owing to the difficulty of extracting aluminum metal from its ore. It was in that year that Charles Martin Hall and Paul Heroult, independently, devised a practical process for the electrolytic production of aluminum. Still used today, the Hall-Heroult process consumes enormous amounts of electrical energy. In fact, electrolytic production of aluminum is the largest single consumer of electricity in the United States today.

The colossal costs of electrolytically producing aluminum provide a major impetus for its recycling. The energy costs of recycling aluminum metal, by shredding, melting, and casting, are 5–10% of those for producing metal from the ore. Recycling aluminum also would lead to a reduction of the amount that finds its way into landfills. The resistance to corrosion that aluminum is so prized for causes it to exist in landfills for many years; it is estimated that a discarded aluminum can has an average "lifetime" in the environment of greater than 100 years.

This experiment will examine the chemical, as opposed to mechanical, recycling of aluminum. The recycling of aluminum in this experiment is recycling in the general sense, to adapt to a new use or function, not the conversion of aluminum metal in one shape to aluminum metal in another. In this experiment aluminum will be converted into alum [$KAl(SO_4)_2 \cdot 12H_2O$]. *Alum* actually is a common term for a class of compounds with the general formula $MM'(SO_4)_2 \cdot 12H_2O$, where M is a monovalent cation, such as K^+, Na^+, NH_4^+, etc., and M' is a trivalent cation, such as Al^{3+}, Cr^{3+}, etc. In this experiment alum will be taken to specifically denote potassium aluminum sulfate dodecahydrate [$KAl(SO_4)_2 \cdot 12H_2O$].

Alum has many diverse industrial uses. The pulp and paper industry consumes 70% of the more than 1 million tons of alum produced annually in the United States. It is used primarily to "size" paper, a chemical process that, among other things, increases the strength and resistance to oxidative breakdown of paper. Alum is used to reduce turbidity in water destined for both human and industrial consumption. The production of soaps, greases, fire-extinguisher compounds, textiles, leather, synthetic rubber, drugs, cosmetics, cement, plastics, and pickles all use alum. Alum even can be found in the spice aisle of many grocery stores.

It needs mentioning that alum is not prepared industrially via the procedure given here. Alum is made industrially from clay because of the lower expense of such a procedure.

In this experiment you will cut a small piece of aluminum from a soft-drink can. The metal will be placed in an aqueous solution of potassium hydroxide (KOH) to yield the tetrahydroxoaluminate(III) anion [$Al(OH)_4^-$]:

$$2\ Al(s) + 2\ KOH(aq) + 6\ H_2O(l) \rightarrow 2\ K^+(aq) + 2\ Al(OH)_4^-(aq) + 3\ H_2(g) \tag{1}$$

Hydrogen gas (H_2) is also a product of the reaction. The reaction is considered to be complete when no more gas is evolved.

A small amount of sulfuric acid (H_2SO_4) then is combined with the aqueous tetrahydroxoaluminate(III) anion, producing a white, gelatinous precipitate, aluminum hydroxide [$Al(OH)_3$]:

$$Al(OH)_4^-(aq) + H^+(aq) \rightarrow Al(OH)_3(s) + H_2O(l) \tag{2}$$

When additional sulfuric acid is added to the aluminum hydroxide the soluble aluminum(III) ion (Al^{3+}) is formed:

$$Al(OH)_3(s) + 3\ H^+(aq) \rightarrow Al^{3+}(aq) + 3\ H_2O(l) \tag{3}$$

Reactions (2) and (3) release heat. As the solution is cooled, aluminum(III) ion reacts with the potassium (K^+) and sulfate (SO_4^{2-}) ions remaining in solution, and the hydrated crystalline solid alum precipitates:

$$K^+(aq) + Al^{3+}(aq) + 2\ SO_4^{2-}(aq) + 12\ H_2O(l) \rightarrow KAl(SO_4)_2 \cdot 12H_2O(s) \tag{4}$$

The net of Reactions (1)–(4) is given by the reaction

$$2\ Al(s) + 2\ KOH(aq) + 4\ H_2SO_4(aq) + 22\ H_2O(l) \rightarrow 2\ KAl(SO_4)_2 \cdot 12H_2O(s) + 3\ H_2(g) \tag{5}$$

The solid alum is washed with an ethanol-water mixture to help remove excess water and facilitate air drying of the crystals.

Your objective is to perform an experiment that will have a theoretical yield of approximately 18 g of alum. Maximizing the actual yield of alum produced via your procedure is a priority. To measure the effectiveness with which this objective is met, you will calculate the percent yield, where

$$\text{Percent yield} = \frac{\text{actual yield}}{\text{theoretical yield}} \times 100 \tag{6}$$

It is also desirous to produce alum free of contaminants. The purity of your alum crystals will be determined by performing a melting-point determination. For a pure substance melting typically occurs at a specific, characteristic temperature. The melting point of pure alum is 92.5°C. For an impure substance, however, melting will occur over a range of a few degrees, the so-called melting range. Not only will the presence of an impurity broaden the melting point of the primary component, but it also will lower the temperature at which melting begins. The nearer the experimentally determined melting point is to 92.5°C and the smaller the melting range, the higher the purity of the alum produced.

In the following you will find a procedure for the recycling of aluminum into alum. However, this procedure is for the recycling of a much greater amount of aluminum than what you will need to produce about 18 g of alum. It will be necessary for you to scale down the reaction. Moreover, the aqueous reagents you will have access to in the lab are of different concentrations than those given in the procedure. It will be necessary for you to account for this as well.

ADDITIONAL READING

Read the sections in the "Laboratory Techniques" chapter at the beginning of this Lab Manual on cleaning glassware, handling chemicals, weighing, heating liquids and solutions, crystallization, melting-point determinations, and gravity and vacuum filtration prior to performing this experiment.

 Safety Precautions:

- Protective eyewear approved by your institution must be worn at all times while you are in the laboratory.

- Flammable H_2 gas is formed in this experiment. Make sure that there are no open flames in the laboratory. Work in the fume hood. In the event of a fire, notify your instructor. A fire in a small container can be smothered by covering the vessel with a watchglass. Use a fire extinguisher to eliminate a larger fire. If the fire is burning over too large an area to be extinguished easily, evacuate the area and activate the fire alarm. If your clothing should catch on fire, use the safety shower or a fire blanket to extinguish the flames.

- Use gloves while obtaining and pouring 10.0 M H_2SO_4. Chemical burns can result when 10.0 M H_2SO_4 comes in contact with skin. If you spill 10.0 M H_2SO_4 on your skin, immediately wash the affected area with water. Continue washing with water for 15 minutes. Have a classmate notify your instructor.

- 95% ethanol is not safe for human consumption.

EXPERIMENT

You are to revise the procedure below in order to prepare a theoretical yield of 18 g of alum. The procedure below allows for the preparation of a theoretical yield much greater than 18 g of alum. It will be necessary for you to scale down the reaction. Moreover, the aqueous reagents you will have access to in the lab are of different concentrations than those given in the procedure. It will be necessary for you to account for this as well. Answering the prelaboratory questions will guide you through this process. The reagents available to you are listed in the "Equipment and Reagents" section of this experiment.

PROCEDURE

Week 1

Cut an aluminum can into one or more squares weighing a total of about 12.0 g. Cut the metal into very small strips. Determine and record the mass of the aluminum. Place the metal strips in a 600 mL beaker. In the fume hood, add 420 mL of 2.0 M KOH to the beaker, and heat gently on a hotplate until gas production ceases (about 30 minutes).

WARNING: Flammable H_2 gas is formed in this reaction. Make sure that there are no open flames in the laboratory. Work in the fume hood.

If an observable amount of water has been lost through evaporation, replace *no more* than half the evaporated water with distilled water. Use gravity filtration to remove any solid residue in the solution (see Figure 8.1).

Figure 8.1
Gravity filtration.

Rinse the residue twice with small (60 mL or less) portions of distilled water. Discard the residue. Obtain 216 mL of 10.0 M H_2SO_4 in a graduated cylinder.

CAUTION: Use gloves while obtaining and pouring 10.0 M H_2SO_4. Chemical burns can result when 10.0 M H_2SO_4 comes in contact with skin. If you spill 10.0 M H_2SO_4 on your skin, immediately wash the affected area with water. Continue washing with water for 15 minutes. Have a classmate notify your instructor.

Add about 54 mL of acid slowly, with stirring, to the clear solution in the beaker. Gently warm the solution while continuing to stir. Add the remaining acid slowly and incrementally. Gravity filter to remove any solids. Discard the residue. Pour the clear filtrate and washings into a clean flask and allow to stand without stirring until the solution gradually reaches room temperature.

Fill a large beaker with a mixture of ice and water. Chill 300 mL of a 50/50 ethanol–water mixture in this ice-bath for later use.

Fill another large beaker with a mixture of ice and water. Place the room-temperature flask containing alum solution in the ice-bath until the solution is thoroughly chilled and alum crystals form (about 10 minutes). If crystals fail to form add a seed crystal of alum or scratch the inside of the flask wall with a glass stirring rod. Crystallization is not instantaneous; wait 5–10 minutes for crystals to form. If nothing happens try adding a few drops of the chilled ethanol–water mixture. As a last resort, gently evaporate some liquid and cool the solution again.

Write your initials on the bottom of a piece of filter paper to be used in a vacuum filtration apparatus. Set the piece of filter paper on a watchglass, and weigh the two. Set up a vacuum filtration apparatus, and filter the alum crystals from solution (see Figure 8.2).

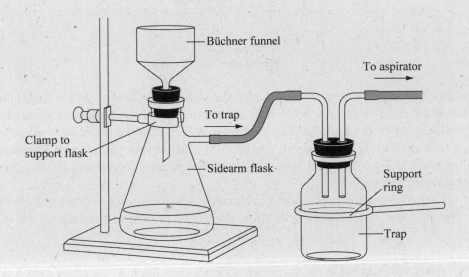

Figure 8.2
Vacuum filtration setup.

It may be necessary to filter the solution several times to effectively separate the solid and liquid mixture components. Use the chilled ethanol–water mixture to rinse remaining crystals and the walls of the beaker into the Büchner funnel containing the alum crystals. Continue to use the aspirator for about 10 minutes, air drying the crystals and filter paper. Place the filter paper and crystals on the watchglass, and place them in your desk until the following lab period.

Week 2

Weigh and record the mass of the crystals, filter paper, and watchglass.

Place alum to a depth of about 0.5 cm in the bottom of a melting-point capillary tube. Mount the capillary tube beside the bulb of a thermometer with a rubberband (see Figure 8.3).

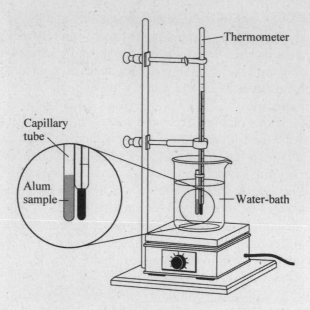

Figure 8.3
Melting point setup.

Assemble a water-bath using a beaker. Slowly heat the water in the beaker. Record the temperature range over which the solid melts. Allow the bath to cool to just below this approximate melting point. Place another sample of alum in another capillary tube. Again, heat the water bath until the solid melts, but at a slower rate this time. Record this approximate melting-point range. Repeat the heating-cooling cycle using a different alum sample each time until reproducibility is achieved. This is the melting point of the sample.

Transfer the remaining alum crystals to a container supplied by your instructor. Label the container with your name, date, and mass of alum, and give it to your instructor.

EQUIPMENT AND REAGENTS

To perform this experiment, you will have access to all the equipment in your lab drawer, one aluminum can, a hotplate, 95% ethanol, 1.4 M KOH, and 9.0 M H_2SO_4.

Waste Disposal. Dispose of all chemical waste as directed by your instructor.

Name: _____   Date: _____

Lab Instructor: _____   Lab Section: _____

EXPERIMENT 8

Recycling Aluminum

PRELABORATORY QUESTIONS

1. In the given procedure for the production of alum, what is the limiting reagent? Since water is the solvent, assume that it is present in excess.

2. How much aluminum is necessary to produce 18 g of alum by way of the given procedure? This is the amount of aluminum you should use in your procedure.

Name: _____   Date: _____

Lab Instructor: _____   Lab Section: _____

3. By what factor must the mole amount of Al in the given procedure be reduced to produce 18 g of alum?

4. What is the result of scaling the number of moles of KOH used in the given procedure by the factor determined in question 3? What volume of 1.4 M KOH contains this number of moles? This is the volume of 1.4 M KOH you should use in your procedure.

5. What is the result of scaling the number of moles of H_2SO_4 used in the given procedure by the factor determined in question 3? What volume of 9.0 M H_2SO_4 contains this number of moles? This is the volume of 9.0 M H_2SO_4 you should use in your procedure.

Name: _____   Date: _____

Lab Instructor: _____   Lab Section: _____

6. In the given procedure, a 54 mL aliquot of 10.0 M H_2SO_4 was used. Scale the moles of sulfuric acid in the 54 mL aliquot of 10.0 M H_2SO_4 by the factor determined in question 3. What volume of 9.0 M H_2SO_4 contains this number of moles? This is the size aliquot of sulfuric acid you should use in your procedure.

7. In the given procedure, 60 mL of distilled water is used to rinse the residue obtained after gravity filtration. What is the result of scaling the 60 mL sample of distilled water by the factor determined in question 3? This is the volume of distilled water you should use in your procedure.

8. What is the result of scaling the volume of the 50/50 ethanol–water mixture used in the given procedure by the factor determined in question 3? This is the volume of this mixture that you should use in your procedure.

9. Look up the Material Safety Data Sheet (MSDS; **http://msds.ehs.cornell.edu/msdssrch.asp**) of either potassium hydroxide or sulfuric acid. (*Hint:* For sulfuric acid, search under for "sulfuric acid, concentrated.") What safety precautions are necessary?

Name: _____   Date: _____

Lab Instructor: _____   Lab Section: _____

EXPERIMENT 8

Recycling Aluminum

RESULTS/OBSERVATIONS

Week 1

1. Mass of aluminum: _____

2. Volume of 1.4 M KOH: _____

3. Volume of distilled water used to rinse the residue: _____

4. Total volume of 9.0 M H_2SO_4 obtained: _____

5. Volume of initial aliquot of 9.0 M H_2SO_4 used: _____

6. Volume of 50/50 ethanol–water mixture: _____

7. Mass of filter paper and watchglass: _____

8. Calculation of the theoretical yield of alum.

Name: _____   Date: _____

Lab Instructor: _____   Lab Section: _____

Week 2

9. Mass of crystals, filter paper, and watchglass: _____

10. Melting temperature range: _____ (Trial #1) _____ (Trial #2)

    _____ (Trial #3) _____ (Trial #4)

11. Mass of crystals: _____

12. Percent yield of alum: _____

Name: _____   Date: _____

Lab Instructor: _____   Lab Section: _____

EXPERIMENT 8

Recycling Aluminum

POSTLABORATORY QUESTIONS

1. Why were gravity filtrations performed in this experiment?

2. Why was vacuum filtration performed in this experiment?

3. Two different students performed this experiment. The first student's percent yield of alum was 87.9%, and the second student's was 76.1%. Which student better achieved the aim of maximizing his or her yield? Explain.

4. Two different students performed this experiment. The first student's alum sample had a melting-point range of 85.3–9.6°C. The second student's alum sample had a melting-point range of 90.8–91.9°C. Which student synthesized alum of higher purity? Explain.

EXPERIMENT 9
Chemical Reactions

OBJECTIVE

To develop observation skills by observing chemical reactions of a variety of types. To use observations of physical changes to construct balanced chemical equations for reactions. To learn some basic laboratory techniques, such as gravity filtration, heating of liquids and solutions, and gas collection via water displacement. To learn some general characteristics of the various classes of chemical reactions.

INTRODUCTION

Chemistry is an experimental science. Performing tests and observing the results are inherent parts of its study. Developing the ability to make keen, careful observations is a necessity.

In this experiment you will observe a large number of chemical reactions. As you perform the reactions, observe carefully what happens, and make a complete record of your observations. Record your observations as you observe them. Don't wait until later; you might forget what happened. Noteworthy observations include whether a gas is produced, a color change occurs, heat is absorbed (testtube cools off) or produced (testtube warms up), or if a precipitate (solid) forms. It is these kinds of changes that signal whether a reaction is occurring—and in some cases what type of reaction is occurring.

In this experiment you will not have time to carry out chemical reactions of every known general type, but there will be time for you to sample a number of reactions representing some of the most important and interesting classes of chemical reactions, including

(Part A) Metathesis (double-displacement) and complexation reactions
(Part B) Decomposition reactions
(Part C) Combination reactions
(Part D) Single-displacement reactions
(Part E) Other oxidation-reduction reactions
(Part F) Dissolution reactions

Your purpose in this experiment is to observe chemical reactions of various types and summarize the chemical changes in terms of balanced chemical equations. There are six parts to this experiment, Parts A–F. Each part deals with a different class of chemical reactions. More specific details about each class of reaction is provided in each section. Depending on the length of your lab period, you will have either two or three weeks to complete this experiment.

ADDITIONAL READING

Read the sections in the "Laboratory Techniques" chapter at the beginning of this Lab Manual on cleaning glassware, handling chemicals, weighing solids, heating liquids and solutions, Bunsen burners, gas collection by water displacement, and decantation prior to performing this experiment.

Part A: Metathesis Reactions and Complexation Reactions

Metathesis reactions, also known as *double-displacement reactions,* are those in which the reactants exchange partners with each other:

$$AB + CD \rightarrow AD + CB \tag{1}$$

Importantly, metathesis reactions occur without oxidation or reduction taking place. Often these reactions occur in aqueous solution. Two of the most common types of metathesis reactions are double-displacement reactions between salts and acid-base neutralization reactions.

Complexation reactions are similar to metathesis reactions in that they do not involve oxidation or reduction, and they often consist of replacement of one species by another. They are, however, fundamentally different from metathesis reactions because they involve a special type of compound called a *complex* or *coordination compound.* The nature of the bonding in complexes is quite different from that in compounds containing ordinary covalent bonds. In a complex, the shared electrons in the covalent bond come from only one of the species involved. To form a complex, a number of anions or neutral molecules (termed *ligands*) combine with a central ion; the ligands provide the electron pairs necessary for bonding. Once formed, a complex can react with other ligands to form new complexes.

More specific characteristics of double-displacement reactions between salts, acid-base neutralization reactions, and complexation reactions follow.

Double-Displacement Reactions between Salts

When double-displacement reactions between salts in aqueous solution occur they often lead to the formation of an insoluble solid called a *precipitate;* because of this, these reactions are often called *precipitation reactions.* For example, aqueous potassium chloride (KCl) and aqueous silver nitrate ($AgNO_3$) react to produce solid silver chloride (AgCl) and aqueous potassium nitrate (KNO_3), which remains as hydrated ions in solution:

$$K^+(aq) + Cl^-(aq) + Ag^+(aq) + NO_3^-(aq) \rightarrow AgCl(s) + K^+(aq) + NO_3^-(aq) \tag{2}$$

The potassium and nitrate ions are not active participants in the reaction; they are called *spectator ions.*

The following solubility rules usually are sufficient for predicting the aqueous solubilities of salts:

1. *A compound is probably soluble if it contains one of the following* cations:

   ■ Group 1A cations: Li^+, Na^+, K^+, Rb^+, Cs^+
   ■ Ammonium ion: NH_4^+

2. *A compound is probably soluble if it contains one of the following anions:*

- Halides: Cl^-, Br^-, I^-, *except Ag^+, Hg_2^{2+}, and Pb^{2+} compounds*
- Nitrate (NO_3^-), perchlorate (ClO_4^-), acetate ($CH_3CO_2^-$), sulfate (SO_4^{2-}), *except Ba^{2+}, Hg_2^{2+}, and Pb^{2+} sulfates.*

A compound that does *not* contain one of the ions listed above is probably *not* soluble.

Acid-Base Neutralization Reactions

An acid-base neutralization reaction leads to the formation of a salt and water. For example, combination of hydrochloric acid (HCl) and sodium hydroxide (NaOH) yields sodium chloride (NaCl) and water:

$$H^+(aq) + Cl^-(aq) + Na^+(aq) + OH^-(aq) \rightarrow Na^+(aq) + Cl^-(aq) + H_2O(l) \tag{3}$$

Protons (H^+) from the HCl vigorously remove virtually all the hydroxide ions (OH^-) from solution by combining to produce water molecules. The sodium (Na^+) and chloride (Cl^-) ions are spectators. Incidentally, it is better to think of H^+ ions as protons with water molecules bound to them than as bare protons. These hydrated protons are called *hydronium ions;* they are frequently written as H_3O^+ in order to emphasize the fact that they are hydrated. Hydrated molecules or ions, such as Na^+, OH^-, and NH_3, are surrounded by water molecules in aqueous solution. Often the water molecules are arranged in a specific geometry around the central ion, forming a complex in which water molecules are the ligands. The Al^{3+} ion, for example, exists as six water molecules attached to the central aluminum ion in an octahedral geometry (see Figure 9.1). Thus Al^{3+} is really better described as $Al(H_2O)_6^{3+}$.

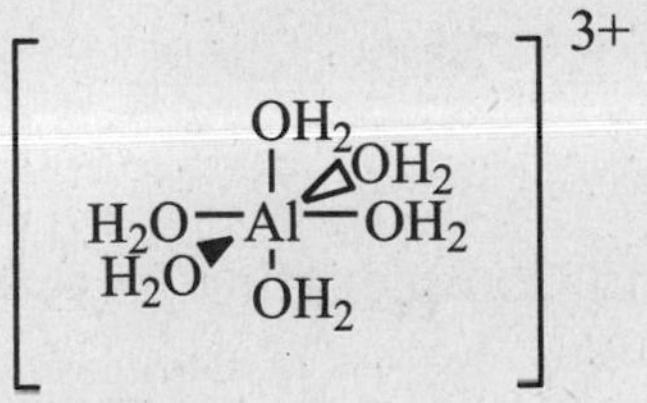

Figure 9.1
The octahedral arrangement of H_2O molecules about Al^{3+}.

Complexation Reactions

Reactions involving complexes are numerous and important. One of the fundamental processes of photography, for example, involves removal of unexposed silver bromide (AgBr) from film or photographic paper by washing it with a solution of sodium thiosulfate ($Na_2S_2O_3$), commonly known as *hypo*. Silver from the solid AgBr combines with thiosulfate ($S_2O_3^{2-}$) to form a soluble complex ion [$Ag(S_2O_3)_2^{3-}$], releasing bromide ions (Br^-) into solution:

$$AgBr(s) + 4\,Na^+(aq) + 2\,S_2O_3^{2-}(aq) \rightarrow Ag(S_2O_3)_2^{3-}(aq) + 4\,Na^+(aq) + Br^-(aq) \tag{4}$$

In this example, in which the ligands ($S_2O_3{}^{2-}$) are anions, the complexation reaction is similar to an ordinary double-displacement metathesis reaction. When neutral ligands such as NH_3 or H_2O are involved, however, the reactions are better described as *single-displacement reactions*. For example, ammonia can displace the chloride ion in solid zinc chloride ($ZnCl_2$) to give a complex ion:

$$ZnCl_2(s) + 4\,NH_3(aq) \rightarrow Zn(NH_3)_4{}^{2+}(aq) + 2\,Cl^-(aq) \qquad (5)$$

 Safety Precautions:

■ Protective eyewear approved by your institution must be worn at all times while you are in the laboratory.

■ $BaCl_2$ is quite toxic. Be sure to wash your hands after handling it.

■ Chemical burns can result when 6 M HCl comes in contact with skin. If you spill 6 M HCl on your skin, immediately wash the affected area with water. Continue washing with water for 15 minutes. Have a classmate notify your instructor.

■ NaOH is especially caustic to the eyes. If NaOH gets in your eyes, immediately wash the affected area with water in an eyewash fountain. Continue washing with water for 15 minutes. Have a classmate notify your instructor.

■ Avoid breathing in too much ammonia vapor. It is quite irritating to mucous membranes. If you spill 6 M NH_3 on your skin, immediately wash the affected area with water. Continue washing with water for 15 minutes. Have a classmate notify your instructor.

PROCEDURE

Part A: Metathesis Reactions and Complexation Reactions

To make the best use of time and materials during Part A of this experiment, different students will begin at different points. Your instructor will tell you which of the following three sequences to follow:

1. Start with A-1; then do A-2, A-3, A-4, A-5, A-6, and A-7.

2. Start with A-3; then do A-4, A-5, A-6, A-7, A-1, and A-2.

3. Start with A-4; then do A-5, A-6, A-7, A-1, A-2, and A-3.

Necessary reagents are available in bottles on the reagent shelves. Be sure to use distilled water, never tap water, throughout the experiment. All reaction products must be disposed of properly. Follow the waste disposal instructions and use the waste containers provided.

As you work on this experiment, please do your part to keep things neat and organized. Return reagent bottles immediately after use to their proper location, or other students will not be able to find what they need.

For each reaction you carry out throughout this experiment you are to record the following:

1. A physical description (color, texture, etc.) of all reactants and products.

2. A description of any changes in color, temperature, or physical state that occur.

A-1: Magnesium Sulfate + Barium Chloride

Barium chloride reacts with magnesium sulfate to give solid barium sulfate, so this reaction is an example of a precipitation reaction. You might know someone who has ingested a "barium milkshake", an emulsion of barium sulfate in water, before being X-rayed. Barium sulfate is used to coat the gastrointestinal tract before giving X-rays of that area because it is opaque to X-rays.

In a small testtube add 5 drops of 0.1 M $MgSO_4$ solution to 5 drops of 0.1 M $BaCl_2$ solution and mix the contents. Record your observations. Explain your observations using balanced chemical equations.

CAUTION: $BaCl_2$ is quite toxic. Be sure to wash your hands after handling it.

Waste Disposal (A-1). Dispose of the products as directed by your instructor.

A-2: Sodium Carbonate + HCl

CAUTION: Chemical burns can result when 6 M HCl comes in contact with skin. If you spill 6 M HCl on your skin, immediately wash the affected area with water. Continue washing with water for 15 minutes. Have a classmate notify your instructor.

The common name of sodium carbonate (Na_2CO_3) is *washing soda.* Added to laundry, it improves the effectiveness of soap or detergent in breaking down grease. Place 2 mL of saturated Na_2CO_3 solution in a small testtube and add 5 drops of 6 M HCl. One of the products is water and one is NaCl. Carbon dioxide (CO_2) is also produced. Record your observations. Explain your observations using balanced chemical equations.

Waste Disposal (A-2). Dispose of the products as directed by your instructor.

A-3: Acetic Acid + Sodium Hydroxide

Acetic acid (CH_3CO_2H) is the main constituent (other than water) of vinegar. Sodium hydroxide, sometimes called *lye,* is an active ingredient (along with some aluminum chunks) in Drano^{TM}. The products of the reaction between acetic acid and NaOH are water and sodium acetate (CH_3CO_2Na).

CAUTION: NaOH is especially caustic to the eyes. If NaOH gets in your eyes, immediately wash the affected area with water in an eyewash fountain. Continue washing with water for 15 minutes. Have a classmate notify your instructor.

Obtain approximately 2 mL of standardized 0.150 M acetic acid solution and about 2 mL of the NaOH solution of unknown molarity in two separate, small, clean testtubes. Clean and rinse with distilled water a third and a fourth small testtube and two Pasteur pipets. You will use the pipets to perform two "minititrations." In the minititrations drops are treated just like you would treat milliliters in a regular titration with a buret. Place 10 drops of acetic acid solution and 2 drops of phenolphthalein acid-base indicator in the third testtube. Note the color of the indicator; it should be colorless. To the acetic acid solution, add the sodium hydroxide solution, one drop at a time, counting each drop, until the indicator changes from colorless to pink. After the addition of each drop you should tap the testtube a bit on the side to mix the contents thoroughly. At the end point (color change) all the H^+ from the acetic acid should be neutralized by the OH^- from the sodium hydroxide, forming water. Note the total number of drops of standardized NaOH solution required.

Place 10 drops of the unknown NaOH solution and 2 drops of phenolphthalein in the fourth testtube. To the NaOH solution, add the standardized acetic acid dropwise until the NaOH has been exactly neutralized (a color change from pink to colorless). Record the number of drops of CH_3CO_2H solution required. Determine the balanced chemical equation. Calculate the molarity of the sodium hydroxide as determined by each minititration.

Waste Disposal (A-3). Dispose of the products as directed by your instructor.

A-4: Sodium Chloride + Silver Nitrate

Sodium chloride is ordinary table salt. Silver nitrate is a mild antiseptic sometimes used to disinfect the eyes of newborn infants.

In a small testtube add 5 drops of 0.1 M NaCl solution to 5 drops of 0.1 M $AgNO_3$ solution, mix the contents by tapping the testtube briskly from the side, and record what happens. Explain your observations using balanced chemical equations.

Save the solid AgCl for Part A-5.

Obtain about 1 mL of 6 M NH_3 (often called 6 M NH_4OH) in a small testtube, and take it to your laboratory station for use in Parts A-5 to A-7.

CAUTION: Avoid breathing in too much ammonia vapor. It is quite irritating to mucous membranes. If you spill 6 M NH_3 on your skin, immediately wash the affected area with water. Continue washing with water for 15 minutes. Have a classmate notify your instructor.

A-5: Silver Chloride + Ammonia

Ammonia gas is used widely as a fertilizer; aqueous ammonia is used as a cleaning solution.

In Part A-4 you made solid AgCl. Silver chloride reacts with aqueous ammonia to form a soluble complex ion $[Ag(NH_3)_2^+]$. Silver does not, however, form a complex with ammonium ion (NH_4^+).

Get 1 mL of 3 M HNO_3 in a small testtube, and take it to your work area. To the solid AgCl prepared in Part A-4, add 6 M NH_3 dropwise, tapping the testtube from the side after each drop to mix the contents, until the AgCl dissolves completely. Then add 3 M HNO_3 dropwise until something happens (*Hint:* $NH_3 + H^+ \rightarrow NH_4^+$). Record your observations. Explain your observations using balanced chemical equations.

Waste Disposal (A-5). Dispose of the products as directed by your instructor.

A-6: Copper Sulfate + Sodium Hydroxide

Copper sulfate, a fungicide and algicide, is used to prevent fungi from attacking grapes and to kill algae growing in ponds.

Using the dropper bottle containing 0.100 M $CuSO_4$ solution, squeeze exactly 10 drops of the $CuSO_4$ solution into a small testtube. Then, to that same testtube, add 10 drops of 0.100 M NaOH solution from the dropper bottle provided, mix for 5 seconds with your stirring rod, wait for 90 seconds, and record what happens. The solid formed is copper hydroxide $[Cu(OH)_2]$. Describe it. Is it gelatinous or crystalline? Is it blue or white? Explain your observations using balanced chemical equations.

Save the copper hydroxide for Part A-7.

A-7: Copper Hydroxide + Ammonia (and Several Related Reactions)

There are three steps for this sequence of reactions:

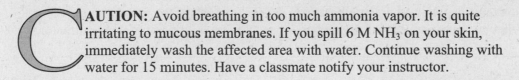

CAUTION: Avoid breathing in too much ammonia vapor. It is quite irritating to mucous membranes. If you spill 6 M NH_3 on your skin, immediately wash the affected area with water. Continue washing with water for 15 minutes. Have a classmate notify your instructor.

Step 1: Using a Pasteur pipet, add 2 drops of 6 M NH_3 to the $Cu(OH)_2$ precipitate from Part A-6 in its testtube, and stir with your glass stirring rod until the precipitate dissolves completely. This should take between 20 seconds and a minute. Explain your observations with balanced chemical equations.

Step 2: Obtain approximately 5 mL of 0.500 M HNO_3 (nitric acid) in your 50 mL beaker. With a clean Pasteur pipet add 4 drops of 0.500 M HNO_3 to the solution in the testtube and stir with your stirring rod for 5 seconds. Add 2 drops of 0.500 M HNO_3 and stir again. Now add 0.500 M HNO_3 one drop at a time, stirring after each drop, until the solution changes color to a cloudy light blue, indicating the presence of a precipitate. (*Hint:* Each drop makes a difference. About 10 drops of 0.500 M HNO_3 may be required.) Record your observations. Explain your observations with balanced chemical equations.

Step 3: Add a few more drops of 0.500 M HNO_3 until the precipitate dissolves; the solution should clear; i.e., there should be no cloudiness. Record your observations. Explain your observations using balanced chemical equations.

Waste Disposal (A-7). Dispose of the products as directed by your instructor.

Part B: Decomposition Reactions

Many compounds decompose when heated to a high enough temperature. Clearly, these are endothermic reactions. The heat (represented by a Δ over the "yields" arrow) provides the energy to necessary for breaking bonds. Sometimes, but not always, the reactions involve oxidation and reduction. Often, gaseous products, such as O_2, H_2O, or CO_2, are generated. For example, mercuric oxide, when heated, decomposes to form mercury metal and oxygen gas. (Incidentally, it was this reaction that led to the discovery of oxygen by Joseph Priestley in 1774.)

$$2\ HgO(s) \xrightarrow{\Delta} 2\ Hg(l) + O2(g)\ 2\ HgO(s) \longrightarrow 2\ Hg(l) + O_2(g) \tag{6}$$

In this reaction, mercury gets reduced from +2 in HgO to zero in metallic mercury, whereas oxygen changes oxidation state from –2 in HgO to zero in O_2.

   Safety Precautions:

■ Protective eyewear approved by your institution must be worn at all times while you are in the laboratory.

■ The unknown solid material you will try to dissolve in Part B-2-iii (Na_2CO_3 or NaOH) is very caustic to the eyes. Do not heat the testtube with your burner because that might cause the solution to be expelled from the tube. If you should spill any of the substance on yourself, immediately remove any affected clothing, rinse with water, and notify your instructor.

To make the best use of time and materials during Part B of this experiment, different students will begin at different points. Your instructor will tell you which of the following two sequences to follow:

1. Start with B-1; then do B-2, B-3, and B-4.

2. Start with B-2; then do B-3, B-1, and B-4.

B-1: Dehydration of Copper Sulfate Pentahydrate

Place about 0.5 g of solid $CuSO_4 \cdot 5H_2O$ in a crucible (just enough to cover the bottom), and support the crucible on a wire triangle on a ring on your ringstand. Gently heat the copper sulfate pentahydrate with your Bunsen burner for about a minute, quickly withdrawing the heat if it starts to pop. Then heat it strongly for a few additional minutes until the bottom of the crucible glows red. Record your observations. Explain your observations using balanced chemical equations.

Once the solid has cooled a bit, add some distilled water dropwise. Record your observations. Explain your observations using balanced chemical equations.

Be sure to wait for everything to cool down before you disassemble the setup lest you burn yourself on the hot crucible or ringstand.

Waste Disposal (B-1). Dispose of the products as directed by your instructor.

B-2: Decomposition of Sodium Bicarbonate

Sodium bicarbonate ($NaHCO_3$) is baking soda. When heated, 2 mol of it decomposes to give 1 mol of CO_2 gas, 1 mol of water vapor, and 1 mol of a white solid. In this part of the experiment, your task is to determine whether that solid is $NaOH$ or Na_2CO_3.

Place about 10 mL of nearly saturated $Ca(OH)_2$ solution in a medium-sized testtube. Support a funnel in a ring attached to a ringstand, and get a piece of qualitative filter paper.

Fold the filter paper into quarters, tear off one corner, and place it in the funnel as shown in Figure 9.2.

(*Normally at this point the next step would be to wet the filter paper with distilled water in order to seal it to the funnel; this step will be skipped, however, because that would dilute the $Ca(OH)_2$ solution.*)

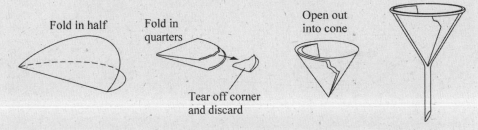

Figure 9.2
Fold the filter paper in half. Fold it in quarters. Tear off a corner. Open the side of the filter paper that has not been torn to make a cone. Fit this cone of filter paper snugly into the funnel.

With one hand, hold an empty small testtube under the funnel, and with your other hand, pour the $Ca(OH)_2$ solution into the filter paper in the funnel. You should be able to collect about 6–7 mL of the filtered solution, which will be used both in the next two steps of Part B-2 and in Part B-3. Transfer one-third of the filtered $Ca(OH)_2$ solution into each of two other clean small testtubes, and place a stopper in each of those other testtubes. The first testtube of $Ca(OH)_2$ will be used in the first experiment of this part (B-2-i), one of the stoppered testtubes will be used in the second experiment of this part (B-2-ii), and the other stoppered testtube will be used in part B-3. The material remaining behind in the filter paper will be discarded. [The reason for filtering the solution is that $Ca(OH)_2$ reacts with CO_2 in the air to form solid $CaCO_3$, which turns the solution cloudy.]

To get rid of the filter paper and contents, being careful not to get caustic $Ca(OH)_2$ on your hands, fill the funnel with tap water, and pour the contents into the your largest beaker. Then remove the filter paper and hold it in your hand to rinse it under the tap into the sink. Finally, squeeze it out and dispose of it in the plastic waste basket.

i.   Obtain an assembly consisting of a one-holed stopper containing a bent piece of glass tubing. Place about 1 g of sodium bicarbonate solid (about 1 cm deep in the bottom of the testtube) in a 16 mm testtube, stopper the testtube with the one-holed stopper containing the piece of glass tubing, and clamp it about halfway up your ring stand (see Figure 9.3). *Note: Do not use a plastic-coated buret clamp. The plastic will melt and catch fire when you heat the testtube with your gas burner.*

Heat the solid $NaHCO_3$ strongly in the direct flame of your Bunsen burner for at least 1 minute, but don't heat it so strongly that the bottom of the testtube melts. As you are heating it, dip the end of the glass tubing into the filtered $Ca(OH)_2$ solution so that any gas produced will bubble through the solution, being careful to withdraw the glass tubing from the $Ca(OH)_2$ solution *before* you stop heating [otherwise, as the testtube cools, the $Ca(OH)_2$ solution will be sucked back into the glass tubing].

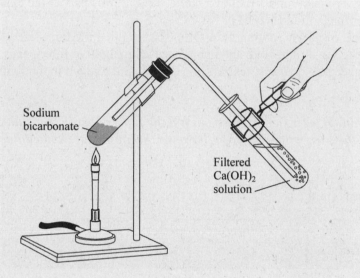

Sodium bicarbonate

Filtered $Ca(OH)_2$ solution

Figure 9.3
Decomposition of sodium bicarbonate setup: The $NaHCO_3$ is decomposed by heating. The gas produced by the decomposition of $NaHCO_3$ is bubbled through a solution of $Ca(OH)_2$.

Record what happens. If a precipitate of $CaCO_3$ forms, the $Ca(OH)_2$ solution will turn cloudy:

$$Ca(OH)_2(aq) + CO_2(g) \rightarrow CaCO_3(s) + H_2O(l) \tag{7}$$

Continuing to bubble CO_2 through the $Ca(OH)_2$ solution eventually should cause the solid $CaCO_3$ to dissolve because CO_2 reacts with water according to the reaction

$$CO_2(g) + H_2O(l) \longrightarrow H^+(aq) + HCO_3^-(aq) \tag{8}$$

and the H^+ so generated should react with solid $CaCO_3$ to produce Ca^{2+} and HCO_3^-. Record all your observations. Explain your observations using balanced chemical equations.

ii. Repeat the experiment you just carried out with a fresh sample of NaHCO₃ and using the stoppered Ca(OH)₂ solution you filtered earlier, except this time add 4 drops of bromothymol blue acid-base indicator to the Ca(OH)₂ solution before heating the testtube containing the solid NaHCO₃. Bromthymol blue is blue to purple in basic, yellow in acidic, and green in neutral solution. Note the initial color of the solution, and also note any color changes that occur while the reaction is underway. Record your observations. Explain your observations using balanced chemical equations.

iii. After the testtube in which you have been heating the NaHCO₃ cools to just warm to the touch, add about 2 mL of distilled water to the testtube, and break up the solid with a stirring rod in order to get some of the solid to dissolve.

CAUTION: The solid material you are trying to dissolve (Na₂CO₃ or NaOH) is very caustic to the eyes. Do not heat the testtube with your burner because that might cause the solution to be expelled from the tube. If you should spill any of the substance on yourself, immediately remove any affected clothing, rinse with water, and notify your instructor.

Add 2 drops of bromothymol blue indicator to the testtube containing the partially dissolved solid. Both Na₂CO₃ and NaOH dissolve to give a basic solution. If you get a result that suggests that the white solid is either NaOH or Na₂CO₃, devise a simple test that will allow you to tell which of the two it is. After checking with your instructor to make sure that your method is safe, carry out the test. Record all your observations. Explain your observations using balanced chemical equations.

Before returning the one-holed stopper containing the bent glass tubing to your instructor, remove any precipitates that are adhering to the tubing. Stubborn precipitates can be dissolved with a few drops of 6 M HCl solution. Be sure to rinse the tubing with distilled water.

Waste Disposal (B-2). Dispose of the products as directed by your instructor.

B-3: Decomposition of Ammonium Carbonate

Ammonium carbonate is sometimes used as smelling salts. When heated it decomposes to give off three different gases, one of them being water:

$$(NH_4)_2CO_3(s) \xrightarrow{\Delta} H_2O(g) + \underline{\hspace{1cm}}(g) + \underline{\hspace{1cm}}(g) \quad\quad (9)$$
(smelling salts)

Your task is to determine the identities of the other two gases.

Place 1.0 g of (NH₄)₂CO₃ in each of two different 16 mm testtubes, and place 1 mL of 0.10 M CuSO₄ in a third testtube. You also will need the testtube containing at least 1 mL of nearly saturated Ca(OH)₂ solution filtered earlier in part B-2.

To one of the tubes containing ammonium carbonate, attach a one-holed stopper containing a bent piece of glass tubing (see Figure 9.3). Dip the glass tubing extending from the testtube containing ammonium carbonate into the CuSO₄ solution, and warm the ammonium carbonate *gently* (just "tickle it" with the flame) with your Bunsen burner. *Be sure to pull the glass tubing out of the CuSO₄ solution before you remove the flame; otherwise, the liquid will be sucked back into the testtube containing the solid (NH₄)₂CO₃ as the testtube cools.* Continue heating the ammonium carbonate until it has decomposed completely; it shouldn't take longer than a few minutes. One of the gases produced in the decomposition should react with the CuSO₄ solution. Because this gas is basic, it will react first to produce OH⁻ ions in the solution, which will react with the CuSO₄ to give solid Cu(OH)₂. Then the gas should continue to react with the copper solution to give a dark-blue complex. (You have seen this copper complex in a previous part of the experiment.) Record your observations. Explain your observations using balanced chemical equations.

Next, rinse the glass tubing with distilled water, and connect it to the second testtube containing ammonium carbonate. Dip the glass tubing into the calcium hydroxide solution filtered in Part B-2. Warm the ammonium carbonate gently, as before, and note what happens in the solution (faint cloudiness would indicate the formation of a precipitate). Again, you should have seen this product before. What gas should react with the Ca(OH)₂ to give a precipitate? Does the precipitate eventually dissolve? Record your observations. Explain your observations using balanced chemical equations.

Waste Disposal (B-3). Dispose of the products as directed by your instructor. Clean the glass tubing and stopper well, and return them to their proper place.

B-4: Decomposition of Hydrogen Peroxide—Preparation of Oxygen

Set up the gas-collection apparatus shown in Figure 9.4.

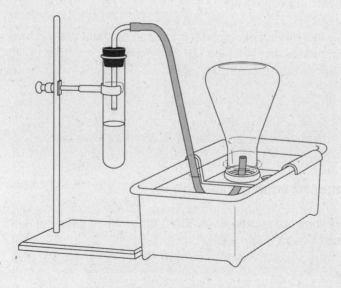

Figure 9.4
Gas collection by water displacement.

You will need a one-holed stopper that fits a 25 mm testtube and a piece of glass tubing shaped as shown in Figure 9.4. The tubing should be inserted into the rubber stopper and the rubber stopper inserted into the 25 mm testtube. You also will need an Erlenmeyer flask, a large stopper that fits the Erlenmeyer flask, and a pneumatic trough, which you should fill with enough water that the water level is above the level of the shelf. Place the overflow spout so that water will drip from the pneumatic trough into the sink. The Erlenmeyer flask should be filled completely with tap water, covered with a large stopper or small watch glass, and placed in an inverted position in the water in the pan. Then the stopper or watch glass should be removed, and the inverted flask should be supported on the shelf by placing the top (the open end) of the flask above the hole in the shelf. Try to keep air from getting into the flask. Finally, the glass tubing connected to the testtube should be placed so that it extends under the shelf and into the Erlenmeyer flask as far as possible.

The apparatus is designed to collect oxygen by water displacement. You will produce oxygen in the testtube. The oxygen, being less dense than water, will push water out of the inverted Erlenmeyer flask. The oxygen will be generated by adding some potassium iodide (KI) to an aqueous 3% H_2O_2 solution in the testtube. The KI will catalyze the decomposition, allowing it to proceed at a faster rate than it would otherwise. Because it is a catalyst, however, KI is not shown as a reactant or product in the balanced equation. Catalysts normally are shown above the "yields" arrow:

$$2 H_2O_2(aq) \xrightarrow{\text{KI}} O_2(g) + 2 H_2O(g) \tag{10}$$

Remove the rubber stopper from the top of the testtube, place 50 mL of 3% H_2O_2 and 2 g of solid KI in it, and replace the rubber stopper. You should see bubbles forming immediately. Soon oxygen should begin displacing water from the Erlenmeyer flask. When nearly all the water has been displaced from the flask remove it from the pan, and immediately seal it with a rubber stopper or glass plate.

Oxygen supports combustion. Putting out a fire by smothering it, in chemical terms, amounts to depriving the reaction of oxygen. The production of oxygen in this reaction will be tested for by observing the effect of oxygen on a combustion reaction.

Obtain a wooden splint. Use a match to light one end of the wooden splint. Blow out the flame, and quickly insert the still glowing splint into the Erlenmeyer flask containing oxygen gas. Observe what happens to the wooden splint.

Waste Disposal (B-4). Dispose of the products as directed by your instructor.

Part C: Combination Reactions

There are two common types of combination reactions, both involving oxidation and reduction: nonmetal plus nonmetal and metal plus nonmetal.

Nonmetal Plus Nonmetal

Carbon (graphite), a nonmetal, reacts with sulfur, another nonmetallic solid, to give carbon disulfide, a liquid, plus heat:

$$C(s) + 2 S(s) \rightarrow CS_2(l) \tag{11}$$

Carbon goes from zero to +4 in oxidation state, and sulfur goes from zero to –2.

Metal Plus Nonmetal

Strontium metal reacts strongly with oxygen to give strontium oxide plus a great deal of heat energy:

$$2 \, Sr(s) + O_2(g) \rightarrow 2 \, SrO(s) \tag{12}$$

Strontium goes from an oxidation state of zero in Sr metal to +2 in SrO. Electrons are transferred from the strontium metal to the oxygen, so strontium is the reducing agent and oxygen is the oxidizing agent. Oxygen goes from an oxidation state of zero in O_2 to –2 in SrO. Oxidation-reduction reactions are also known as *redox reactions* or *electron-transfer reactions*. The bonds formed in SrO are strong enough to break the bonds in both strontium and oxygen and still have excess energy remaining.

It is characteristic of metallic oxides that if they react with water, they do so to give basic solutions. Strontium oxide, for example, reacts with water to give strontium hydroxide:

$$SrO(s) + H_2O(l) \rightarrow Sr^{2+}(aq) + 2 \, OH^-(aq) \tag{13}$$

Nonmetallic oxides, conversely, react with water to give acidic solutions. For example,

$$SO_3(g) + H_2O(l) \rightarrow H_2SO_4(aq) \tag{14}$$

    Safety Precautions:

- Protective eyewear approved by your institution must be worn at all times while you are in the laboratory.

- Protective welder's glasses must be worn while burning magnesium because it gives off a great deal of ultraviolet light that can be quite damaging to the eyes.

Part C: Combination Reactions

C-1: Magnesium + Air (N_2 and O_2)

Obtain a piece of magnesium ribbon about 3 in long. Take it, together with your evaporating dish and your crucible tongs, to the hood where there is a special setup (a black metal box) for burning magnesium. Put on the welder's glasses.

CAUTION: Burning magnesium gives off a great deal of ultraviolet light.

Working inside the black metal box in order to shield your classmates from the light, hold the magnesium ribbon with your crucible tongs a few inches above a clean evaporating dish, and set one end of the ribbon on fire with a burner. You will see that the magnesium burns with a very hot, bright flame. Magnesium gives off so much heat in its reaction with oxygen that nitrogen molecules, which normally are too stable to react, break apart into atoms and combine with the magnesium to form some Mg_3N_2 in addition to the MgO that forms. Record your observations of burning magnesium. Explain your observations using balanced chemical equations.

After the magnesium has reacted, allow the residue, which consists of a mixture of MgO and Mg_3N_2, to drop into the evaporating dish. Carry everything back to your regular laboratory desk. Pour 2 mL of distilled water into the dish, and add 4 drops of bromothymol blue indicator (blue in basic, yellow in acidic, and green in neutral solution). Record your observations. MgO reacts with water to give $Mg(OH)_2$. Mg_3N_2 reacts with water to give NH_3 and $Mg(OH)_2$. Provide balanced chemical equations for these reactions.

Waste Disposal (C-1). Dispose of the products as directed by your instructor.

Part D: Single Displacement Reactions

Aluminum metal reacts with Fe_2O_3, an oxide of iron, to give Al_2O_3 and iron metal, plus a great deal of heat:

$$2\ Al(s) + Fe_2O_3(s) \rightarrow Al_2O_3(s) + 2\ Fe(s) \tag{15}$$

This reaction, called the *thermite reaction*, gives off so much heat that it has been used both to weld railroad tracks together and to detonate bombs. It is a redox reaction. Aluminum is oxidized from zero to +3, and iron is reduced from +3 to zero in oxidation state. A metal that can displace another metal from a compound is said to be more *active* than the other metal. By carrying out a series of displacement reactions it is possible to assemble an *activity series*, a list of metals arranged such that each metal in the list will displace any metal below it from its compounds. On such a list aluminum would be higher than iron.

After completing all the reactions in Part D, rank the four metals examined, Ag, Ca, Cu, and Zn, according to their activities.

Note: Set up reactions D-1, D-2, and D-3 to be run simultaneously. While you are waiting for these three reactions to take place, which will take quite a bit of time, proceed with Parts D-4 to D-6. If reagents for Part D are in use, it might be advisable to skip to Part E and return later to work on Part D.

 Safety Precautions:

■   Protective eyewear approved by your institution must be worn at all times while you are in the laboratory.

■   Chemical burns can result when 6 M HCl comes in contact with skin. If you spill 6 M HCl on your skin, immediately wash the affected area with water. Continue washing with water for 15 minutes. Have a classmate notify your instructor.

D-1: Copper + Silver Nitrate

Sand the surface of a piece of copper wire to remove surface oxides. Coil it around a pencil, remove it from the pencil, and put it in a small testtube. Half-fill the testtube with 0.1 M $AgNO_3$ solution, allow it to sit for 30 minutes, and record your observations. Explain your observations using balanced chemical equations. If it reacts, solid Ag should form on the surface of the wire and $Cu(NO_3)_2$ should be produced in the solution. If you wish, you can test for the formation of $Cu(NO_3)_2$ by adding some 6 M NH_3 to the solution and looking for the characteristic dark-blue color of $Cu(NH_3)_4^{2+}$.

Waste Disposal (D-1). Dispose of the products as directed by your instructor.

D-2: Copper + Zinc Sulfate

Sand, coil, and place another piece of copper wire in a small testtube. Half-fill the testtube with 0.1 M $ZnSO_4$ solution, and allow the mixture to sit for 30 minutes. Record your observations. Explain your observations using balanced chemical equations. If it occurs, a reaction would produce aqueous $CuSO_4$ and Zn metal.

Waste Disposal (D-2). Dispose of the products as directed by your instructor.

D-3: Zinc + Copper Sulfate

Place two pieces of Zn metal (mossy zinc) in a small testtube, and half-fill it with 0.1 M $CuSO_4$ solution. Allow the mixture to sit for 30 minutes, and record your observations. Explain your observations using balanced chemical equations. If it reacts, aqueous $ZnSO_4$ and Cu metal will be produced, the copper depositing on the surface of the zinc.

Waste Disposal (D-3). Dispose of the products as directed by your instructor.

D-4: Zinc + Water/Zinc + HCl

C AUTION: Chemical burns can result when 6 M HCl comes in contact with skin. If you spill 6 M HCl on your skin, immediately wash the affected area with water. Continue washing with water for 15 minutes. Have a classmate notify your instructor.

Get 5 mL of 6 M HCl in a small testtube for use in Parts D-4, D-5, and D-6. Place two pieces of Zn metal in a small testtube, and add 2 mL of water and 2 drops of phenolphthalein acid-base indicator (colorless in acidic, pink in basic solution). Insert a cork (not a rubber stopper) *very loosely* into the top of the testtube; it must be loose enough to allow gas to pass freely past it. After about 30 seconds, note whether there is any color change in the indicator. If aqueous $Zn(OH)_2$ is formed, the indicator should turn pink. Light a splint, remove the cork, bring the lighted splint to the top of the testtube, and note what happens. Record your observations. Explain your observations with a balanced chemical equation. If there is quite a bit of hydrogen present, there should be a small explosion (sounds like a puppy barking) as it reacts with oxygen in the air.

If there appears to have been no reaction between the zinc and water you should try adding HCl, a more vigorous reagent. If you got a good reaction with water it is not necessary to add acid. *If there was no reaction with water* add 1 mL of 6 M HCl to the testtube, replace the cork very loosely, and note any changes (color, reaction speed, etc.). Record your observations. Explain your observations using balanced chemical equations. A reaction would produce aqueous $ZnCl_2$, among other things. Again test for the presence of hydrogen gas with a lighted splint.

Waste Disposal (D-4). Dispose of the products as directed by your instructor.

D-5: Calcium + Water/Calcium + HCl

Place a piece of calcium metal about a quarter the size of a dime in a small testtube. Perform the same tests on calcium as you did on zinc; that is, first check for a reaction with water containing phenolphthalein indicator, and then, if no reaction with water occurred, check whether it reacts with 6 M HCl. As with zinc, place a cork loosely in the top of the testtube, and test for hydrogen with a lighted splint. If calcium reacts with water, $Ca(OH)_2$ also will be produced; if it reacts with HCl, $CaCl_2$ will be a product. Record your observations. Explain your observations using balanced chemical equations.

Waste Disposal (D-5). Dispose of the products as directed by your instructor.

D-6: Copper + Water/Copper + HCl

Sand the surface of a piece of copper wire, coil it around a pen, and place it in a small testtube. Perform the same tests on copper as you did on zinc and calcium; that is, first check to see if it reacts with water containing phenolphthalein indicator, and then, if necessary, test to see if it reacts with 6 M HCl. As you did with the other metals, place a cork loosely in the top of the testtube, and test for the presence of hydrogen gas with a lighted splint. If copper reacts with water, $Cu(OH)_2$ also will be produced; if it reacts with HCl, $CuCl_2$ will be a product.

Waste Disposal (D-6). Dispose of the products as directed by your instructor.

Part E: Other Oxidation-Reduction Reactions

Electron-transfer reactions are pervasive, being especially important in metabolism and other physiologic processes. The reactions involved in fireworks and explosives are also redox reactions. Black powder, used in both explosives and fireworks, is a mixture of potassium nitrate, charcoal, and sulfur that react to give products such as potassium carbonate, potassium sulfate, carbon dioxide, and nitrogen:

$$10 \text{ KNO}_3(s) + 8 \text{ C}(s) + 3 \text{ S}(s) \rightarrow 2 \text{ K}_2\text{CO}_3(s) + 3 \text{ K}_2\text{SO}_4(s) + 6 \text{ CO}_2(g) + 5 \text{ N}_2(g) \qquad (16)$$

Carbon and sulfur are oxidized, nitrogen (in KNO_3) is reduced, and a great deal of heat and light are given off.

 Safety Precautions:

- Protective eyewear approved by your institution must be worn at all times while you are in the laboratory.

- The experiment in Part *E-1* must be carried out in the hood, behind a protective sheet of safety glass.

- Potassium permanganate is a powerful oxidizing agent. If you get it on your clothing, the clothing should be removed. Immediately wash off any spills with plenty of water, and notify your instructor.

E-1: Potassium Chlorate + Glucose ($C_6H_{12}O_6$)

Potassium chlorate ($KClO_3$) is a powerful oxidizing agent used in both matches and fireworks. To make it react it is mixed with one or more reducing agents, such as glucose (grape sugar), carbon (graphite), or sulfur. In the reaction of potassium chlorate and glucose ($C_6H_{12}O_6$) the main products are solid KCl, water vapor, and CO_2. To initiate the reaction, a drop of concentrated H_2SO_4 is added to the mixture. The acid reacts very strongly with the glucose, giving off a great deal of heat, enough heat to initiate the reaction between $KClO_3$ and glucose.

Note: This reaction will be carried out, not by each student, but by small groups of students under the supervision of the instructor.

C AUTION: This experiment must be carried out in the hood, behind a protective sheet of safety glass.

Obtain 1 g of $KClO_3$. Weigh out 0.5 g of glucose into a dry beaker. Take the $KClO_3$ and glucose to the hood, and pour them into the bottle labeled "Mixing Bottle for $KClO_3$ and Glucose." Also take a wash bottle to the hood. Place the $KClO_3$ and glucose in the mixing bottle, replace the cover, and mix thoroughly by inverting the bottle and shaking it. *Do not mix them with a stirring rod; that could be dangerous.* After the solids have been mixed, pour them in a heap into an evaporating dish. Place the evaporating dish on the transite sheet behind the safety glass. There should be gloves and a long dropper there, as well as a container of concentrated H_2SO_4. Put on the gloves. Reaching around the safety glass, draw some H_2SO_4 into the dropper, squeeze a drop or two of it onto the pile of $KClO_3$ and glucose, and pull your arm out. You should observe lots of smoke and fire. If little happens add a drop or two more of H_2SO_4. If nothing happens, even after adding more H_2SO_4, fill the evaporating dish halfway with water, take it to the sink, and rinse the mixture down the drain with plenty of water. (Do *not* put the mixture into a waste container because this could cause a fire.) Observe what happens, especially taking note of the color. Record your observations. Explain your observations using balanced chemical equations. (*Note:* You do not have to know an equation for the reaction between glucose and sulfuric acid, which is complex and gives a mixture of products. The sulfuric acid is just added to provide enough heat to initiate the main reaction.)

Waste Disposal (E-1). Dispose of the products as directed by your instructor. Return the empty mixing bottle to its proper location in the hood.

E-2: Potassium permanganate + Ferrous Sulfate

$$H_2SO_4 + KMnO_4 + FeSO_4 \rightarrow Fe_2(SO_4)_3 + MnSO_4 + K_2SO_4 + H_2O \text{ (unbalanced)} \qquad (17)$$

Potassium permanganate ($KMnO_4$) is a widely used oxidizing agent. It is sometimes used to clear up algae growth in aquariums. Ferrous sulfate ($FeSO_4$) is used commonly to treat or prevent iron-deficiency anemia in humans.

CAUTION: Potassium dichromate ($K_2Cr_2O_7$) is a powerful oxidizing agent and is quite toxic. If you get it on your clothing, the clothing should be removed. Immediately wash off any spills with plenty of water, and notify your instructor.

Obtain about 2 mL of 0.10 M $KMnO_4$ solution and about 2 mL of 0.25 M $FeSO_4$ (in 3 M H_2SO_4) solution, placing each solution in a different small testtube. Using a clean Pasteur pipet, place 10 drops of the $FeSO_4$ solution in a third testtube. Then add the $KMnO_4$ dropwise to the acidic $FeSO_4$ solution, counting and recording the number of drops, until you see a slight pinkish coloration of the solution, indicating that a small excess of $KMnO_4$ has been added. Tap the side of the testtube briskly after every couple of drops to mix the contents. (As the concentration of Fe^{3+} builds up the solution will become somewhat yellowish, but you still should be able to see the color change at the end point; Mn^{2+}, the other main product of the reaction, is nearly colorless.) Record your observations. Explain your observations using balanced chemical equations.

Next, you will carry out the reaction in the opposite way; i.e., you will add the acidic $FeSO_4$ to the $KMnO_4$. Place 10 drops of the 0.10 M $KMnO_4$ solution in a clean testtube. Then add 0.25 M $FeSO_4$ dropwise, again counting the number of drops, until the dark color of the solution abruptly lightens. During the early portion of this minititration, you should observe a brown solid, MnO_2, forming in the solution. Toward the end, however, as the acidity of the solution increases, owing to the H_2SO_4 in the $FeSO_4$ solution, the solid MnO_2 will be reduced to aqueous Mn^{2+}. You should not have seen the formation of MnO_2 in the first reaction (i.e., adding $KMnO_4$ to $FeSO_4$ in 3 M H_2SO_4) because the solution was too acidic for it to form. Record the number of drops of $FeSO_4$ added to reach the point at which the solution lightens in color. Record your observations. Explain your observations using balanced chemical equations.

Waste Disposal (E-2). Dispose of the products as directed by your instructor.

E-3: Sodium Sulfite + Potassium Dichromate

$$H^+(aq) + Cr_2O_7^{2-}(aq) + SO_3^{2-}(aq) \rightarrow Cr^{3+}(aq) + HSO_4^-(aq) + H_2O(l) \quad \text{(unbalanced)} \tag{18}$$

CAUTION: Potassium dichromate is a powerful oxidizing agent and is quite toxic. If you get it on your clothing, the clothing should be removed. Immediately wash off any spills with plenty of water, and notify your instructor.

Obtain about 1 mL of the acidic 0.1 M $K_2Cr_2O_7$ solution and about one-fourth of a gram of Na_2SO_3, placing each reagent in a different small testtube. Combine the contents of the two testtubes, and record your observations. Explain your observations using balanced chemical equations.

Waste Disposal (E-3). Dispose of the products as directed by your instructor.

Part F: Dissolution Reactions

Whether a solute dissolving in a solvent is considered to be a chemical reaction is a matter of semantics in that while a change of some sort certainly occurs during solution formation, the number of particles that interact with each other in the formation of a solution is indeterminate; it isn't a case of x molecules of substance A reacting with y molecules of substance B, as in a normal chemical reaction. Moreover, in forming solutions, the attractions that are broken and formed often are considerably weaker than the bonds that break and form in a normal chemical reaction.

There are many types of attractive forces between particles of solute and solvent, including ionic (electrostatic) attractions, dipolar attractions, and hydrogen bonds. When a solute is mixed with a solvent, in order for dissolution to take place, attractions between the solute and solvent must overcome attractions both between the solvent particles, such as hydrogen bonding, and between the solute particles, such as electrostatic forces between ions. As the solution process takes place, depending on the relative magnitudes of the different attractive forces that must be overcome and those that develop between the solvent and the solute, heat can be either released or absorbed. If a solution warms up when a solute is dissolved in a solvent, it is an indication that the energy released when the solute dissolves in the solvent is great enough to overcome the attractive forces between both solute particles and solvent molecules. If cooling accompanies solution formation, it indicates that the sum of the solvent–solvent attractions and the solute–solute attractions exceeds the attractions created when the solute–solvent attractions are established.

Only those dissolution reactions in which water is the solvent will be considered in this experiment. Water is an excellent solvent for ionic and polar solutes, especially if it can form hydrogen bonds with the solutes. When water dissolves an ionic solute, the slightly negative end of the water dipole is attracted to the positive ions (cations), and the slightly positive end of the water dipole is attracted to the negative ions (anions).

   Safety Precautions:

- Protective eyewear approved by your institution must be worn at all times while you are in the laboratory.

- Avoid breathing in too much ammonia vapor. It is quite irritating to mucous membranes. If you spill concentrated NH_3 on your skin immediately wash the affected area with water. Continue washing with water for 15 minutes. Have a classmate notify your instructor.

F-1: Ammonia Gas + Water

When ammonia gas dissolves in water, the ammonia forms hydrogen bonds with the water molecules, giving aqueous ammonia:

$$NH_3(g) + H_2O(l) \rightarrow NH_3(aq) \tag{19}$$

Although hydrogen bonds typically are about 10 to a 100 times weaker than ordinary covalent bonds, they are much stronger than most other intermolecular attractions. Hydrogen bonding can occur between molecules whenever the molecules contain hydrogen atoms that are bonded to either nitrogen, oxygen, or fluorine. It can occur between molecules of the same substance, or it can occur between molecules of different substances. For example, there is hydrogen bonding between water molecules in pure water; likewise, there is hydrogen bonding between ammonia molecules in pure ammonia. In aqueous ammonia there are hydrogen bonds not only between water molecules and between ammonia molecules but also between water and ammonia molecules (see Figure 9.5).

Hydrogen bond

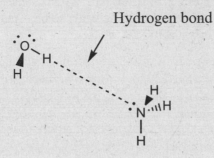

Figure 9.5
Hydrogen bonding between water and ammonia.

In some ways, hydrated ammonia is similar to any other complex, such as $Al(H_2O)_6^{3+}$, but in the case of aqueous ammonia, the number of water molecules associated with a given ammonia molecule is not well defined. Thus it is more representative of the reaction to write

$$NH_3(g) + n\ H_2O(l) \rightarrow NH_3(aq) \tag{20}$$

where n is an indeterminate number of water molecules. Actually, the more usual convention is just to place H_2O above the "yields" arrow:

$$NH_3(g) \xrightarrow{\ H_2O\ } NH_3(aq) \tag{21}$$

implying that water is involved in the reaction but in an indeterminate amount.

A very small percentage of the hydrated ammonia molecules reacts (in a normal chemical reaction) with the water to produce ammonium ions and hydroxide ions, giving a mildly basic solution:

$$NH_3(aq) + H_2O(l) \rightarrow NH_4^+(aq) + OH^-(aq) \tag{22}$$

It is important to note that by far the main reaction that occurs when ammonia gas dissolves in water is the simple hydration of the ammonia molecules; Reaction (22) occurs to a very slight extent.

Obtain two Beral pipets, one having a larger diameter shaft than the other. Obtain about 4 mL of concentrated (15 M) NH_3 in a small testtube from the bottle in the hood.

CAUTION: Avoid breathing in too much ammonia vapor. It is quite irritating to mucous membranes. If you spill concentrated NH_3 on your skin, immediately wash the affected area with water. Continue washing with water for 15 minutes. Have a classmate notify your instructor.

Using a pair of scissors, cut the bottom portion of the plastic pipet having the smaller diameter shaft off about ¼ in below the bulb. Invert the piece that is shaft only and insert it, tip-end first, into the remaining short shaft on the bulb, as shown in Figure 9.6. You will be collecting ammonia gas in this pipet. The larger plastic pipet will be used in generating the ammonia gas by heating concentrated aqueous ammonia.

Squeeze the bulb of the larger Beral pipet, draw at least 2 mL of the concentrated NH_3 solution into the bulb (the bulb should be about one-third full), and invert the pipet so that the bulb is at the bottom.

In the fume hood, place about 100 mL of water in a 200 mL beaker, set it on a hotplate, and heat the water until it boils. While you are waiting for the water to boil, place 20 mL of room-temperature water and 4 drops of phenolphthalein in a 100 mL beaker.

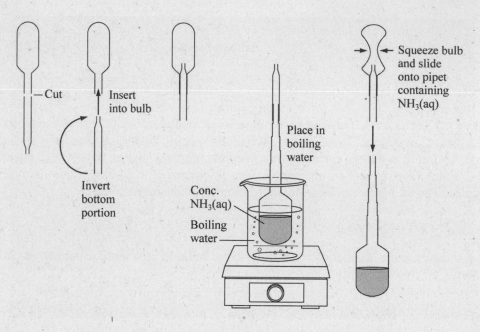

Figure 9.6
Setup for collecting ammonia gas.

Once the water is boiling, shut off the hotplate. Immerse the bulb containing the aqueous ammonia into the boiling water. Warming the concentrated ammonia solution will cause it to release ammonia gas, which you will collect in the upper bulb. Wait about 12–15 seconds, until you see some bubbles of ammonia gas forming and leaving the solution, then squeeze the upper pipet bulb as tightly as you can in order to get the air out, and while it is still compressed, slide the shaft of the upper pipet firmly over the tip of the bottom pipet, as shown in Figure 9.6. Slowly release the pressure on the bulb (take at least 5 seconds) in order to allow it to fill with ammonia gas being released from the bottom bulb. After doing this one time, there still will be quite a bit of air in the upper bulb, so after sliding the upper pipet shaft off the tip of the lower pipet, squeeze the upper bulb again, reattach it, and repeat the filling procedure. Repeat a third time. Now the upper bulb should be completely filled with nearly pure ammonia gas.

Next, separate the upper pipet from the bottom pipet and quickly immerse the lower end of the shaft into the solution of water and phenolphthalein in the 100 mL beaker, as shown in Figure 9.7. Squeeze the bulb a little bit in order to get a bubble or two of the ammonia down into the water, and then release the bulb. You will have to wait at this point until some ammonia dissolves in the water.

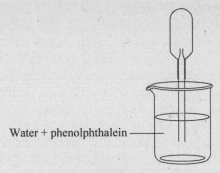

Figure 9.7
Setup for ammonia fountain.

When the ammonia gas dissolves in the water, the pressure in the bulb will drop below the pressure of the atmosphere acting on the surface of the water. This will cause water to be pushed up into the pipet bulb. Record your observations. Explain your observations using balanced chemical equations.

Waste Disposal (F-1). Dispose of the products as directed by your instructor.

F-2: Ammonium Chloride + Water

When solid NH_4Cl dissolves in water both ionic attractions between NH_4^+ cations and Cl^- anions in the solid and hydrogen bonds and dipole–dipole attractions between the water molecules must be overcome. Overcoming these forces requires energy (endothermic). Also, as dissolution occurs, the chloride ions become surrounded by the slightly positive (δ^+) ends of the dipolar water molecules, and the ammonium ions become surrounded by the slightly negative (δ^-) ends. In addition, hydrogen bonds form between the ammonium ions and the water molecules. These interactions all release energy (exothermic). See Figure 9.8.

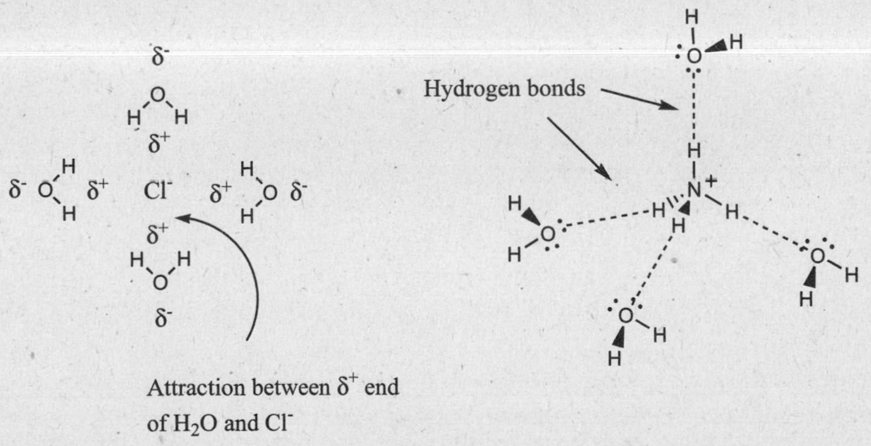

Hydrogen bonds

Attraction between δ^+ end
of H_2O and Cl^-

Figure 9.8
The attractive forces responsible for the dissolution of NH_4Cl in H_2O.

To test whether the endothermic or exothermic processes dominate, dissolve about 1 g of ammonium chloride in 2 mL of water, and note whether the testtube warms up or cools off. Record your observations. Explain your observations using balanced chemical equations.

Waste Disposal (F-2). Dispose of the products as directed by your instructor.

F-3: Calcium Chloride Dihydrate + Water

The case of calcium chloride dihydrate dissolving in water is simpler to analyze than that of ammonium chloride because, in this situation, no hydrogen bonds are formed. Here, the endothermic process involves breaking apart the calcium chloride crystal lattice (as well as a few hydrogen bonds between water molecules), and the exothermic part involves the slightly positive ends and slightly negative ends of water molecules associating themselves with the calcium ions and chloride ions.

Dissolve about 1 g of solid calcium chloride dihydrate in about 2 mL of water, and note whether heat is absorbed or released. Record your observations. Explain your observations using balanced chemical equations.

Waste Disposal (F-3). Dispose of the products as directed by your instructor.

Name: _____   Date: _____

Lab Instructor: _____   Lab Section: _____

EXPERIMENT 9

Chemical Reactions

PRELABORATORY QUESTIONS

1. Define each of the following general types of reactions, and provide an example of each.

    (a) Metathesis (double-displacement) reactions

    (b) Complexation reactions

    (c) Decomposition reactions

    (d) Combination reactions

    (e) Single-displacement reactions

    (f) Other oxidation-reduction reactions

    (g) Dissolution reactions

Name: _____   Date: _____

Lab Instructor: _____   Lab Section: _____

EXPERIMENT 9

Chemical Reactions

RESULTS/OBSERVATIONS

Part A: Metathesis Reactions and Complexation Reactions

A-1: Magnesium Sulfate + Barium Chloride

1. Observations on adding 5 drops of 0.1 M $MgSO_4$ solution to 5 drops of 0.1 M $BaCl_2$ solution:

2. Balanced chemical equation:

A-2: Sodium Carbonate + HCl

3. Observations on adding 5 drops of 6 M HCl to 2 mL of saturated Na_2CO_3 solution:

4. Balanced chemical equation:

A-3: Acetic Acid + Sodium Hydroxide

5. Drops of NaOH solution necessary to completely neutralize 10 drops of 0.150 M acetic acid:

6. Drops of 0.150 M acetic acid necessary to completely neutralize 10 drops of NaOH solution:

7. Balanced chemical equation:

8. Calculation of NaOH solution concentration using the data from question 5:

Name: _____   Date: _____

Lab Instructor: _____   Lab Section: _____

9. Calculation of NaOH solution concentration using the data from question 6:

10. Average NaOH solution concentration: _____

A-4: Sodium Chloride + Silver Nitrate

11. Observations on adding 5 drops of 0.1 M NaCl solution to 5 drops of 0.1 M $AgNO_3$ solution:

12. Balanced chemical equation:

A-5: Silver Chloride + Ammonia

13. Observations on adding 6 M NH_3 dropwise to solid AgCl:

14. Observations on adding 3 M HNO_3 dropwise after the addition of 6 M NH_3:

15. Balanced chemical equations:

A-6: Copper Sulfate + Sodium Hydroxide

16. Observations on combining 10 drops of 0.100 M $CuSO_4$ solution with 10 drops of 0.100 M NaOH solution:

17. Is solid $Cu(OH)_2$ gelatinous or crystalline? Is it blue or white? _____

18. Balanced chemical equation:

Name: _____   Date: _____

Lab Instructor: _____   Lab Section: _____

A-7: Copper Hydroxide + Ammonia (and Several Related Reactions)

Step 1

19. Observations on adding 2 drops of 6 M NH_3 to copper hydroxide:

20. Balanced chemical equation:

Step 2

21. Observations on adding 0.500 M HNO_3 to the solution from step 1:

22. Balanced chemical equation:

Step 3

23. Observations on adding a few more drops of 0.500 M HNO_3 to the solution from step 2:

24. Balanced chemical equation:

Part B: Decomposition Reactions

B-1: Dehydration of Copper Sulfate Pentahydrate

25. (a) Observations on heating solid $CuSO_4 \cdot 5H_2O$ in a crucible:

    (b) Observations on adding a few drops of water the previously heated solid:

26. Balanced chemical equations:

Name: _____   Date: _____

Lab Instructor: _____   Lab Section: _____

B-2: Decomposition of Sodium Bicarbonate

27. Observation on filtering the $Ca(OH)_2$ solution:

28. Observations on heating $NaHCO_3$ and bubbling the gas produced by decomposition through the $Ca(OH)_2$ solution:

29. Balanced chemical equations:

30. Observations on heating $NaHCO_3$ and bubbling the gas produced by decomposition through the $Ca(OH)_2$ solution in the presence of bromothymol blue acid-base indicator solution:

31. Balanced chemical equations:

32. Observations on adding bromothymol blue to the partially dissolved unknown solid (Na_2CO_3 or NaOH?) in the testtube:

Name: _____   Date: _____

Lab Instructor: _____   Lab Section: _____

33. Description of test to determine the identity of the partially dissolved unknown solid:

34. Observations on performing the test to determine the identity of the partially dissolved unknown solid:

35. Balanced chemical equation(s), if applicable, associated with the test to determine the identity of the partially dissolved unknown solid:

36. The unknown solid is _____ .

B-3: Decomposition of Ammonium Carbonate

37. Observations on heating $(NH_4)_2CO_3$ and bubbling the gas produced by decomposition through the $CuSO_4$ solution:

38. Balanced chemical equations:

39. Observations on heating $(NH_4)_2CO_3$ and bubbling the gas produced by decomposition through the $Cu(OH)_2$ solution:

40. Balanced chemical equations:

Name: _____   Date: _____

Lab Instructor: _____   Lab Section: _____

41. Balance Reaction (9):

$$(NH_4)_2CO_3(s) \xrightarrow{\Delta} H_2O(g) + \underline{\hspace{1cm}}(g) + \underline{\hspace{1cm}}(g)$$

B-4: Decomposition of Hydrogen Peroxide—Preparation of Oxygen

42. Observations on adding 2 g of solid KI to 50 mL of 3% H_2O_2:

43. Balanced chemical equation:

44. Observation on inserting a glowing splint into the Erlenmeyer flask containing oxygen gas:

Part C: Combination Reactions

C-1: Magnesium + Air (N_2 and O_2)

45. Observations on burning magnesium metal:

46. Balanced chemical equations:

47. Observations on adding 2 mL of distilled water and 4 drops of bromothymol blue indicator to the residue from burning magnesium metal:

48. Balanced chemical equations:

Name: _____   Date: _____

Lab Instructor: _____   Lab Section: _____

Part D: Single-Displacement Reactions

D-1: Copper + Silver Nitrate

49. Observations on combining Cu metal and 0.1 M $AgNO_3$ solution:

50. Balanced chemical equation:

D-2: Copper + Zinc Sulfate

51. Observations on combining Cu metal with 0.1 M $ZnSO_4$ solution:

52. Balanced chemical equation:

D-3: Zinc + Copper Sulfate

53. Observations on combining Mg metal and 0.1 M $CuSO_4$ solution:

54. Balanced chemical equation:

D-4: Zinc + Water/Zinc + HCl

55. Observation on combining Zn metal, 2 mL of water, and 2 drops of phenolphthalein acid-base indicator:

56. Balanced chemical equation:

57. Observation on adding 6 M HCl to the testtube containing Zn metal:

Name: _____   Date: _____

Lab Instructor: _____   Lab Section: _____

58. If applicable, observation on placing a glowing splint in the gas produced:

59. Balanced chemical equation(s):

D-5: Calcium + Water/Calcium + HCl

60. Observation on combining Ca metal, 2 mL of water, and 2 drops of phenolphthalein acid-base indicator:

61. Balanced chemical equation:

62. Observation on adding 6 M HCl to the testtube containing Ca metal:

63. If applicable, observation on placing a glowing splint in the gas produced:

64. Balanced chemical equation(s):

D-6: Copper + Water/Copper + HCl

65. Observation on combining Cu metal, 2 mL of water, and 2 drops of phenolphthalein acid-base indicator:

66. Balanced chemical equation:

67. Observations on adding 6 M HCl to the testtube containing Cu metal:

68. Balanced chemical equation:

69. Rank the four metals according to their activity: _____ > _____ > _____ > _____

Name: _____   Date: _____

Lab Instructor: _____   Lab Section: _____

Part E: Other Oxidation-Reduction Reactions

E-1: Potassium Chlorate + Glucose ($C_6H_{12}O_6$)

70. Observations on reacting potassium chlorate and glucose:

71. Balanced chemical equation:

E-2: Potassium permanganate + Ferrous Sulfate

72. Number of drops of $KMnO_4$ solution necessary to react with 10 drops of the $FeSO_4$ solution:

73. Observations on adding $KMnO_4$ solution $FeSO_4$ solution:

74. Balanced chemical equation:

75. Number of drops of $FeSO_4$ solution necessary to react with 10 drops of $KMnO_4$ solution:

76. Observations on adding $FeSO_4$ solution to $KMnO_4$ solution:

77. Balanced chemical equations:

E-3: Sodium Sulfite + Potassium Dichromate

78. Observations on combining the acidic 0.1 M $K_2Cr_2O_7$ solution and about one-fourth of a gram of Na_2SO_3:

Name: _____   Date: _____

Lab Instructor: _____   Lab Section: _____

79. Balanced chemical equation:

Part F: Dissolution Reactions

F-1: Ammonia Gas + Water

80. Observations:

81. Balanced chemical equation:

F-2: Ammonium Chloride + Water

82. Observations on dissolving solid NH_4Cl in water:

83. Is the dissolution endo- or exothermic?

84. Balanced chemical equation:

F-3: Calcium Chloride Dihydrate + Water

85. Observations on dissolving solid $CaCl_2 \cdot 2H_2O$ in water:

86. Is the dissolution endo- or exothermic?

87. Balanced chemical equation:

Name: _____   Date: _____

Lab Instructor: _____   Lab Section: _____

EXPERIMENT 9

Chemical Reactions

POSTLABORATORY QUESTIONS

1. What is the general purpose of gravity filtration?

2. When aqueous solutions of $Pb(NO_3)_2$ and NaI are combined a yellow solid results. What is the chemical identity of the yellow solid? Write a balanced chemical equation consistent with this observation. To which general class does this chemical reaction belong?

3. When heated to sufficiently high temperatures solid calcium carbonate ($CaCO_3$) produces solid calcium oxide (CaO) and a gas. What is the chemical identity of the gas? Write a balanced chemical equation consistent with this observation. To which general class does this chemical reaction belong?

4. Dissolution of solid ammonium nitrate (NH_4NO_3) in water results in a measurable temperature decrease. Write a balanced chemical equation consistent with this observation. Did this reaction absorb or release heat? To which general class does this chemical reaction belong?

5. If barium ion is toxic, how can it be that barium sulfate is nontoxic?

Name: _____   Date: _____

Lab Instructor: _____   Lab Section: _____

6. Write a balanced chemical equation for the reaction between hydrobromic acid (HBr) and aqueous potassium hydroxide (KOH). To which general class does this chemical reaction belong? How many drops of 2.0 M HBr are necessary to completely neutralize 10 drops of 1.0 M KOH?

7. Write a balanced chemical equation for the combination reaction between aluminum metal and oxygen gas.

8. Addition of hydrochloric acid to Ni metal results in the production of a flammable gas. What is the chemical identity of the gas? Write a balanced chemical equation consistent with this observation. To which general class does this chemical reaction belong?

9. Balance the chemical equation below. To which general class does this chemical reaction belong?

$$KHSO_3(aq) + KMnO_4(aq) + H_2SO_4(aq) \rightarrow KHSO_4(aq) + MnSO_4(aq) + K_2SO_4(aq) + H_2O(l)$$

Name: _____   Date: _____

Lab Instructor: _____   Lab Section: _____

10. A student has 12 mL of a solution of what is either magnesium chloride or barium chloride. Despite not being able to recall the chemical identity of this solution, the student does remember that the solution has a concentration of 0.10 M. Explain how the student can determine the chemical identity of this solution. The student may use any instrument that you used in the experiment Chemical Reactions and any amount of the following chemicals: 0.10 M KNO_3, 0.10 M HCl, 0.10 M HNO_3, 0.10 M K_2SO_4, phenolphthalein indicator solution, bromothymol blue indicator solution, solid $NaHCO_3$, and solid $BaSO_4$. Justify your answer with a balanced chemical equation.

Ten Unknown Solutions

OBJECTIVE

To determine the identities of 10 unknown solutions by performing reactions and observing the results.

INTRODUCTION

In this experiment you will use what you learned about chemical reactions in Experiment 9: Chemical Reactions to solve a puzzle. You will be issued 10 testtubes that have been filled with 10 different chemical solutions of unknown identity. Your task will be to determine what is in each testtube. You are to do this by carrying out reactions between the 10 unknown solutions and the reagents provided. All the chemicals will be among those examined in Experiment 9. The unknown chemicals in your set of 10 testtubes have been selected from 15 possible solutions:

0.1 M $Cu(NO_3)_2$	0.1 M $AgNO_3$	3 M H_2SO_4
0.1 M $CuSO_4$	0.25 M $FeSO_4$ (in 3 M H_2SO_4)	0.1 M $MgSO_4$
6 M NH_3	0.1 M $BaCl_2$	Saturated Na_2CO_3
1 M NaOH	0.1 M NaCl	0.1 M $K_2Cr_2O_7$
3 M HCl	0.1 M $KMnO_4$	3 M CH_3CO_2H

There will be labeled bottles of each of these solutions, as well as litmus paper and some solid sodium sulfite (Na_2SO_3), available in the laboratory for you to use in carrying out tests on and reactions with the unknown solutions.

ADDITIONAL READING

Read the sections in the "Laboratory Techniques" chapter at the beginning of this Lab Manual on cleaning glassware, handling chemicals, weighing solids, heating liquids and solutions, and decantation prior to performing this experiment.

Safety Precautions:

- Protective eyewear approved by your institution must be worn at all times while you are in the laboratory.

- Since you don't know its identity, you will need to observe all these precautions when dealing with any unknown.

- Chemical burns can result when 3 M HCl or 3 M H_2SO_4 comes in contact with skin. If you spill 3 M HCl or 3 M H_2SO_4 on your skin, immediately wash the affected area with water. Continue washing with water for 15 minutes. Have a classmate notify your instructor.

- Avoid breathing in too much ammonia vapor. It is quite irritating to mucous membranes. If you spill 6 M NH_3 on your skin, immediately wash the affected area with water. Continue washing with water for 15 minutes. Have a classmate notify your instructor.

- Potassium permanganate is a powerful oxidizing agent. If you get it on your clothing, the clothing should be removed. Immediately wash off any spills with plenty of water, and notify your instructor.

- Potassium dichromate is a powerful oxidizing agent and is quite toxic. If you get it on your clothing, the clothing should be removed. Immediately wash off any spills with plenty of water, and notify your instructor.

EXPERIMENT

From having carried out the reactions in Experiment 9, you should be able to determine a strategy for identifying each of your 10 unknown solutions. Before coming to lab, think about how you can identify each solution by reacting it with one or more of the other available test solutions. You will be issued only 4 mL of each of the 10 unknown solutions. Be careful; there are *no* refills. Do not test reactions on the entire 4 mL sample. Use just a few drops for each reaction.

     Record your observations as usual.
     To avoid waste, take no more than 2 mL of any test solution you wish to use.
     Litmus turns red in acid and blue in base. To test a solution with litmus paper place a piece of litmus paper on a dry watch glass, then place your stirring rod in the solution, withdraw a drop, and place the drop on the litmus paper.

Waste Disposal. Dispose of all chemical waste as directed by your instructor.

Name: _____   Date: _____

Lab Instructor: _____   Lab Section: _____

EXPERIMENT 10

Ten Unknown Solutions

PRELABORATORY QUESTIONS

1. State three examples of physical changes that can be used to identify a chemical reaction.

2. Why is it unwise to test a reaction on the entire unknown sample?

Name: _____   Date: _____

Lab Instructor: _____   Lab Section: _____

EXPERIMENT 10

Ten Unknown Solutions

RESULTS/OBSERVATIONS

Testtube A:

Test	Observation

Testtube B:

Test	Observation

Name: _____   Date: _____

Lab Instructor: _____   Lab Section: _____

Testtube C:

Test	Observation

Testtube D:

Test	Observation

Testtube E:

Test	Observation

Name: _____   Date: _____

Lab Instructor: _____   Lab Section: _____

Testtube F:

Test	Observation

Testtube G:

Test	Observation

Testtube H:

Test	Observation

Name: _____   Date: _____

Lab Instructor: _____   Lab Section: _____

Testtube I:

Test	Observation

Testtube J:

Test	Observation

Testtube	Solution Identity
A	
B	
C	
D	
E	
F	
G	
H	
I	
J	

Name: _____   Date: _____

Lab Instructor: _____   Lab Section: _____

EXPERIMENT 10

Ten Unknown Solutions

POSTLABORATORY QUESTIONS

1. Use the observations below to identify the unknown compound in each case.

(a)

Test	Observation
The unknown was added to 0.1 M NaCl.	A white precipitate resulted.
6 M NH_3 was added dropwise to the precipitate.	The precipitate was destroyed, a clear solution forming in its place.

(b)

Test	Observation
A drop of unknown was placed on litmus paper.	Litmus turned red.
1 M NaOH was added dropwise to 10 drops of the unknown and 1 drop of phenolphthalein.	30 drops of 1 M NaOH were required to reach the end-point.
The unknown was added dropwise to a sample of 0.1 M $AgNO_3$.	A white precipitate formed.

(c)

Test	Observation
Visual appearance.	An orange solution.
Solid Na_2SO_3 was added to a small portion of the unknown solution.	The initially orange solution is converted to a green solution.

Copper Reactions

OBJECTIVE

To perform and observe a sequence of reactions that begin and end with copper metal. The percent recovery of copper metal will be determined, the balanced chemical equation for each reaction in the sequence will be determined, and each reaction in the sequence will be classified as belonging to one of four types: precipitation reactions, acid-base reactions, oxidation-reduction reactions, or decomposition reactions.

INTRODUCTION

When first encountered, the variety and number of chemical reactions seem overwhelming. Literally millions of compounds are known to exist, each of which is involved in one or more reactions. How is one to make sense of such a massive amount of information? How does one lend order to these myriad data? A classification scheme. Despite their enormous number, chemical reactions can be classified into a relatively small number of categories. Throughout time, a number of schemes for classifying chemical reactions have been applied successfully. Each has its merits. The classification scheme presented here is one of the simpler ones. It involves classifying reactions as one of four types: precipitation reactions, acid-base reactions, oxidation-reduction reactions, or decomposition reactions.

Precipitation Reactions

Precipitation reactions occur when soluble reactants yield an insoluble solid product, the precipitate. Most precipitation reactions involve soluble ionic compounds that switch partners to form a precipitate. For example, both calcium chloride ($CaCl_2$) and sodium carbonate (Na_2CO_3) are soluble salts, but when solutions of the two are combined together a precipitate of calcium carbonate ($CaCO_3$) results:

$$Ca^{2+}(aq) + 2\ Cl^-(aq) + 2\ Na^+(aq) + CO_3^{2-}(aq) \rightarrow 2\ Na^+(aq) + 2\ Cl^-(aq) + CaCO_3(s) \qquad (1)$$

Acid-Base Reactions

An acid is a substance that reacts with water to form hydronium ions (H_3O^+). For example, hydrochloric acid (HCl):

$$HCl(aq) + H_2O(l) \rightarrow H_3O^+(aq) + Cl^-(aq) \qquad (2)$$

An acid also can be defined as an H^+ donor. As is clear from an examination of Reaction (2), the acid HCl donates an H^+ to H_2O, forming hydronium and chloride ions.

A base is a substance that forms hydroxide ions (OH^-) when dissolved in water. Sodium hydroxide (NaOH) is a common example:

$$NaOH(aq) \rightarrow Na^+(aq) + OH^-(aq) \tag{3}$$

A typical acid-base reaction leads to the formation of a salt and water. The reaction of hydrobromic acid (HBr) and potassium hydroxide (KOH) is a representative example:

$$HBr(aq) + KOH(aq) \rightarrow KBr(aq) + H_2O(l) \tag{4}$$

Again, the acid acts as an H^+ donor, donating H^+ this time to OH^- and forming water and potassium bromide.

Oxidation-Reduction Reactions

Oxidation-reduction reactions, often referred to as *redox reactions*, involve the transfer of electrons from one reactant to another. As a simple example, consider the reaction of elemental calcium (Ca) with molecular bromine (Br_2) to form calcium bromide ($CaBr_2$):

$$Ca(s) + Br_2(l) \rightarrow CaBr_2(s) \tag{5}$$

The product is an ionic solid: Forming the product involved the net transfer of electrons from calcium to bromine atoms.

But it isn't always so easy to recognize oxidation-reduction reactions. Consider the reaction between sulfur dioxide (SO_2) and oxygen (O_2) to form sulfur trioxide (SO_3):

$$SO_2(g) + O_2(g) \rightarrow 2\ SO_3(g) \tag{6}$$

No ions are formed in this reaction, nor are any of the reactants ions, but Reaction (6) is an oxidation-reduction reaction. How was this designation made? By assigning and comparing oxidation numbers. An oxidation number is a measure of whether an atom is neutral, electron-rich, or electron-poor. Changes in an atom's oxidation number from the reactant to the product state indicate a gain or loss of electrons.

Oxidation numbers can be assigned using the following rules:

1. *An atom in its elemental state has an oxidation number of 0.* For example, Na, Cl_2, and P_4 all have oxidation numbers of 0.

2. *For monatomic ions the oxidation number is equal to the charge.* For example, Na^+ has an oxidation number of +1; Ca^{2+} has an oxidation number of +2; and S^{2-} has an oxidation number of –2.

3. *Oxygen has an oxidation number of –2, except in peroxides, where it is –1.* For example, the oxidation number of oxygen in H_2O, CO_2, and MnO_4^- is –2; the oxidation number of oxygen in hydrogen peroxide (H_2O_2) is –1.

4. *Hydrogen has an oxidation number of +1, except in hydrides, where it is –1.* For example, the oxidation number of hydrogen in H_2O, HCl, and H_2O_2 is +1; the oxidation number of hydrogen in sodium hydride (NaH) is –1.

5. *The sum of the oxidation numbers is equal to the charge of the molecule or ion.* For example, in CO_2, the oxidation number of each oxygen is –2, the molecule has no charge, so the oxidation number of carbon must be +4; in HSO_4^-, each oxygen has an oxidation number of –2, hydrogen has an oxidation number of +1, and the ion has a –1 charge, so sulfur must have an oxidation number of +6.

Now to assign oxidation numbers to the atoms in Reaction (6):

$$SO_2(g) + O_2(g) \rightarrow 2\ SO_3(g) \tag{6}$$

$$+4\ -2 \quad 0 \quad +6 \quad -2$$

Sulfur changed from an oxidation number of +4 to +6; it lost electrons. Oxygen changed its oxidation number from 0 to –2; it gained electrons. Hence its designation as an oxidation-reduction reaction.

Decomposition Reactions

Most reactions can be identified as belonging to one of the three previous groupings. But there are some reactions that cannot be described as precipitation, acid-base, or redox. For example, when calcium carbonate ($CaCO_3$) is heated it breaks down into calcium oxide (CaO) and carbon dioxide (CO_2):

$$CaCO_3(s) \xrightarrow{\Delta} CaO(s) + CO_2(g) \tag{7}$$

No precipitate forms because of this reaction. It cannot be described as an acid-base reaction. The oxidation numbers of the atoms don't change. This reaction is classified as a decomposition reaction. Decomposition reactions tend to involve a more complex compound falling apart into simpler species, often owing to heating.

In this experiment you will perform and observe reactions of a variety of types. The common feature of these reactions is that they all involve copper. You will begin with a sample of copper metal. You will convert that piece of metal to a series of copper compounds, one after another, and finally, in the last reaction, you will try to regenerate as much of the copper metal as possible. You will be responsible for balancing and classifying each reaction. There are a total of five reactions involving copper, plus a sixth reaction, a side reaction involving magnesium. The *unbalanced* equations for the reactions are as follows:

(Part A)　　$Cu(s) + HNO_3(aq) \rightarrow Cu(NO_3)_2(aq) + NO_2(g) + H_2O(l)$ 　　　(8)

(Part B)　　$Cu(NO_3)_2(aq) + OH^-(aq) \rightarrow Cu(OH)_2(s) + NaNO_3(aq)$ 　　　(9)

(Part C)　　$Cu(OH)_2(s) \rightarrow CuO(s) + H_2O(l)$ 　　　(10)

(Part D)　　$CuO(s) + H_2SO_4(aq) \rightarrow CuSO_4(aq) + H_2O(l)$ 　　　(11)

(Part E)　　$CuSO_4(aq) + Mg(s) \rightarrow Cu(s) + MgSO_4(aq)$ 　　　(12)

(Part E)　　$Mg(s) + HCl(aq) \rightarrow MgCl_2(aq) + H_2(g)$ 　　　(13)

ADDITIONAL READING

Read the sections in the "Laboratory Techniques" chapter at the beginning of this Lab Manual on cleaning glassware, handling chemicals, weighing solids, heating liquids and solutions, decantation, and vacuum filtration prior to performing this experiment.

Safety Precautions:

- Protective eyewear approved by your institution must be worn at all times while you are in the laboratory.

- Nitric acid is a powerful acid and oxidizing agent. Avoid spilling it on yourself, but, if you do, rinse it thoroughly with water immediately. If you spill it on your clothing, remove the affected clothing before rinsing with water. Notify your lab instructor, who will assist you. Also, the brown gas, NO_2, produced when nitric acid acts on copper is very toxic. Be sure that you are working in the hood while NO_2 is being produced. If you should accidentally inhale any of this gas, notify your instructor immediately.

- Sodium hydroxide is corrosive, especially to the eyes. Thoroughly wash off any sodium hydroxide spilled on your skin. If you get it into your eyes, immediately flush your eyes with water for at least 10 minutes. If you spill it on your clothing, remove the affected clothing before rinsing with water. Notify your lab instructor, who will assist you.

- Sulfuric acid is corrosive and must be kept off skin and clothing. If you get it into your eyes, immediately flush them with water for at least 10 minutes. If you spill it on your clothing, remove the affected clothing before rinsing with water. Notify your lab instructor, who will assist you.

- Make sure that there are no flames in the lab as you carry out the reaction in Part E: Recovery of Copper because highly flammable hydrogen gas is produced. In the event of a fire, notify your instructor. A fire in a small container can be smothered by covering the vessel with a watchglass. Use a fire extinguisher to eliminate a larger fire. If the fire is burning over too large an area to be extinguished easily, evacuate the area, and activate the fire alarm. If your clothing should catch on fire, use the safety shower or a fire blanket to extinguish the flames.

PROCEDURE

Part A: Conversion of Copper Metal to Copper(II) Nitrate

Obtain a piece of copper wire having a mass of approximately 0.5 g. Clean the surface of the wire by rubbing it with sandpaper to remove any surface oxides or tarnish that might be present. Note and record the color of the wire before and after cleaning it. Determine and record the mass of the clean wire.

C AUTION: Concentrated nitric acid is a powerful acid and oxidizing agent. Avoid spilling it on yourself, but if you do, rinse it thoroughly with water immediately. If you spill it on your clothing, remove the affected clothing before rinsing with water. Notify your lab instructor, who will assist you. Also, the brown gas, NO_2, produced when nitric acid acts on copper is very toxic. Be sure that you are working in the hood while NO_2 is being produced. If you should accidentally inhale any of this gas, notify your instructor immediately.

In the fume hood, support a funnel on a ringstand (see Figure 11.1), and use the funnel to carefully pour 5 mL of concentrated nitric acid (HNO_3) into a graduated cylinder. If you exceed 5 mL, don't try to correct this; just use the amount you poured.

Lay the copper wire flat in the bottom of a 200 mL beaker, carry the beaker to the hood, set the beaker on the hood floor at least 6 in behind the sash, and then slowly add the 5 mL of concentrated

HNO_3 to the copper. Note your observations. If the copper does not dissolve completely, add more concentrated HNO_3, and stir the mixture with a glass stirring rod.

When the reaction is complete, dilute the solution with distilled water to a total volume of about 75 mL. Record your observations.

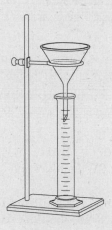

Figure 11.1
A funnel supported on a ringstand for pouring concentrated HNO_3 into a graduated cylinder.

Part B: Formation of Copper(II) Hydroxide

CAUTION: Sodium hydroxide is corrosive, especially to the eyes. Thoroughly wash off any sodium hydroxide spilled on your skin. If you get it into your eyes, immediately flush your eyes with water for at least 10 minutes. If you spill it on your clothing, remove the affected clothing before rinsing with water. Notify your lab instructor, who will assist you.

Take the 200 mL beaker out of the fume hood and to your lab bench. Slowly add 20 mL of 3 M NaOH, stirring continuously with a glass stirring rod. Record your observations. Test the resulting solution with litmus paper to make certain that it is basic (litmus is blue in basic and red in acidic environments). Do this by dipping a glass stirring rod into the solution and touching a drop of the solution to a piece of litmus paper on a watch glass. If necessary, add more NaOH until a drop of the solution turns a piece of red litmus paper blue.

Part C: Conversion of Copper(II) Hydroxide to Copper(II) Oxide

Add two boiling chips to the solution from Part B, and place the beaker on a hotplate. Slowly bring the solution to a gentle boil, stirring constantly. If bumping (intermittent violent boiling) becomes too severe transfer the solution to a larger beaker and return to heating. Gently boil the mixture until all the solid has changed color. Note and record the color change. Turn the hotplate off, and let the beaker cool somewhat before setting it aside and allowing the solid to settle.

In a 500 mL Florence flask, heat to boiling roughly 150 mL of distilled water to be used in rinsing the solid. Assemble a vacuum filtration setup. Place a piece of filter paper in the funnel, turn on the aspirator, and wet the paper with distilled water to seal it to the funnel. Carefully decant about two-thirds of the liquid into the Büchner funnel, losing as little solid as possible. Add about 100 mL of hot distilled water to the beaker, swirl, and transfer the contents to the funnel. Pour it through slowly so that the solid does not seep under the filter paper. Transfer any solid remaining in the beaker or on the stirring rod to the Büchner funnel with short bursts of distilled water from your wash bottle. Wash the solid on the filter paper with the rest of the hot distilled water.

Part D: Conversion of Copper(II) Oxide to Copper(II) Sulfate

Turn off the water aspirator to stop the suction. Using a stirring rod to first loosen and fold up the edges, lift the filter paper and solid from the Büchner funnel, and place it flat in the bottom of a 600 mL beaker. Clean the filter flask and Büchner funnel. (Caustic solutions left in a funnel or filter flask might cause injury to the next person to use the apparatus.)

CAUTION: Sulfuric acid is corrosive and must be kept off skin and clothing. If you get it into your eyes, immediately flush them with water for at least 10 minutes. If you spill it on your clothing, remove the affected clothing before rinsing with water. Notify your lab instructor, who will assist you.

Dissolve the solid by pouring 10 mL of 2 M H_2SO_4 into the beaker, swirling the solution gently. (It will take several minutes for the CuO to dissolve completely. While you are waiting, weigh out the magnesium turnings necessary for Part E.) When the solid CuO dissolves the solution should be transparent. Record your observations.

After the CuO has dissolved completely, use a stirring rod to hold the filter paper down, and decant the solution into a 200 mL beaker. Rinse the filter paper with four 5 mL portions of distilled water, combining each rinse with the copper solution in the 200 mL beaker. Dispose of the filter paper by rinsing it under the tap, wringing it out, and placing it in the plastic waste basket.

Part E: Recovery of Copper

WARNING: Make sure that there are no flames in the lab as you carry out this reaction because highly flammable hydrogen gas is produced.

Slowly add 0.6 g of magnesium turnings to the copper solution and stir. Solid copper should collect gradually at the bottom of the beaker. The solution should be colorless when all the Cu^{2+} has been converted to solid Cu. Thus, if all the magnesium disappears before the solution is colorless, more magnesium should be added.

CAUTION: Concentrated hydrochloric acid is quite corrosive and must be kept off skin and clothing. If you get it into your eyes, immediately flush them with water for at least 10 minutes. If you spill it on your clothing, remove the affected clothing before rinsing with water. Notify your lab instructor, who will assist you.

When the bubbling stops, which should occur about 20 to 30 minutes after adding the magnesium, add 2 mL of concentrated HCl to dissolve any excess magnesium. Allow the copper to settle to the bottom of the beaker, and decant as much of the supernatant liquid as possible into the sink without loss of copper. Transfer all the copper into a clean, weighed evaporating dish. Record the mass of the empty evaporating dish. Rinse the copper *at least three times* with small portions of distilled water, being careful not to lose any solid when decanting the water. With a Pasteur pipet remove as much of the remaining water as possible. Place the dish in a drying oven set at 120°C for 15 minutes. Allow the dish to cool to room temperature, and weigh it on the electronic balance. Dry in the oven again for 10 minutes. Determine and record the mass. Continue to dry the copper in the oven, for 10 minutes at a time, until two successive mass determinations differ by less than 5 mg. Record the appearance of the sample.

Show your sample of recovered copper to your laboratory instructor, and then place it in the "Recovered Copper" container provided.

Waste Disposal. Dispose of all chemical waste as directed by your instructor.

Name: _____   Date: _____

Lab Instructor: _____   Lab Section: _____

EXPERIMENT 11

Copper Reactions

PRELABORATORY QUESTIONS

1. Verify that Reaction (7) is not an oxidation-reduction reaction by determining the oxidation numbers of each of the atoms and showing that they do not change from reactants to products.

$$CaCO_3(s) \xrightarrow{\Delta} CaO(s) + CO_2(g) \tag{7}$$

2. Classify each of the reactions below as either precipitation, acid-base, redox, or decomposition.

   (a) $H_2SO_4(aq) + Ca(OH)_2(aq) \rightarrow CaSO_4(aq) + 2\ H_2O(l)$

   (b) $AgNO_3(aq) + NaCl(aq) \rightarrow AgCl(s) + NaNO_3(aq)$

   (c) $2\ Fe_2O_3(s) + 3\ C(s) \rightarrow 4\ Fe(s) + 3\ CO_2(g)$

   (d) $Cu(s) + 2\ AgNO_3(aq) \rightarrow Cu(NO_3)_2(aq) + Ag(s)$

   (e) $C_6H_{12}O_6 + 6\ O_2 \rightarrow 6\ CO_2 + 6\ H_2O$

   (f) $2\ NaHCO_3(s) \xrightarrow{\Delta} Na_2CO_3(s) + CO_2(g) + H_2O(g)$

3. Why is it necessary to clean the surface of the copper wire with sandpaper prior to reacting it with nitric acid?

Name: _____   Date: _____

Lab Instructor: _____   Lab Section: _____

4. What safety precautions must be used when carrying out the reaction between copper and nitric acid? Why must these precautions be observed?

5. How will the copper oxide be separated from aqueous sodium nitrate?

6. How will the excess magnesium solid be removed after the copper solid has been completely recovered?

Name: _____   Date: _____

Lab Instructor: _____   Lab Section: _____

EXPERIMENT 11

Copper Reactions

RESULTS/OBSERVATIONS

Part A: Conversion of Copper Metal to Copper(II) Nitrate

1. Color of the copper wire before cleaning: _____

2. Color of the copper wire after cleaning: _____

3. Mass of the copper wire: _____

4. Observations on adding concentrated HNO_3: _____

5. Observations on diluting with water: _____

Part B: Formation of Copper(II) Hydroxide

6. Observations on adding 3 M NaOH: _____

Part C: Conversion of Copper(II) Hydroxide to Copper(II) Oxide

7. Observation of precipitate color change: _____

Part D: Conversion of Copper(II) Oxide to Copper(II) Sulfate

8. Observation on adding 10 mL of 2 M H_2SO_4: _____

Part E: Recovery of Copper

9. Observation on adding magnesium turnings: _____

10. Mass of the evaporating dish: _____

11. Mass of the evaporating dish and copper after drying in the oven: _____ (Trial #1)

    _____ (Trial #2) _____ (Trial #3) _____ (Trial #4) _____ (Trial #5)

12. Appearance of the copper sample: _____

Name: _____   Date: _____

Lab Instructor: _____   Lab Section: _____

EXPERIMENT 11

Copper Reactions

POSTLABORATORY QUESTIONS

1. Balanced equation for Part A. This is an example of a _____ reaction.

2. Balanced equation for Part B. This is an example of a _____ reaction.

3. Balanced equation for Part C. This is an example of a _____ reaction.

4. Balanced equation for Part D. This is an example of a _____ reaction.

5. Balanced equations for Part E. The first is an example of a _____ reaction. The second is an example of a _____ reaction.

6. Calculation of the percent recovery of copper, where percent recovery is defined as

$$\text{Percent recovery} = \frac{\text{mass of recovered Cu}}{\text{initial mass of Cu}} \times 100$$

7. Suggest reasons why the percent recovery might be greater or less than 100%.

EXPERIMENT 12

Standardization of a Sodium Hydroxide Solution

OBJECTIVE

To standardize a solution of NaOH by titrating samples containing known masses of KHP.

INTRODUCTION

Solutions of sodium hydroxide (NaOH) can be prepared by either dissolving solid NaOH pellets in water or by diluting a concentrated solution of NaOH. However, an accurate concentration of the solution prepared by these methods cannot be calculated from the weighed mass or using the dilution equation for two reasons. First, solid NaOH is hygroscopic; pellets of NaOH exposed to air will absorb water, so the actual mass of pure NaOH is not known accurately. Second, NaOH in solution reacts with carbonic acid (H_2CO_3), i.e.,

$$H_2CO_3(aq) + NaOH(aq) \rightarrow H_2O(l) + NaHCO_3(aq) \tag{1}$$

causing its concentration to decrease with time. Carbonic acid is formed when small amounts of carbon dioxide gas (CO_2), which is always present in air, dissolves in solution:

$$CO_2(g) + H_2O(l) \rightarrow H_2CO_3(aq) \tag{2}$$

The water used to make the NaOH solution can be boiled to expel the dissolved CO_2, but because of the time involved this is often not possible in a short laboratory period. A stock solution of NaOH can be made in advance with boiled water, but it will reabsorb CO_2 over a period of time unless stored in an airtight container. To determine the concentration accurately of a freshly prepared NaOH solution or one that has been allowed to stand in air for some time it is necessary to standardize it and to determine its concentration by titrating it with a known mass of a primary standard acid.

A *primary standard* is a substance that is used to determine the concentration of a solution. An effective primary standard is available in very high purity at a reasonable cost, has a high molecular weight to minimize weighing errors, is stable at room temperature, is easy to dry, and should not readily absorb water when exposed to air.

Potassium hydrogen phthalate ($KC_8H_4O_4H$), often denoted by the shorthand notation KHP, is the primary standard used commonly to standardize NaOH. It is a monoprotic acid with a molecular weight of 204.22 g/mol.

The KHP is normally heated to 110°C for 1 hour to remove any loosely bound waters of hydration and then cooled in a desiccator prior to use. The mass of dried KHP is determined by weighing. The KHP then is dissolved in water, and NaOH is added incrementally until just enough NaOH has been added to completely react with the acid. The point at which this occurs, the end-point, is often denoted by the color change of an indicator solution. The indicator solution used in this experiment, phenolphthalein, is colorless in acidic solutions and pink in basic ones. The solution containing KHP will remain colorless as long as any KHP is present. Once all the KHP has reacted the solution will turn pink with 1 drop of added base.

At the end-point of this titration experiment the moles of NaOH added is stoichiometrically equivalent to the moles of KHP present initially. This information can be used in conjunction with the balanced chemical equation for the reaction between KHP and NaOH,

$$KC_8H_4O_4H(aq) + NaOH(aq) \rightarrow KC_8H_4O_4Na(aq) + H_2O(l) \tag{3}$$

to determine the moles of NaOH added during the titration. The volume of NaOH used in the titration can be read off the buret. With these two data the concentration of NaOH can be determined.

In this experiment you will standardize a solution of NaOH by titrating samples containing known masses of KHP.

ADDITIONAL READING

Read the sections in the "Laboratory Techniques" chapter at the beginning of this Lab Manual on cleaning glassware, handling chemicals, weighing, and burets prior to performing this experiment.

Safety Precautions:

■ Protective eyewear approved by your institution must be worn at all times while you are in the laboratory.

■ Sodium hydroxide is corrosive, especially to the eyes. Thoroughly wash off any sodium hydroxide spilled on your skin. If you get it into your eyes, immediately flush your eyes with water for at least 10 minutes. If you spill it on your clothing, remove the affected clothing before rinsing with water. Notify your lab instructor, who will assist you.

PROCEDURE

Measure and record the mass of a 125 mL Erlenmeyer flask as precisely as possible. Add approximately 3 g of KHP to the flask. Measure and record this combined mass as precisely as possible. Determine the mass of KHP in the beaker by subtraction.

Dissolve the KHP in the Erlenmeyer flask in 50–75 mL of distilled water. Add 2–3 drops of phenolphthalein indicator solution to this mixture.

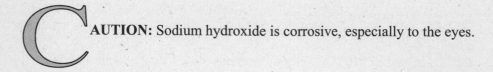

CAUTION: Sodium hydroxide is corrosive, especially to the eyes.

Fill a clean buret with the dilute NaOH solution [see Figure 12.1(a)]. Allow a few milliliters of NaOH to drain through the stopcock into a waste beaker in order to flush out any air bubbles trapped in the tip. Touch the wall of the waste beaker to the buret tip in order to remove the drop of NaOH likely hanging from the buret tip. Record the initial volume of NaOH.

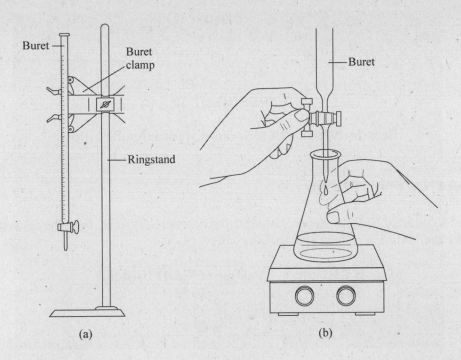

Figure 12.1
Using a buret: (a) support a buret with a buret clamp attached to a ringstand; (b) a buret
being used in a titration experiment.

Place a clean magnetic stirbar in the flask containing the KHP solution, and set the flask on a stirplate. Turn the stirplate control knob until the solution is churning gently. Position the ringstand so that the buret tip is in the neck of the Erlenmeyer flask [see Figure 12.1(b)]. Add NaOH from the buret slowly to the KHP solution. Periodically stop the addition of NaOH and wash the wall of the Erlenmeyer flask with distilled water from a washbottle. There will be momentary flashes of pink color as the NaOH is added, and as the end-point nears these flashes will be visible for longer periods of time. These flashes of pink occur because the solution is momentarily basic in the local region where the drop touches the solution. Mixing homogenizes the solution, removing the locally basic region. The end-point of the titration is reached when a *very faint pink* color persists throughout the solution for more than 30 seconds (i.e., the solution is no longer locally basic; it is wholly basic). Record the final volume of NaOH. (If the solution is dark pink the end-point has been overshot by one or two drops of NaOH.) Determine the volume of NaOH used by subtraction.

Repeat the preceding steps until at least three very close determinations are made.

Waste Disposal. Dispose of all chemical waste as directed by your instructor.

Name: _____   Date: _____

Lab Instructor: _____   Lab Section: _____

EXPERIMENT 12

Standardization of a Sodium Hydroxide Solution

PRELABORATORY QUESTIONS

1. A diluted solution of NaOH was used to titrate three samples of KHP. Find the concentration of NaOH for each titration and the average concentration:

Trial #	Mass of KHP (g)	Volume of NaOH Used (mL)
1	3.182	30.02
2	2.891	28.32
3	2.699	24.89

2. What is the purpose of the indicator solution in this experiment?

3. What signifies the end-point of the titration in this experiment?

Name: _____   Date: _____

Lab Instructor: _____   Lab Section: _____

EXPERIMENT 12

Standardization of a Sodium Hydroxide Solution

RESULTS/OBSERVATIONS

1. Titration data:

	Trial #1	Trial #2	Trial #3	Trial #4
Mass of 125 mL Erlenmeyer flask (g)				
Combined mass of 125 mL Erlenmeyer flask and KHP (g)				
Mass KHP (g)				
Initial NaOH volume in buret (mL)				
Final NaOH volume in buret (mL)				
Volume of NaOH (mL)				

2. Show in detail the calculation of NaOH concentration using the data collected in trial #_____ (choose a trial):

3. Titration results:

	Trial #1	Trial #2	Trial #3	Trial #4
Moles of KHP (mol)				
Concentration of NaOH (M)				

4. Average NaOH concentration: _____

Name: _____   Date: _____

Lab Instructor: _____   Lab Section: _____

EXPERIMENT 12

Standardization of a Sodium Hydroxide Solution

POSTLABORATORY QUESTIONS

1. If the KHP used in this experiment was not dried prior to use, how would this affect the calculated concentration of the standardized NaOH solution?

2. If the 125 mL Erlenmeyer flasks used in this experiment were washed and rinsed with distilled water but not dried prior to use, how would this affect the calculated concentration of the standardized NaOH solution?

3. If a student performing this experiment failed to remove a drop of NaOH hanging from the buret tip before titrating a solution of KHP and the drop fell into the flask when the titration began, how would this affect the calculated concentration of the standardized NaOH solution?

4. If air bubbles were not removed from the tip of the buret prior to titrating a solution of KHP, how would this affect the calculated concentration of the standardized NaOH solution?

5. A student performing this experiment forgot to add phenolphthalein until after several milliliters of NaOH had been added in the titration. Trying to salvage the experiment, the student added two drops of phenolphthalein indicator solution and noticed that the solution turned dark pink. How would this affect the calculated concentration of the standardized NaOH solution?

EXPERIMENT 13

Molar Mass
of a Volatile Liquid

OBJECTIVE

To determine the molar mass of a liquid by the Dumas method. The Dumas method requires volatilizing the liquid in a container with a measurable volume. The determined volume, combined with the measured pressure and temperature, can be substituted into the ideal gas law to determine the moles of gas. The mass of the gas divided by the moles equals the molar mass.

INTRODUCTION

During the 1800s the French chemist J. B. A. Dumas devised a simple technique for determining the molar mass of a volatile liquid. The Dumas method involves placing a small amount of the liquid in a preweighed flask of known volume. A cover containing a tiny pinhole is placed over the flask, and the flask is partially immersed in a hot-water bath. All the liquid will volatilize, forcing air out as it fills the flask. Some of the vapor also will leave the flask via the pinhole. This loss of vapor will stop when the pressure inside the flask is equal to the external pressure (the barometric pressure). The temperature of the hot-water bath is recorded, the flask is removed, and the vapor filling it is allowed to condense. The flask then is reweighed to determine the mass of the liquid, which is equal to the mass of the gas that occupied the flask at the temperature of the hot-water bath and barometric pressure.

This procedure leads to direct measurement of all the variables necessary for determining the molar mass of the volatile liquid. Specifically, the ideal gas law allows the moles of gas to be calculated:

$$PV = nRT \tag{1}$$

and the mass of the gas divided by the number of moles yields the molar mass:

$$\text{Molar mass} = \frac{m_{gas}}{n_{gas}} \tag{2}$$

Your purpose in this experiment is to determine the molar mass of a volatile liquid of unknown identity using the Dumas method.

ADDITIONAL READING

Read the sections in the "Laboratory Techniques" chapter at the beginning of this Lab Manual on cleaning glassware, handling chemicals, weighing, and heating liquids and solutions prior to performing this experiment.

 Safety Precautions:

- Protective eyewear approved by your institution must be worn at all times while you are in the laboratory.

- Many of the liquid unknowns used in this experiment are flammable liquids. No open flames should be used during this experiment. In the event of a fire, notify your instructor. A fire in a small container can be smothered by covering the vessel with a watch glass. Use a fire extinguisher to eliminate a larger fire. If the fire is burning over too large an area to be extinguished easily, evacuate the area, and activate the fire alarm. If your clothing should catch on fire, use the safety shower or a fire blanket to extinguish the flames.

PROCEDURE

Securely crimp a piece of aluminum foil over the mouth of a 125 mL Erlenmeyer flask. Punch a tiny pinhole (as small as possible) in the foil to allow air and excess vapor to escape. Determine and record the mass of the flask and cap.

Obtain an unknown volatile liquid sample from your instructor. Record the unknown identification number. Place about 6 mL of the liquid in the flask, and resecure the cap.

Fill a 600 mL beaker about one-half full, and, working in the fume hood, heat it to boiling on a stirrer-hotplate. Attach a clamp to a ringstand, and tighten the clamp around the neck of the flask. Partially immerse the flask in the hot-water bath. Place a split one-hole rubber stopper around a thermometer, and support it on the ringstand using a second clamp. See Figure 13.1 for the experimental setup. Measure and record the temperature of the boiling water and the barometric pressure.

Heat the flask in the hot-water bath until all the liquid has vaporized. This sometimes can be difficult to observe. Watch the light reflected from the surface of the liquid pool and the light refracted by the jet of vapor escaping through the pinhole. When both of these disappear all the liquid has vaporized. At this point, the flask should be removed from the hot-water bath by unfastening the clamp from the ringstand and lifting the flask out, holding it by the clamp. Set the flask aside to cool, removing the clamp.

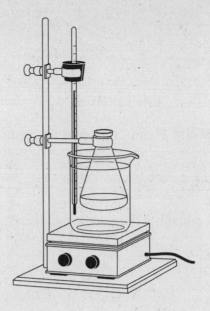

Figure 13.1
The Dumas method setup.

After the flask is cool, examine it and the cap for water droplets. Wipe it dry if necessary. Weigh the flask, cap, and unknown liquid. Record this mass. Determine the mass of the liquid by difference.

Determine and record the mass of a 400 mL beaker. Fill the flask completely full of water. Pour the water into the beaker. Determine and record the mass of the water and the beaker. Calculate the volume of water, which is equal to the volume of the flask, using the water mass and density (1.00 g/mL).

Repeat the procedure.

Waste Disposal. Dispose of all chemical waste as directed by your instructor.

Name: _____   Date: _____

Lab Instructor: _____   Lab Section: _____

EXPERIMENT 13

Molar Mass of a Volatile Liquid

PRELABORATORY QUESTIONS

1. Why is it necessary to place a pinhole in the aluminum foil cap?

2. What is the molar mass of 0.527 g of a gas that occupies 252.03 mL at 0.942 atm and 99.8°C?

3. What safety precaution is necessary in this experiment?

Name: _____   Date: _____

Lab Instructor: _____   Lab Section: _____

EXPERIMENT 13

Molar Mass of a Volatile Liquid

RESULTS/OBSERVATIONS

1. Mass of flask and cap: _____

2. Unknown identification number: _____

3. Temperature of the boiling water: _____

4. Barometric pressure: _____

5. Mass of flask, cap, and unknown liquid: _____ (Trial #1) _____ (Trial #2)

6. Mass of liquid: _____ (Trial #1) _____ (Trial #2)

7. Mass of the 400 mL beaker: _____

8. Mass of the 400 mL beaker and water: _____

9. Mass of water: _____

10. Calculation of the volume of the flask from the water mass and density:

11. Calculation of the molar mass of the unknown compound using the data from Trial #_____
    (choose a trial):

12. Average molar mass: _____

Name: _____   Date: _____

Lab Instructor: _____   Lab Section: _____

EXPERIMENT 13

Molar Mass of a Volatile Liquid

POSTLABORATORY QUESTIONS

1. What does the flask contain at the following points during the experiment?

   (a) Before adding the volatile liquid

   (b) At the point when the volatile liquid has completely vaporized

   (c) After the flask has been removed from the hot-water bath and allowed to cool to room temperature

2. Would the Dumas method of molar mass determination work well when applied to a liquid that is volatile at room temperature? Explain.

3. Why was the volume of the flask calculated by weighing the water rather than measuring its volume in a graduated cylinder?

Experiment 14

Stoichiometry and the Ideal Gas Law

OBJECTIVE

To devise and implement a procedure capable of determining the identity of the unknown nitrite salt. The procedure must involve the reaction between sulfamic acid and a nitrite salt [Reaction (2)] and the ideal gas law.

INTRODUCTION

Numerous experiments performed in the seventeenth, eighteenth, and nineteenth centuries led to an empirical description of the physical properties of gases in terms of their pressure (P), volume (V), number of moles (n), and temperature (T):

$$PV = nRT \tag{1}$$

where P is usually in atmospheres (atm), V is typically in liters (L), T is expressed in units of kelvin (K), and R is a constant of proportionality termed the *universal gas constant* (0.08206 atm · L/mol · K; it has different values with different units, but this is the value pertinent to this experiment). Equation (1) is termed the *ideal gas law*. As the name suggests, strict adherence to the ideal gas law is the "ideal" case. Such ideal behavior is observed most commonly under conditions of low pressure and high temperature, but even for conditions of 1 atm and 25°C, approximately those of this experiment, deviations from ideal behavior are negligible for most gases, and use of the ideal gas law is justified.

Your purpose in this experiment is to determine the identity of a nitrite salt of the general formula MNO_2, where M^+ is an alkali metal cation: Li^+, Na^+, K^+, or Rb^+. You will determine the identity of the nitrite salt by dissolving it in water and reacting it with excess sulfamic acid (HSO_3NH_2):

$$MNO_2(aq) + HSO_3NH_2(aq) \rightarrow MHSO_4(aq) + H_2O(l) + N_2(g) \tag{2}$$

The moles of nitrogen gas produced as a consequence of this reaction are stoichiometrically related to the moles of MNO_2 originally present. From the moles MNO_2 originally present and the mass of the MNO_2 sample it is possible to identify the nitrite salt.

The mass of the nitrogen gas produced in Reaction (2) could readily be used to determine the number of moles of nitrogen gas. But because it is a gas, a direct measurement of the mass of nitrogen produced is difficult to achieve. However, since the conditions of this experiment are such that use of the ideal gas law is justified, measurement of the volume, temperature, and pressure of the nitrogen gas produced allows for the moles of nitrogen to be calculated.

The apparatus shown in Figure 14.1 will be used in this experiment. The reaction between aqueous MNO_2 and sulfamic acid will occur in the 250 mL Erlenmeyer flask. The volume of nitrogen gas produced in the 250 mL Erlenmeyer flask will equal the volume of water displaced from the 500 mL

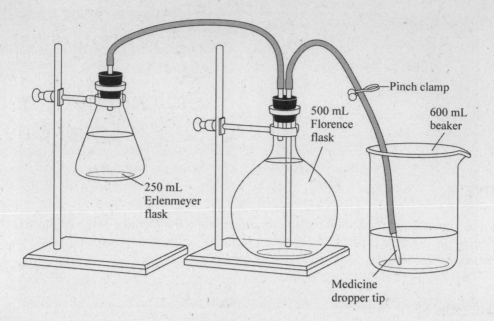

Figure 14.1
Water displacement setup.

Florence flask. It will be assumed that the temperature of the nitrogen gas is the same as the water temperature in the Erlenmeyer flask. Measuring the pressure is a bit more complicated, however, because the nitrogen gas produced in this reaction will be saturated with water vapor. According to Dalton's law of partial pressures, the total pressure of the gas will be equal to the partial pressure of nitrogen plus the partial pressure of water vapor:

$$P_{Total} = P_{H_2O} + P_{N_2} \qquad (3)$$

In this experiment the total pressure of the gas inside the apparatus will be made equal to the pressure in the laboratory, i.e., the barometric pressure. The barometric pressure will be determined by reading a barometer. The partial pressure of water vapor (P_{H_2O}) is a function of temperature and can be obtained directly from a table, which has been provided for you (Appendix B).

 Safety Precautions:

■ Protective eyewear approved by your institution must be worn at all times while you are in the laboratory.

■ Use the proper technique and great care when inserting glass tubing through a rubber stopper. Serious cuts to the hand can occur quite easily if this task is performed incorrectly. Be sure that the stopper hole is lubricated, and wrap the glass tube in a towel or lab apron before attempting to insert. Grasp the glass rod near the point of insertion to minimize torque on the glass.

ADDITIONAL READING

Read the sections in the "Laboratory Techniques" chapter at the beginning of this Lab Manual on cleaning glassware, handling chemicals, inserting glass tubing through a rubber stopper, and weighing prior to performing this experiment.

EXPERIMENT

You are to use the procedure below to determine the identity of the unknown nitrite salt. *Note: The procedure below only describes how to assemble and use the apparatus shown in Figure 14.1 properly. You are responsible for deciding what additional steps are necessary to determine the identity of the nitrite salt.*

PROCEDURE

Assemble the apparatus shown in Figure 14.1.

CAUTION: Use the proper technique and great care when inserting glass tubing through a rubber stopper. Serious cuts to the hand can occur quite easily if this task is performed incorrectly.

Be sure that the longer piece of glass tubing going into the 500 mL Florence flask extends nearly to the bottom. Clean the Erlenmeyer flask thoroughly, and rinse it with distilled water. It is not necessary that it be dry. The Florence flask will serve as a water reservoir. Nitrogen will be produced in the Erlenmeyer flask and then travel through the rubber tubing into the Florence flask, pushing water out of the Florence flask and into the 600 mL beaker. The volume of water collected in the beaker will be equal to the volume of nitrogen gas produced. A medicine dropper tip should be inserted into one end of the rubber tubing; this will help keep water from dripping out of the rubber tubing into the beaker.

Clean and dry the 600 mL beaker. Fill the 500 mL Florence flask nearly full of water. Remove the stopper from the Erlenmeyer flask, and attach a pipet bulb to the glass tubing extending through the stopper.

Fill a 400 mL beaker about two-thirds full of water. With the pinch clamp off the rubber tubing and with the medicine dropper tip below the surface of the water in the 400 mL beaker, squeeze the pipet bulb, forcing water from the flask through the rubber tubing and out into the beaker. Pinch the rubber tubing leading from the Florence flask to the 400 mL beaker to dislodge any bubbles of air. Remove the pipet bulb. Verify that there are no air bubbles in the glass or rubber tubing from the Florence flask to the 400 mL beaker. If air bubbles are present, dislodge them by pinching or forcing water through the tubing by squeezing the pipet bulb.

(The presence of air bubbles in the glass and rubber tubing connecting the Florence flask to the beaker will detrimentally affect the accuracy of this experiment. As nitrogen gas is produced in the Erlenmeyer flask, it exerts a pressure on the surface of the water in the Florence flask. If there are no air bubbles in the tubing, this pressure causes a displacement of water equal in volume to the nitrogen produced. However, if air bubbles are present, the increased pressure first compresses the air bubbles and then displaces water, causing the volume of water displaced to be less than the volume of nitrogen produced.)

Obtain a 10 × 75 mm culture tube from your lab instructor. Test the fit of the culture tube inside your Erlenmeyer flask while both are empty. The culture tube should be too long to lay fully across the bottom of the flask. With it bottom resting on the bottom of the flask, the culture tube opening should be above the 100 mL mark on the Erlenmeyer flask. If it isn't, try a different Erlenmeyer flask or culture tube. Once a proper fit has been established, remove the culture tube from the flask.

Add 25.0 mL of sulfamic acid solution (80.0 g/L of solution) to the Erlenmeyer flask. Add an additional 65–70 mL of distilled water to the Erlenmeyer flask.

Obtain your unknown nitrite salt from your lab instructor. Place the entirety of your nitrite salt sample in the culture tube. Add about 3 mL of distilled water to the culture tube, and, being careful not to splash, tap the tube briskly on the side until the nitrite salt is at least partially dissolved. Tilt the Erlenmeyer flask at an angle, and slide the culture tube down into it, being careful not to spill any of the contents of the culture tube into the solution in the flask (see Figure 14.2).

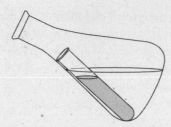

Figure 14.2
Tilt the Erlenmeyer flask at an angle, and slide the culture tube down into it, being careful not to spill any of the contents of the culture tube into the solution in the flask.

Set the Erlenmeyer flask down. Raise the 400 mL beaker in order to siphon enough water back into the Florence flask that the water level is at least halfway up the neck of the Florence flask. Then quickly stopper the Erlenmeyer flask securely with the rubber stopper.

An accurate determination of the amount of gas produced requires that the system be airtight. Test for leaks by lifting the dropper tip barely out of the water in the 400 mL beaker. If there are no leaks in the apparatus, almost no water should drip out of the tip. If there are leaks lower the dropper tip into the water and check the snugness of all the connections. Do not proceed until the system is airtight.

Close the pinch clamp on the rubber tubing. Carefully remove the rubber tubing and dropper tip from the 400 mL beaker. Gently place the rubber tubing and dropper tip inside the dry 600 mL beaker. The tip should nearly reach the bottom of the 600 mL beaker.

Open the pinch clamp on the rubber tubing. Unclamp the Erlenmeyer flask from the ring stand, and tilt the flask enough to get a little bit of the solution in the culture tube to mix with the sulfamic acid solution in the flask. *Do not mix too fast, or the nitrogen gas may be generated so quickly that it will blow the stopper out of the flask.* Water should start flowing into the 600 mL beaker as nitrogen is produced.

After about a minute, tip the flask again to add some more nitrite salt solution, and swirl the contents. Continue doing this until the entirety of the contents of the culture tube have been mixed with the solution in the flask and no more nitrogen gas is produced. Toward the end, tip the flask as much as you can, without getting liquid up into the glass tubing at the top, in order to make sure that all the nitrite salt solution in the culture tube has reacted.

When all the nitrite salt solution has reacted, adjust the pressure of the gas inside the apparatus to make it the same as the pressure in the laboratory. To do this, raise or lower the Florence flask and beaker until the upper water surfaces in each container are at the same level, i.e., until the water levels have been equalized. Close the pinch clamp.

EQUIPMENT AND REAGENTS

To perform this experiment, you will have access to all the equipment in your lab drawer, a pipet bulb, the electronic balances, a 10 × 75 mm culture tube, and the additional items necessary to construct the apparatus shown in Figure 14.1.

Waste Disposal. Dispose of all chemical waste as directed by your instructor.

Name: _____   Date: _____

Lab Instructor: _____   Lab Section: _____

EXPERIMENT 14

Stoichiometry and the Ideal Gas Law

PRELABORATORY QUESTIONS

1. Why is it necessary to remove any air bubbles from the glass and rubber tubing connecting the Florence flask to the beaker before initiating the reaction?

2. Why must the apparatus in Figure 14.1 be airtight prior to initiating the reaction?

3. Why is it important to react the nitrite salt solution in the vial with the sulfamic acid solution in small amounts?

4. How will you determine the temperature of the nitrogen gas produced in this experiment?

Name: _____   Date: _____

Lab Instructor: _____   Lab Section: _____

5. If the barometric pressure in the lab is 755 mm Hg and the temperature of the nitrogen gas is 23.5°C, what is the pressure of the nitrogen gas?

6. There are two ways in which the volume of nitrogen gas produced in this experiment can be determined. One is by direct measurement of the volume of displaced water in the 600 mL beaker. The other is by determining the mass of the displaced water; this mass divided by the density of water gives the volume of water displaced. Which of these methods is more precise? Explain your reasoning. You should use the more precise method in your experimental procedure.

7. Why must the mass of the nitrite salt be determined in this experiment? How will you determine this mass?

Name: _____   Date: _____

Lab Instructor: _____   Lab Section: _____

EXPERIMENT 14

Stoichiometry and the Ideal Gas Law

RESULTS/OBSERVATIONS

1. Unknown nitrite salt identification number: _____

2. Write any other pertinent data and calculations in the space below. Clearly indicate both the property and the amount.

3. The identity of your unknown nitrite salt: _____.

Name: _____   Date: _____

Lab Instructor: _____   Lab Section: _____

EXPERIMENT 14

Stoichiometry and the Ideal Gas Law

POSTLABORATORY QUESTIONS

1. Hydrogen gas can be prepared by reaction of zinc metal with aqueous hydrochloric acid (HCl):

    $$Zn(s) + 2\ HCl(aq) \rightarrow ZnCl_2(aq) + H_2(g)$$

    (a) How many liters of hydrogen gas would produced at 752 mm Hg and 21.0°C if 12.5 g of zinc was reacted with 300.00 mL of 1.50 M HCl?

    (b) How many grams of zinc are necessary to prepare 10.00 L of hydrogen gas at 27.5°C and 1.10 atm?

    (c) An excess of zinc metal was combined with 100.00 mL of aqueous HCl, and the resulting reaction yielded 3.82 L of gaseous hydrogen at 24.8°C and 759 torr. What was the molarity of the HCl solution?

VSEPR and Molecular Shape

OBJECTIVE

To generate Lewis structures for a number of molecules and ions. The Lewis structures then will be used according to the VSEPR model to determine the molecular geometry and molecular polarity of each molecule or ion.

INTRODUCTION

The three-dimensional shape of a molecule has a profound effect on its properties. A number of striking examples can be found among biological molecules. For example, slight changes in the structure of a biomolecule can render it useless to a cell.

There are a number of methods by which one could determine the molecular geometry of a molecule or ion. The model used in this experiment is a simple one called the *valence shell electron-pair repulsion* (VSEPR) *model,* which, despite its simplicity, does very well at predicting the approximate shape of a molecule. The fundamental postulate of the VSEPR model is that the geometry around a given atom is that which minimizes electron-pair repulsions. The pairs of valence electrons surrounding an atom, be they bonding or lone pairs, repel one another. These valence pairs of electrons around an atom adopt an orientation that maximizes their separation, thus minimizing their repulsions.

Of course, in order to use the VSEPR model it is necessary to know how many pairs of valence electrons a molecule or ion possesses. Once this is known and the VSEPR model applied to determine the molecular geometry a number of other interesting characteristics of the molecule can be determined, such as information about the nature of the bonding and polarity of the molecule. The following sections describe how all these characteristics can be determined.

Lewis Structures

The number of bonding and lone pairs of valence electrons is derived from the Lewis structure of the molecule or ion. The following rules can be used to determine the Lewis structure:

1. *Arrange the atoms usually with the least electronegative atom in the center. (Hydrogen is never the central atom.) If the species is an ion, bracket it and indicate the charge as a right superscript.* To represent the method, the Lewis structures of CH_4 and OCN^- will be determined:

H

H   C   H
$$\left[\; O \quad C \quad N \;\right]^{-}$$

H

2. *Determine the total number of valence electrons available. For polyatomic ions, add 1 e⁻ for each negative charge of the ion or subtract 1 e⁻ for each positive charge. For CH₄ and OCN⁻:*

	1 O atom: 6 e⁻
	1 C atom: 4 e⁻
1 C atom: 4 e⁻	1 N atom: 5 e⁻
4 H atoms: 4(1 e⁻)	−1 charge: 1 e⁻
CH₄ valence: 8 e⁻	OCN⁻ valence: 16 e⁻

3. *Connect each surrounding atom to the central atom by a single bond, subtracting two valence electrons from the total for each bond. (Note: A dash represents a pair of bonding electrons.)*

H
|
H—C—H
|
H

$$\left[\; O{-}C{-}N \;\right]^{-}$$

CH₄ valence remaining: 8 e⁻ − 8 e⁻ = 0 e⁻   OCN⁻ valence remaining: 16 e⁻ − 4 e⁻ = 12 e⁻

4. *Distribute the remaining electrons in pairs so that each atom has 8 valence electrons (2 for hydrogen). Do this for the surrounding (more electronegative) atoms first.*

All the atoms in CH₄ have an octet (or duet, in the case of hydrogen). Only oxygen and nitrogen have octets in OCN⁻.

5. *If at this point the central atom does not have an octet of electrons change a lone pair from one of the surrounding atoms into a bonding pair with the central atom.*

H
|
H—C—H
|
H

$$\left[\; \ddot{O}{=}C{=}\ddot{N} \;\right]^{-}$$

Now all the atoms in both CH_4 and OCN^- satisfy the octet rule.

There are some molecules and ions that do not satisfy the octet rule. These exceptions involve central atoms from Groups 2 and 3, which can have fewer than 8 valence electrons, e.g., $BeCl_2$ and BF_3:

Other exceptions involve expanded octets, which have more than 8 valence electrons, e.g., PCl_5 and XeF_4,

Only elements of the third row of the periodic table or below will display expanded valence shells; only these elements possess energetically accessible unfilled d orbitals that can be used in bonding.

VSEPR Model

Once the Lewis structure has been determined, the number of electron groups about a central atom is determined by counting: Carbon in CH_4 has 4 e^- groups, carbon in OCN^- has two e^- groups (double bonds and triple bonds are only considered one e^- group), beryllium in $BeCl_2$ has 2 e^- groups, boron in BF_3 has 3 e^- groups, phosphorus in PCl_5 has 5 e^- groups, and xenon in XeF_4 has 6 e^- groups. There are five different electron group geometries possible, each corresponding to a different number of e^- groups: linear (2 e^- groups), trigonal planar (3 e^- groups), tetrahedral (4 e^- groups), trigonal bipyramidal (5 e^- groups), and octahedral (6 e^- groups). Each electron group geometry maximizes the separation between electron groups.

The molecular geometries are related but not necessarily equivalent to the electron group geometries. This is because the molecular geometry is defined by the locations of the nuclei: Lone pairs affect the molecular geometry, but they are not part of it. Table 15.1 summarizes the relationships between the number of e^- groups, the e^- group geometry, and the molecular geometry.

Table 15.1

No. of e^- Groups	e^- Group Geometry	Perspective Drawing	Molecular Geometry	Approximate Bond Angles
2	Linear	X——A——X	Linear	180°
3	Trigonal planar		Trigonal planar	120°
			Bent	120°
4	Tetrahedral		Tetrahedral	109.5°
			Trigonal pyramidal	109.5°
			Bent	109.5°
5	Trigonal bipyramidal		Trigonal bipyramidal	90 & 120°
			See-saw	90° & 120°

Table 15.1 (Continued)

No. of e^- Groups	e^- Group Geometry	Perspective Drawing	Molecular Geometry	Approximate Bond Angles
			T-shaped	90°
			Linear	180°
6	Octahedral		Octahedral	90°
			Square pyramidal	90°
			Square planar	90°

Once the Lewis structure of a molecule or ion has been determined and the number of e^- groups it contains counted, Table 15.1 can be used to determine the molecular geometry.

Notice that Table 15.1 includes perspective drawings of each of the molecular geometries. It is common practice to use perspective drawings to represent the three-dimensional shapes of molecules. In these perspective drawings, a solid line represents a bond in the plane of the paper, a wedge indicates a bond coming out of the plane of the paper toward the observer, and a dashed line indicates a bond going out of the plane of the paper away from the observer.

The approximate bond angles associated with each molecular geometry are also listed in Table 15.1.

Valence Bond Theory and Hybridization

Valence bond theory describes a covalent bond as the overlapping of two half-filled valence orbitals. For example, the formation of H_2 is described by the overlap of two half-filled $1s$ orbitals:

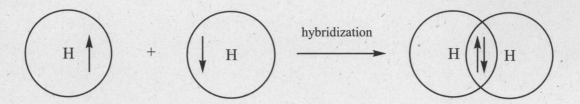

But it isn't possible to overlap simple atomic orbitals and achieve all the known molecular geometries. Take CH_4 as an example. It has 4 e^- groups, a tetrahedral e^- group geometry, and a tetrahedral molecular geometry with 109.5° bond angles. It isn't possible to overlap hydrogen $1s$ orbitals and carbon $2s$ or $2p$ orbitals and achieve a bond angle of 109.5°. In such cases, the valence orbitals of the central atom are said to *hybridize,* to mathematically mix together and form an equivalent number of hybrid orbitals, orbitals separated by angles different than those of the original orbitals. In the case of CH_4 the hybridization involves the $2s$ and the three $2p$ orbitals to form four hybrid orbitals displaying angles of 109.5°; the hybrid orbitals in this case are termed sp^3 hybrid orbitals because they are the result of combining one s and three p orbitals:

$$2s \uparrow\downarrow \; 2p \uparrow \; \uparrow \xrightarrow{\text{hybridization}} sp^3 \uparrow \; \uparrow \; \uparrow \; \uparrow$$

Since the sp^3 hybrid orbitals have 109.5° angles they are appropriate anytime a molecule or ion has a tetrahedral e^- group geometry. Other types of hybridization correspond to other e^- group geometries. The hybridization type necessary to achieve a given e^- group geometry is summarized in Table 15.2. The total number of hybridized orbitals is equal to the number of e^- groups about the central atom.

Table 15.2

e^- Group Geometry	Hybridization
Linear	sp
Trigonal planar	sp^2
Tetrahedral	sp^3
Trigonal bipyramidal	sp^3d
Octahedral	sp^3d^2

Molecular Polarity

The polarity of an individual chemical bond is determined by the relative electronegativities of the two atoms that make up that bond. For example,

$$\overset{\longrightarrow}{H\!-\!\ddot{\underset{\cdot\cdot}{Cl}}\!:}\qquad\overset{\longrightarrow}{:C\!\equiv\!O:}\qquad\text{(No bond dipole)}\quad :\!\ddot{\underset{\cdot\cdot}{Br}}\!-\!\ddot{\underset{\cdot\cdot}{Br}}\!:$$

(The head of the dipole arrow points toward the more electronegative atom of the bond.) The polarity of a molecule is given by the vector sum of the individual bond dipoles. This sum, termed the *molecular dipole moment* (μ), depends on the molecular shape as well as the individual bond polarities. For example, each of the carbon–oxygen bonds in carbon dioxide is polar covalent, i.e.,

$$:\ddot{O}\!=\!C\!=\!\ddot{O}: \qquad \mu = 0$$

but because carbon dioxide is a linear molecule, the two bond polarities are of the same magnitude and directed oppositely to one another, causing the net polarity of the molecule to be zero. Contrast carbon dioxide with water, i.e.,

$$\mu = \uparrow$$

which also has two identical polar covalent bonds, but owing to the bent shape of water these two bond polarities do not sum to zero, instead leading to a nonzero molecular dipole moment, the direction of which is indicated above.

PROCEDURE

In this experiment you will use the VSEPR model to determine the molecular geometries of a number of molecules and ions (see the following pages). You then will use the molecular geometries and the information used to derive them to determine the hybridization and the molecular dipole moment of the molecule or ion. The schematic below outlines the process:

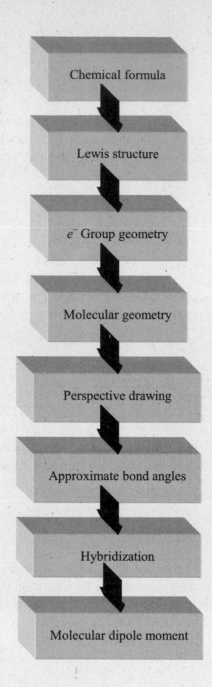

Name: _____   Date: _____

Lab Instructor: _____   Lab Section: _____

EXPERIMENT 15

VSEPR and Molecular Shape

PRELABORATORY QUESTIONS

1. Determine the Lewis structures of ClO_3^- and ClO_4^-.

2. How does the electron group geometry differ from the molecular geometry?

3. Give perspective drawings of ClO_3^- and ClO_4^-.

4. What are the hybridizations of ClO_3^- and ClO_4^-?

5. Explain why ClO_3^- has a molecular dipole moment, but ClO_4^- doesn't.

Name: _____   Date: _____

Lab Instructor: _____   Lab Section: _____

EXPERIMENT 15

VSEPR and Molecular Shape

RESULTS/OBSERVATIONS

Formula:	Lewis Structure:	Electron Group Geometry:	Perspective Drawing:	Approximate Bond Angles:
$BeCl_2$	$:\overset{..}{\underset{..}{Cl}}$—Be—$\overset{..}{\underset{..}{Cl}}:$	Linear Molecular Geometry: Linear	Cl——Be——Cl	$180°$ **Hybridization:** *sp* **Molecular Dipole Moment:** $\mu = 0$
CCl_4	Lewis Structure:	Electron Group Geometry: Molecular Geometry:	Perspective Drawing:	Approximate Bond Angles: Hybridization: Molecular Dipole Moment: $\mu =$
CS_2	Lewis Structure:	Electron Group Geometry: Molecular Geometry:	Perspective Drawing:	Approximate Bond Angles: Hybridization: Molecular Dipole Moment: $\mu =$
SF_6	Lewis Structure:	Electron Group Geometry: Molecular Geometry:	Perspective Drawing:	Approximate Bond Angles: Hybridization: Molecular Dipole Moment: $\mu =$
H_3O^+	Lewis Structure:	Electron Group Geometry: Molecular Geometry:	Perspective Drawing:	Approximate Bond Angles: Hybridization: Molecular Dipole Moment: $\mu =$

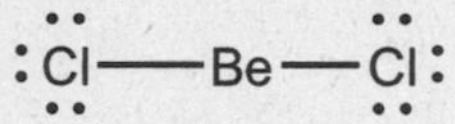

Name: _____   Date: _____

Lab Instructor: _____   Lab Section: _____

Formula:	Lewis Structure:	Electron Group Geometry: Molecular Geometry:	Perspective Drawing:	Approximate Bond Angles: Hybridization: Molecular Dipole Moment: $\mu =$
HCN				
PF_3		Electron Group Geometry: Molecular Geometry:	Perspective Drawing:	Approximate Bond Angles: Hybridization: Molecular Dipole Moment: $\mu =$
PF_5	Lewis Structure:	Electron Group Geometry: Molecular Geometry:	Perspective Drawing:	Approximate Bond Angles: Hybridization: Molecular Dipole Moment: $\mu =$
NH_4^+	Lewis Structure:	Electron Group Geometry: Molecular Geometry:	Perspective Drawing:	Approximate Bond Angles: Hybridization: Molecular Dipole Moment: $\mu =$
H_2CO	Lewis Structure:	Electron Group Geometry: Molecular Geometry:	Perspective Drawing:	Approximate Bond Angles: Hybridization: Molecular Dipole Moment: $\mu =$
NO_2^+	Lewis Structure:	Electron Group Geometry: Molecular Geometry:	Perspective Drawing:	Approximate Bond Angles: Hybridization: Molecular Dipole Moment: $\mu =$

Name: _____   Date: _____

Lab Instructor: _____   Lab Section: _____

Formula: C_2H_4	Lewis Structure:	Electron Group Geometry: Molecular Geometry:	Perspective Drawing:	Approximate Bond Angles: Hybridization: Molecular Dipole Moment: $\mu =$
Formula: NH_2^-	Lewis Structure:	Electron Group Geometry: Molecular Geometry:	Perspective Drawing:	Approximate Bond Angles: Hybridization: Molecular Dipole Moment: $\mu =$
Formula: C_2H_2	Lewis Structure:	Electron Group Geometry: Molecular Geometry:	Perspective Drawing:	Approximate Bond Angles: Hybridization: Molecular Dipole Moment: $\mu =$
Formula: BrF_3	Lewis Structure:	Electron Group Geometry: Molecular Geometry:	Perspective Drawing:	Approximate Bond Angles: Hybridization: Molecular Dipole Moment: $\mu =$
Formula: BrF_4^-	Lewis Structure:	Electron Group Geometry: Molecular Geometry:	Perspective Drawing:	Approximate Bond Angles: Hybridization: Molecular Dipole Moment: $\mu =$
Formula: I_3^-	Lewis Structure:	Electron Group Geometry: Molecular Geometry:	Perspective Drawing:	Approximate Bond Angles: Hybridization: Molecular Dipole Moment: $\mu =$

Name: _____   Date: _____

Lab Instructor: _____   Lab Section: _____

Formula:	Lewis Structure:	Electron Group Geometry: Molecular Geometry:	Perspective Drawing:	Approximate Bond Angles: Hybridization: Molecular Dipole Moment: $\mu =$
BrF_5				
Formula: CH_2Cl_2	Lewis Structure:	Electron Group Geometry: Molecular Geometry:	Perspective Drawing:	Approximate Bond Angles: Hybridization: Molecular Dipole Moment: $\mu =$
Formula: SF_4	Lewis Structure:	Electron Group Geometry: Molecular Geometry:	Perspective Drawing:	Approximate Bond Angles: Hybridization: Molecular Dipole Moment: $\mu =$

Name: _____   Date: _____

Lab Instructor: _____   Lab Section: _____

EXPERIMENT 15

VSEPR and Molecular Shape

POSTLABORATORY QUESTIONS

1. A previously ignored feature of the VSEPR model is the differential repulsion exerted by different groups of electrons: Nonbonding electron pairs exert more repulsion than bonding pairs of electrons, and multiple bonds exert more repulsion than single bonds. These differential repulsions affect the shapes of molecules, causing some bond angles to be larger than expected and others to be smaller. Which of the 20 molecules in the Results/Observations section will have bond angles different from those originally predicted because of these effects? For each of the affected molecules, indicate how the bond angles change.

Molecular Shape
and Polarity

OBJECTIVE

To determine the shapes and polarities of a number of molecules using the VSEPR model. If possible, the shapes and polarities of these same molecules also will be determined via a software program that will perform the necessary quantum mechanical calculations. The agreement between the VSEPR model and the quantum mechanical calculations will be examined. Following this, a drug molecule that will interact favorably with a known receptor site will be created.

INTRODUCTION

Molecular shape and polarity are factors that influence both the chemical and physical properties of compounds. Chemical reactivity, especially in biological settings, is often especially sensitive to molecular shape and polarity. Physical properties such as solubility and melting and boiling points are also strongly dependent on molecular shape and polarity, as is the lattice arrangement in molecular solids.

Molecular shape is determined by energy considerations; a molecule assumes the geometry that gives it the lowest possible potential energy. The most accurate means of determining the geometry of lowest energy for a particular molecule or ion involves sophisticated quantum mechanical calculations that consider numerous possible geometric arrangements for a molecule, calculate the total energy of the molecule for each arrangement, and identify the arrangement with the lowest potential energy. The valence shell electron-pair repulsion (VSEPR) model accomplishes this qualitatively by arranging electron groups about a central atom in such a manner as to maximize their separation, a geometry that should be approximately that of minimum potential energy.

Molecular polarity is determined by summing the individual bond polarities according to the known shape of a molecule. For example, each of the carbon–oxygen bonds in carbon dioxide is polar covalent, but because carbon dioxide is a linear molecule the two bond polarities are of the same magnitude and directed oppositely to one another, causing the net polarity of the molecule, the molecular dipole moment (μ), to be zero (see Figure 16.1).

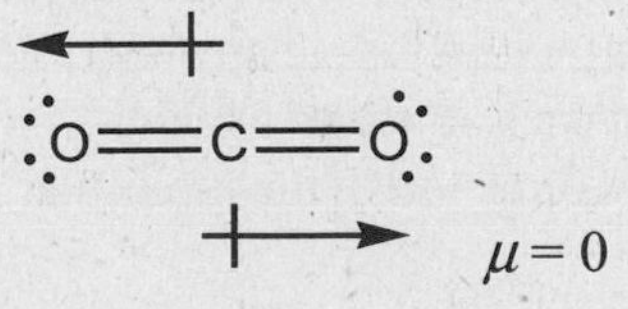

Figure 16.1
Despite its polar bonds, the linear geometry of carbon dioxide causes it to have a molecular dipole moment of zero.

Contrast carbon dioxide with water, which also has two identical polar covalent bonds, but owing to the bent shape of water these two bond polarities do not sum to zero, instead leading to a nonzero molecular dipole moment, the direction of which is indicated below (see Figure 16.2).

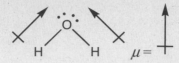

Figure 16.2
The bent geometry and polar bonds of water cause it to have a non-zero molecular dipole moment.

In this experiment you will determine the shapes and polarities of a number of molecules using the VSEPR model. You also might determine the shapes and polarities of a number of molecules via a software program that will perform the necessary quantum mechanical calculations for you. Following this, you will apply your knowledge of molecular shape and polarity to build and examine a drug molecule that you believe will interact with a known receptor site.

Drug activity is a complex process influenced by many different chemical factors: drug ingestion, metabolism, the rate at which the drug passes through the cell membrane and into the cell proper, and the specific shape and atomic composition of the receptor site. The focus here will be on the interaction between the receptor site and the drug itself. While the detailed mechanism of these interactions is still being vigorously debated and actively researched, two defining aspects of this interaction are clear: (1) the receptor site and drug must be attracted to one another, typically through some electrostatic interactions, and (2) the shape of the drug must be complementary to the geometry of the receptor site. This is the lock-and-key model, in which the receptor is the lock into which the key, i.e., the drug, must fit.

The experiment will be performed in one of two ways. Method I uses the VSEPR model and hand-held molecular model kits to determine the shapes and polarities of a number of molecules and ions. Following this, you will use the hand-held molecular model kits to design a drug molecule that you believe will bind with a hypothetical receptor site (see Figure 16.3). You then will compare your drug molecule with that designed by another student. Method II uses a software program to determine the shapes and polarities of a number of molecules and ions. Afterward, you will use the software program to design a drug molecule that you believe will bind with a hypothetical receptor site (see Figure 16.3). You then will compare your drug molecule with that designed by another student. Your instructor will inform you of which method you will be using in this experiment.

EXPERIMENT

Method I: (Molecular Modeling)

Part A: VSEPR and Polarity

Determine the shapes and polarities of the following molecules and ions: PCl_5, ClF_3, $SbCl_5^{2-}$, IF_6^+, ClO_3^-, and IF_4^+. Provide a Lewis structure and a perspective drawing of each molecule or ion, indicating the direction of the molecular dipole moment with a dipole arrow. Molecular models will be available for you to use. You might find it helpful to build each structure using the molecular models before creating the perspective drawing and determining the direction of the molecular dipole moment.

Part B: Pharmaceutical Drug Design

You are to build and examine a drug molecule that you believe will interact favorably with the hypothetical receptor site depicted in Figure 16.3. Assume that the better the complementary fit of and the stronger the electrostatic attraction between the drug and the receptor, the more potent is the drug action.

Start by constructing a scale model of the hypothetical receptor site shown in Figure 16.3. To create the scale model, first use the molecular models to form a C—C bond. Measure the distance from the center of one C atom to the other using a ruler. The average C—C bond length is 1.54 Å. Use this value and your measured C–C bond length to set the scale; i.e., if the C—C distance was measured by ruler to be 1.54 in, then the scale is set so that 1 inch equals 1Å. On paper, draw Figure 16.3 to scale. This might require more than one sheet of paper. If necessary, tape two or more sheets of paper together. Once Figure 16.3 has been drawn to scale, use the molecular models to create the three molecular fragments: —NH_3, —OH, and =O. Place the molecular models of the molecular fragments on the scale model of Figure 16.3.

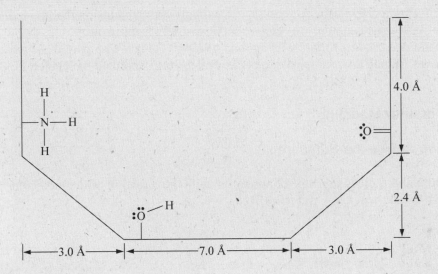

Figure 16.3
Hypothetical drug receptor site. In this figure the solid heavy lines represent the receptor surface. Assume the receptor surface to be nonpolar, composed exclusively of carbon. Note: The single and double bonds shown in this figure are not drawn to scale, but relative placement of the single bonds attached to the receptor surface is.

When creating your drug molecule, assume that it is an organic compound, as is most generally the case. Organic compounds have a very specific definition in chemistry; all organic compounds contain carbon, nearly always bonded to itself and to hydrogen, and to a lesser extent with oxygen and nitrogen, and on even more rare occasions to sulfur, phosphorous, and the halogens. To simplify your drug search, limit the composition of your molecule to these atoms.

Start with a simple organic molecule, such as pentane (C_5H_{12}), shown below in Figure 16.4. By adding, deleting, and replacing atoms with other groups create a drug molecule.

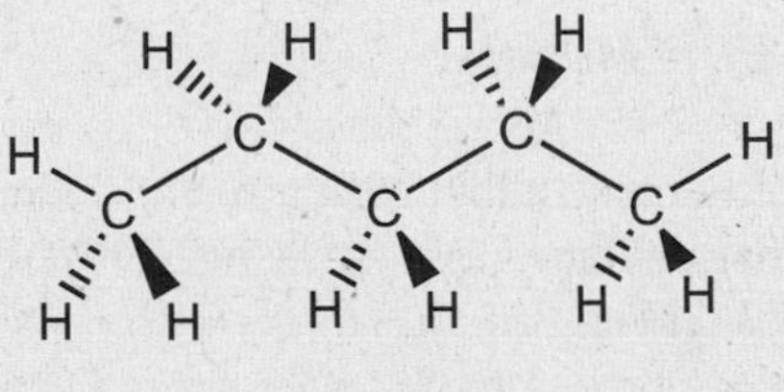

Figure 16.4
Pentane.

For example, by deleting two hydrogen atoms and replacing them with an O= group, 2-pentanone ($C_5H_{10}O$) can be created (see Figure 16.5).

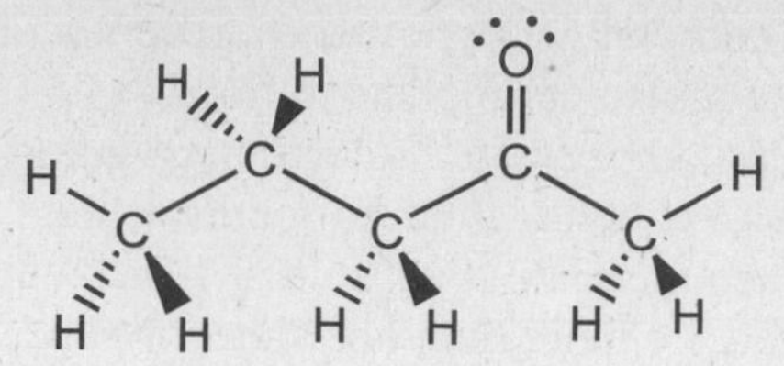

Figure 16.5
2-Pentanone.

Record the information you believe necessary to support your conclusion that the molecule you have created is a likely drug candidate to react with the desired receptor site.

Swap drug candidate molecule data with another group in the lab. Compare these two drug candidate molecules. Which do you believe to be the better drug candidate molecule?

Method II: (Computer Modeling)

Part A: Learning to Use the Software

The Lewis structure and molecular shape of ammonia, with the direction of the molecular dipole moment indicated, are shown below in Figure 16.6.

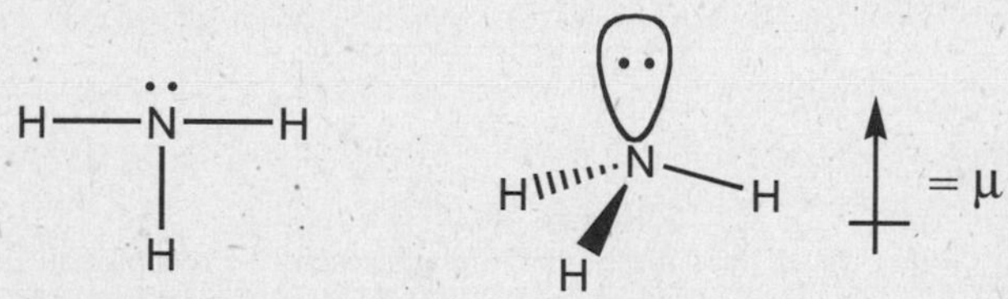

Figure 16.6
The Lewis structure and molecular shape of ammonia, with the direction of the molecular dipole moment indicated.

Use the software program to build a model of NH_3. Your instructor will provide you with instructions for using the software program. Measure and record the bond lengths and bond angles for this initial structure. Use the software program to determine the most energetically stable structure of NH_3. Measure and record the bond lengths and bond angles. The N—H bond lengths should be approximately 1.01 Å. The H—N—H bond angles should be about 107°. Determine the dipole moment of NH_3. It should point in the direction indicated above.

Part B: Verification of the VSEPR Model

Build and determine the energetically most stable structures of the molecules you predicted Lewis structures, molecular shapes, and molecular dipole moments for in the prelab assignment. Record enough information, bond lengths, bond angles, etc. to definitively support or contradict your answers to the prelab questions. Based on your results in lab, does the VSEPR model predict the structures and polarities of molecules well?

Part C: Pharmaceutical Drug Design

You are to use the software program to build and examine a drug molecule that you believe will interact favorably with the hypothetical receptor site in Figure 16.3. Assume that the better the complementary fit of and the stronger the electrostatic attraction between the drug and the receptor, the more potent the drug action.

When creating your drug molecule, assume that it is an organic compound, as is most generally the case. Organic compounds have a very specific definition in chemistry; all organic compounds contain carbon, nearly always bonded to itself and to hydrogen, and to a lesser extent with oxygen and nitrogen, and on even more rare occasions to sulfur, phosphorous, and the halogens. To simplify your drug search, limit the composition of your molecule to these atoms.

Start with a simple organic molecule, such as pentane (C_5H_{12}) shown in Figure 16.4. By adding, deleting, and replacing atoms with other groups create a drug molecule. For example, by deleting two hydrogen atoms and replacing them with an O= group, 2-pentanone ($C_5H_{10}O$) can be created (see Figure 16.5).

Record the information you believe necessary to support your conclusion that the molecule you have created is a likely drug candidate to react with the desired receptor site.

Swap drug candidate molecule data with another group in the lab. Compare these two drug candidate molecules. Which do you believe to be the better drug candidate molecule?

Table 16.1: Average Bond Lengths

Single Bonds

Bond	Length (Å)	Bond	Length (Å)	Bond	Length (Å)
H—H	0.74	N—H	1.01	S—H	1.34
H—F	0.92	N—N	1.46	S—P	2.10
H—Cl	1.27	N—P	1.77	S—S	2.04
H—Br	1.41	N—O	1.44	S—F	1.58
H—I	1.61	N—S	1.68	S—Cl	2.01
		N—F	1.39	S—Br	2.25
C—H	1.09	N—Cl	1.91	S—I	2.34
C—C	1.54	N—Br	2.14		
C—N	1.47	N—I	2.22	F—F	1.43
C—O	1.43			F—Cl	1.66
C—P	1.87	O—H	0.96	F—Br	1.78
C—S	1.81	O—P	1.60	F—I	1.87
C—F	1.33	O—O	1.48		
C—Cl	1.77	O—S	1.51	Cl—Cl	1.99
C—Br	1.94	O—F	1.42	Cl—Br	2.14
C—I	2.13	O—Cl	1.64	Cl—I	2.43
		O—Br	1.72		
P—H	1.42	O—I	1.94	Br—Br	2.28
P—P	2.21			Br—I	2.48
P—F	1.56				
P—Cl	2.04			I—I	2.66
P—Br	2.22				
P—I	2.43				

Multiple Bonds

Bond	Length (Å)	Bond	Length (Å)	Bond	Length (Å)
C=C	1.34	N=N	1.22	C≡C	1.21
C=N	1.27	N=O	1.20	C≡N	1.15
C=O	1.23	O=O	1.21	C≡O	1.13
				N≡N	1.10
				N≡O	1.06

Name: _____   Date: _____

Lab Instructor: _____   Lab Section: _____

EXPERIMENT 16

Molecular Shape and Polarity

PRELABORATORY QUESTIONS

1. Determine the Lewis structures and molecular shapes of $CHCl_3$, SO_3^{2-}, and PF_5. Do any of these molecules have nonzero molecular dipole moments? If so, which and in what direction do they point?

    (a) $CHCl_3$

    (b) SO_3^{2-}

    (c) PF_5

2. Determine the formal charges on each atom in Figure 16.3. For those molecular fragments that lack formal charges indicate polarity with partial charges ($\delta+$ and $\delta-$).

3. Use the Table 16.1 and Figure 16.3 to determine the approximate size (length and width) your drug molecule needs to be.

Name: _____   Date: _____

Lab Instructor: _____   Lab Section: _____

EXPERIMENT 16

Molecular Shape and Polarity

RESULTS/OBSERVATIONS

Method I: (Molecular Modeling)

Part A: VSEPR and Polarity

1. Lewis structures and molecular dipole moments:

Formula: PCl$_5$	Lewis Structure:	Perspective Drawing:	Molecular Geometry: Molecular Dipole Moment: $\mu =$
Formula: ClF$_3$	Lewis Structure:	Perspective Drawing:	Molecular Geometry: Molecular Dipole Moment: $\mu =$
Formula: SbCl$_5^{2-}$	Lewis Structure:	Perspective Drawing:	Molecular Geometry: Molecular Dipole Moment: $\mu =$

Name: _____   Date: _____

Lab Instructor: _____   Lab Section: _____

Formula: IF_6^+	Lewis Structure:	Perspective Drawing:	Molecular Geometry: Molecular Dipole Moment: $\mu =$
Formula: ClO_3^-	Lewis Structure:	Perspective Drawing:	Molecular Geometry: Molecular Dipole Moment: $\mu =$
Formula: IF_4^+	Lewis Structure:	Perspective Drawing:	Molecular Geometry: Molecular Dipole Moment: $\mu =$

Part B: Pharmaceutical Drug Design

2. Measured C—C length: _____

3. Scale setting: _____ = ___1.54 Å__

4. Write any pertinent data in the space below. Clearly indicate both the property and the amount.

Name: _____   Date: _____

Lab Instructor: _____   Lab Section: _____

5. Record the data collected by another student in the space below. Clearly indicate both the property and the amount.

6. Compare these two drug candidate molecules. Which do you believe to be the better drug candidate molecule? Explain.

Name: _____   Date: _____

Lab Instructor: _____   Lab Section: _____

Method II: (Computer Modeling)

Part A: Learning to Use the Software

1. N—H bond lengths in the initial structure: _____ _____ _____

2. H—N—H bond angles in the initial structure: _____ _____ _____

3. N—H bond lengths in the most energetically stable structure: _____ _____

   _____

4. H—N—H bond angles in the most energetically stable structure: _____ _____

   _____

5. Indicate the direction of the NH_3 molecular dipole moment relative to the structure below:

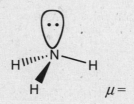

$\mu =$

Part B: Verification of the VSEPR Model

6. Write any pertinent data in the space below. Clearly indicate both the property and the amount.

Name: _____   Date: _____

Lab Instructor: _____   Lab Section: _____

7. Based on your results in lab, does the VSEPR model predict the structures and polarities of molecules well? Explain.

Part C: Pharmaceutical Drug Design

8. Write any pertinent data in the space below. Clearly indicate both the property and the amount.

Name: _____   Date: _____

Lab Instructor: _____   Lab Section: _____

9. Record the data collected by another student in the space below. Clearly indicate both the property and the amount.

10. Compare these two drug candidate molecules. Which do you believe to be the better drug candidate molecule? Explain.

Name: _____   Date: _____

Lab Instructor: _____   Lab Section: _____

EXPERIMENT 16

Molecular Shape and Polarity

POSTLABORATORY QUESTIONS

Method I: (Molecular Modeling)

1. Explain why ClO_3^- has a molecular dipole moment, but ClO_4^- doesn't.

2. Assuming the same distance of separation, which of the two attractions below, that between the hydrogen of NH_3 and the oxygen of H_2O or that between the hydrogen of NH_3 and the sulfur of H_2S, is stronger? Explain. Why is it important when making this determination to assume that the distance of separation is the same in both cases?

Name: _____   Date: _____

Lab Instructor: _____   Lab Section: _____

EXPERIMENT 16

Molecular Shape and Polarity

POSTLABORATORY QUESTIONS

Method II: (Computer Modeling)

1. Assuming the same distance of separation, which of the two attractions below, that between the hydrogen of NH_3 and the oxygen of H_2O or that between the hydrogen of NH_3 and the sulfur of H_2S, is stronger? Explain. Why is it important when making this determination to assume that the distance of separation is the same in both cases?

$$H—\overset{\displaystyle ..}{\underset{\displaystyle |}{N}}—H \qquad :\overset{\displaystyle ..}{\underset{\displaystyle |}{O}}—H \qquad H—\overset{\displaystyle ..}{\underset{\displaystyle |}{N}}—H \qquad :\overset{\displaystyle ..}{\underset{\displaystyle |}{S}}—H$$
$$\quad H \qquad\qquad H \qquad\qquad\quad H \qquad\qquad H$$

2. What advantage(s) does the VSEPR model have over quantum mechanical calculations? What advantage(s) do quantum mechanical calculations have over the VSEPR model?

EXPERIMENT 17

Enthalpy of Formation of an Ammonium Salt

OBJECTIVE

To determine the heat of formation of solid $(NH_4)_2SO_4$. The heat of formation of solid $(NH_4)_2SO_4$ will be determined by combining measurements of enthalpy changes with known heats of formation using Hess's law. The measured enthalpy changes will be determined by constant pressure calorimetry.

INTRODUCTION

The enthalpy of formation per mole of a compound is the enthalpy change (ΔH) accompanying the formation of 1 mole of the compound from its elements in their standard states. Direct measurement of enthalpies of formation is often difficult experimentally, so indirect methods involving enthalpies of reaction are used frequently instead. In this experiment the heat of formation of solid ammonium sulfate $[(NH_4)_2SO_4]$ will be determined by combining measurements of the enthalpy change of the neutralization reaction between aqueous ammonia (NH_3) and sulfuric acid (H_2SO_4),

$$2\,NH_3(aq) + H_2SO_4(aq) \rightarrow (NH_4)_2SO_4(aq) \qquad\qquad \Delta H_{neut} \qquad\qquad (1)$$

and the enthalpy change for the dissolution of solid $(NH_4)_2SO_4$ in water,

$$(NH_4)_2SO_4(s) \xrightarrow{\ \ H_2O\ \ } (NH_4)_2SO_4(aq) \qquad\qquad \Delta H_{diss} \qquad\qquad (2)$$

with the known heats of formation of aqueous NH_3 [-81.2 kJ/mol for 1.5 M $NH_3(aq)$] and aqueous H_2SO_4.

Hess's law states that the change in a thermodynamic property that is a state function, such as enthalpy, depends only on the initial and final states, independent of the particular combination of steps taken between them. In other words, adding several consecutive reactions gives a net reaction for which the ΔH is simply the sum of the ΔHs of the component reactions. For example, the ΔH of formation of aqueous NH_3 is equal to the sum of the ΔH of formation of gaseous NH_3 and the ΔH that accompanies dissolving gaseous NH_3 in water:

$$1/2\,N_2(g) + 3/2\,H_2(g) \rightarrow NH_3(g) \qquad\qquad \Delta H \text{ of formation of } NH_3(g) = \Delta H_1 \qquad\qquad (3)$$

$$NH_3(g) \xrightarrow{\ \ H_2O\ \ } NH_3(aq) \qquad\qquad \Delta H \text{ of dissolution of } NH_3(g) = \Delta H_2 \qquad\qquad (4)$$

_____   _____

$$1/2\,N_2(g) + 3/2\,H_2(g) \rightarrow NH_3(aq) \qquad\qquad \Delta H \text{ of formation of } NH_3(aq) = \Delta H_1 + \Delta H_2 \qquad\qquad (5)$$

Given that ΔH_1 is –45.8 kJ/mol and ΔH_2 is –35.4 kJ/mol, the ΔH of formation of aqueous NH_3 is calculated to be –81.2 kJ/mol. A similar calculation can be set up to get the ΔH of formation of aqueous H_2SO_4.

The enthalpy changes for Reactions (1) and (2) will be measured in this experiment using a calorimeter. A calorimeter is a device used to measure the amount of heat transferred during a chemical reaction. A calorimeter is just an insulated container equipped with a stirrer, a thermometer, and a loose-fitting lid (to maintain the contents at atmospheric pressure). The reaction is carried out inside the container, and the heat evolved or absorbed is calculated from the measured temperature change. Under conditions of constant pressure, the heat is equivalent to the enthalpy change,

$$q = \Delta H \tag{6}$$

The temperature change (ΔT) of an object is proportional to the heat transferred. Each object has a particular capacity for absorbing heat, its own heat capacity (C). The heat capacity of an object is the constant of proportionality in the defining relationship between the temperature change of an object and the heat absorbed or released by it:

$$q = C \cdot \Delta T \tag{7}$$

The heat capacity is the amount of heat necessary to change the temperature of an object by 1°C. A related property is the specific heat capacity (s), the quantity of heat required to change the temperature of 1 g of a substance by 1°C:

$$q = m \cdot s \cdot \Delta T \tag{8}$$

When Reactions (1) and (2) occur they exchange heat with those objects in close proximity; in this case the aqueous mixture and the calorimeter. Consequently, the following relationship holds:

$$-q_{rxn} = q_{calorimeter} + q_{mixture} \tag{9}$$

Any heat lost (or gained) by the reaction is gained (or lost) by the calorimeter and the aqueous mixture. Substituting Equations (6), (7), and (8) into Equation (9) yields

$$-\Delta H_{rxn} = C_{calorimeter} \cdot \Delta T_{calorimeter} + m_{mixture} \cdot s_{mixture} \cdot \Delta T_{mixture} \tag{10}$$

The mass of the aqueous mixture can be measured easily in the laboratory using an electronic balance. Measuring the temperature changes of the calorimeter and the aqueous mixture, which are equivalent on account of their thermal contact, is accomplished easily by observing the initial and final thermometer values. The specific heat capacity of the aqueous mixture is assumed to be equivalent to that of water, 4.184 J/g · °C. Therefore, if the heat capacity of the calorimeter is known or can be determined, the heat of solution can be simply calculated.

In this experiment the heat capacity of the calorimeter will be determined by mixing known quantities of hot and cold water in the calorimeter and measuring the resulting temperature change. The hot water will lose heat to the cold water and the calorimeter, i.e.,

$$-q_{hot} = q_{cold} + q_{calorimeter} \tag{11}$$

Substituting Equations (7) and (8) into Equation (11) gives the relationship

$$-m_{hot} \cdot s_{hot} \cdot \Delta T_{hot} = m_{cold} \cdot s_{cold} \cdot \Delta T_{cold} + C_{calorimeter} \cdot \Delta T_{calorimeter} \tag{12}$$

Of course, the specific heat capacities of the hot and cold water are the same, 4.184 J/g · °C. The masses of the hot and cold water are not necessarily the same, but these values are determined easily by weighing. The temperature changes will be measured using a thermometer. This provides enough information to use Equation (12) to calculate the heat capacity of the calorimeter.

In the preceding description of calorimetry, it was assumed that all the heat transfer was isolated to the calorimeter and its contents. However, a small but measurable amount of heat is transferred to (or from) entities outside the calorimeter. It is possible to correct for this heat. You will do so in this experiment by monitoring the rate at which the temperature of the calorimeter and its contents decreases (or increases) as a function of time. Assuming this loss (or gain) of heat from entities outside the calorimeter to be linear, it is possible to estimate the actual initial and final temperatures by extrapolation. An example will better convey this idea.

In Part I of this experiment the initial temperatures of the hot and cold water will be measured over a period of time (since the cold water will be inside the calorimeter, it and the calorimeter will have the same initial temperature). Before mixing, the temperature of the hot water will be dropping gradually at a constant rate, whereas the temperature of the cold water should be nearly constant or increasing very gradually in a linear manner. Immediately after the hot water has been added to the cold water, the temperature of the contents of the calorimeter will increase rapidly. Eventually, the thermometer readings will peak and decrease slowly as the calorimeter and its contents lose heat. The loss of heat by the calorimeter and its contents should occur at a constant rate, causing these data points to lie along a straight line. These data then will be plotted (see Figure 17.1).

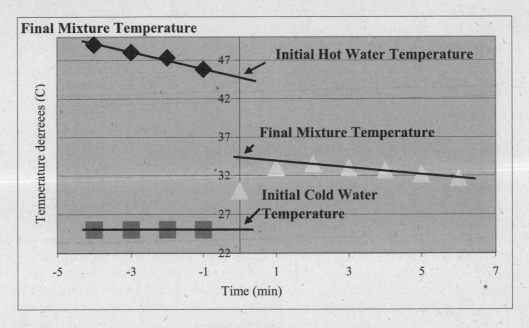

Figure 17.1
Typical temperature as a function of time data collected to determine the heat capacity of the calorimeter.

Draw a vertical line representing the time of mixing on the graph. Draw the best straight line through the hot-water data, extrapolating to the vertical line representing the time of mixing. The temperature at which the lines intersect is the actual initial temperature of the hot water. Repeat for the cold-water data. The temperature at which the lines intersect is the actual initial temperature of the cold water and the calorimeter. For the data after the time of mixing draw the best straight line through the points corresponding to decreasing temperature readings, extrapolating to the vertical line representing the time of mixing. The temperature at which the lines intersect is the actual final temperature of the hot water, cold water, and calorimeter.

Similar plots should be made for the data collected in Parts II and III.

Your purpose in this experiment is to determine the heat of formation of solid $(NH_4)_2SO_4$. The heat of formation of solid $(NH_4)_2SO_4$ will be determined by combining measurements of the enthalpy change for Reactions (1) and (2) with the known heats of formation of aqueous NH_3 and aqueous H_2SO_4 using Hess's law. This experiment has three parts. You will be working in pairs throughout the experiment. In Part I you will determine the heat capacity of the calorimeter, and in Parts II and III you will measure the heat produced or absorbed by Reactions (1) and (2).

ADDITIONAL READING

Read the sections in the "Laboratory Techniques" chapter at the beginning of this Lab Manual on cleaning glassware, handling chemicals, heating liquids and solutions, and weighing prior to performing this experiment.

 Safety Precautions:

- Protective eyewear approved by your institution must be worn at all times while you are in the laboratory.

- Sulfuric acid is corrosive and must be kept off skin and clothing. If you get it into your eyes, immediately flush them with water for at least 10 minutes. If you spill it on your clothing, remove the affected clothing before rinsing with water. Notify your lab instructor, who will assist you.

PROCEDURE

Part I: Determination of the Heat Capacity of the Calorimeter

Place 50 mL of distilled water in a 250 mL Erlenmeyer flask, and heat it gently on a hotplate until the temperature has increased about 30–40° above its initial temperature. Allow it to cool until the temperature is about 20° above room temperature.

Weigh a lidded Styrofoam cup with a stirbar in it. This, in addition to a thermometer, is the calorimeter. This same calorimeter will be used in all subsequent parts of the experiment. To avoid errors caused by the magnetic field of the stirbar interacting with the electronic balance, first place an inverted Nalgene™ beaker on the balance pan. Next, tare the balance. Then set the lidded Styrofoam cup containing the stirbar on top of the inverted beaker. Record the mass. *Note: In subsequent weighings, whenever a stirbar is in a cup being weighed, always place an inverted Nalgene™ beaker on the balance, tare, and set the cup containing the stirbar on top of the inverted beaker.*

Put 50 mL of room-temperature ("cold") distilled water in the cup containing the stirbar. Weigh the lidded cup containing the cold water and stirbar on the balance. Record the mass.

Use a split one-hole rubber stopper and a clamp to support the thermometer on a ringstand. Set the cup on a stirrer-hotplate. (Make sure that the hotplate control is *off*; you don't want to melt the Styrofoam cup.) Put the thermometer through the lid of the cup. The thermometer should be positioned about ½ in from the center so that when the liquid is swirling the tip of the thermometer will remain immersed (see Figure 17.2).

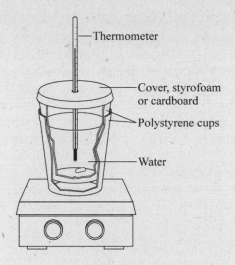

Figure 17.2
Coffee cup calorimeter.

CAUTION: Don't let the stirbar come into contact with the thermometer. It is quite fragile.

Turn on the stirrer, and adjust the stirring speed until the bar is turning at a moderate rate. Check the temperature of the hot water. When it is about 20°C above the temperature of the cold water, proceed with the following.

Take temperature readings of the cold water for a period of 1 minute, measuring and recording data every 15 seconds. *Simultaneously,* your lab partner should use his or her thermometer to measure the temperature of the hot water, collecting and recording data every 15 seconds. Being very careful not to splash, quickly pour the hot water into the calorimeter, and affix the lid. Continue reading the temperature every 15 seconds until the temperature begins decreasing linearly for at least four successive readings (e.g., it may take 45 seconds for a 0.1°C drop in temperature; this rate of drop should be observed for at least four successive 45-s intervals). This might take quite a few minutes. One student should keep track of the elapsed time and record the data while the other reads the temperatures.

Turn off the stirrer, and weigh the calorimeter and its contents. Record the mass. Determine the mass of the hot water by difference. Empty and dry the cup, stirbar, and thermometer before use in subsequent parts of the experiment.

Part II: Neutralization

CAUTION: Sulfuric acid is corrosive and must be kept off skin and clothing. If you get it into your eyes, immediately flush them with water for at least 10 minutes. If you spill it on your clothing, remove the affected clothing before rinsing with water. Notify your lab instructor, who will assist you.

Weigh the lidded Styrofoam cup with the stirbar in it. Record this mass. Place 50 mL of 1.50 M NH_3 in the Styrofoam cup. Next, obtain the volume of 1.50 M H_2SO_4 required to neutralize the NH_3 in the cup (you calculated this volume as part of the prelab assignment), and place that volume of acid in a beaker. Set the cup on a stirrer-hotplate. Place the thermometer in the Styrofoam cup containing the NH_3, turn on the stirrer, and collect temperature-versus-time readings at 15 s intervals for 1 minute prior to mixing. Being careful not to splash, immediately following the last temperature-time reading, add the entirety of the acid solution to the NH_3 solution in the Styrofoam cup, and affix the lid. Continue taking temperature-time data for at least 5 minutes. (Continue for a longer time if needed to observe a linear drop in the temperature.) Turn off the stirrer, and weigh the calorimeter and its contents. Record the mass. Determine the mass of the aqueous mixture by difference. Rinse and dry the cup, stirbar, and thermometer.

Repeat to obtain a second set of data.

Part III: Dissolution

Weigh out a mass of solid $(NH_4)_2SO_4$ equivalent to that produced in the neutralization reaction of Part II (you calculated this mass as part of the prelab assignment) into a clean, dry beaker. Weigh the lidded Styrofoam cup with the stirbar in it. Record this mass. Place a volume of distilled water equal to the final volume of solution from Part II in the Styrofoam cup. Set the cup on a stirrer-hotplate. Place the thermometer in the Styrofoam cup containing the water, turn on the stirrer, and collect temperature-versus-time readings at 15 s intervals for 1 minute prior to mixing. Being careful not to splash, immediately following the last temperature-time reading add the entirety of the solid $(NH_4)_2SO_4$ to the water in the Styrofoam cup. Continue taking temperature-time data for at least 5 minutes. (Continue for a longer time if needed to observe a linear drop in the temperature.) Turn off the stirrer, and weigh the calorimeter and its contents. Record the mass. Determine the mass of the aqueous mixture by difference. Rinse and dry the cup, stirbar, and thermometer.

Repeat the procedure to obtain a second set of data.

Waste Disposal. Dispose of all chemical waste as directed by your instructor.

Name: _____   Date: _____

Lab Instructor: _____   Lab Section: _____

EXPERIMENT 17

Enthalpy of Formation of an Ammonium Salt

PRELABORATORY QUESTIONS

1. Calculate the volume of 1.50 M H_2SO_4 required in Part II of the experimental procedure.

2. Calculate the mass of salt, $(NH_4)_2SO_4$, produced in the neutralization reaction of Part II.

3. Use the following data to calculate the heat of formation of $H_2SO_4(aq)$: ΔH of formation of $H_2SO_4(l) = -814.0$ kJ/mol and ΔH of dissolution of $H_2SO_4(l) = -79.5$ kJ/mol.

Name: _____ Date: _____

Lab Instructor: _____ Lab Section: _____

4. Why is it necessary to determine the heat capacity of the calorimeter?

5. In calculating the heat capacity of the calorimeter, which temperature difference should be larger in magnitude, ΔT_{hot} or ΔT_{cold}? Explain your answer.

6. Why is it necessary to extrapolate the temperature-versus-time data to the time of mixing in order to estimate the actual initial and final temperatures?

7. What action should you take if you spill 1.50 M H_2SO_4 on your hand?

Name: _____   Date: _____

Lab Instructor: _____   Lab Section: _____

EXPERIMENT 17

Enthalpy of Formation of an Ammonium Salt

RESULTS/OBSERVATIONS

Part I: Determination of the Heat Capacity of the Calorimeter

1. Mass of lidded Styrofoam cup and stirbar: _____

2. Mass of lidded Styrofoam cup, stirbar, and cold water: _____

3. Mass of cold water (subtract 1 from 2): _____

4. Indicate in the data table below when mixing occurred. (*Note:* The data in the column T_{cold} become the temperature of the mixture after mixing.)

Elapsed Time (s)	T_{hot} (°C)	T_{cold} (°C)	Elapsed Time (s)	T_{cold} (°C)	Elapsed Time (s)	T_{cold} (°C)	Elapsed Time (s)	T_{cold} (°C)

Name: _____   Date: _____

Lab Instructor: _____   Lab Section: _____

5. Mass of lidded Styrofoam cup, stirbar, hot water, and cold water: _____

6. Mass of hot water (subtract 2 from 5): _____

7. On the graph provided, plot the temperature-versus-time data in order to estimate the initial
   temperatures of the hot and cold water and the final temperature of the mixture.

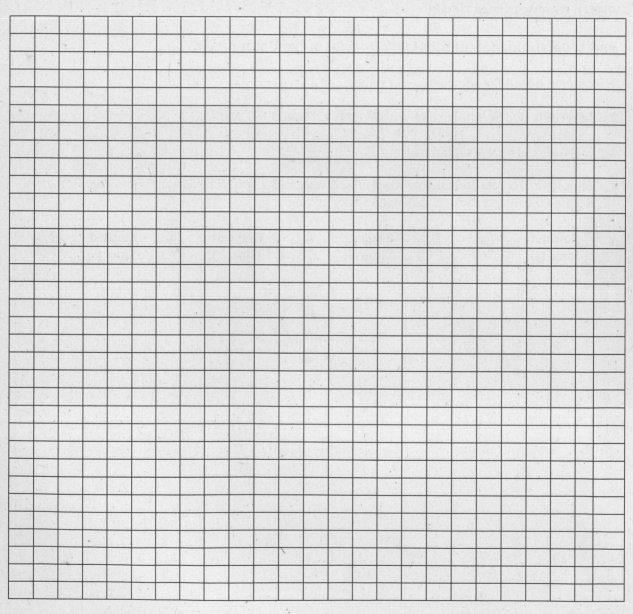

Name: _____   Date: _____

Lab Instructor: _____   Lab Section: _____

8. Calculation of the heat capacity of the calorimeter:

Part II: Neutralization

9. Mass of lidded Styrofoam cup and stirbar: _____

10. Trial #1. Indicate in the data table below when mixing occurred.

Elapsed Time (s)	T (°C)	Elapsed Time (s)	T (°C)	Elapsed Time (s)	T (°C)	Elapsed Time (s)	T (°C)

Name: _____   Date: _____

Lab Instructor: _____   Lab Section: _____

11. Mass of lidded Styrofoam cup, stirbar, and aqueous mixture: _____

12. Mass of aqueous mixture (subtract 9 from 11): _____

13. Mass of lidded Styrofoam cup and stirbar: _____

14. Trial #2. Indicate in the data table below when mixing occurred.

Elapsed Time (s)	T (°C)	Elapsed Time (s)	T (°C)	Elapsed Time (s)	T (°C)	Elapsed Time (s)	T (°C)

Name: _____   Date: _____

Lab Instructor: _____   Lab Section: _____

15. Mass of lidded Styrofoam cup, stirbar, and aqueous mixture: _____

16. Mass of aqueous mixture (subtract 13 from 15): _____

17. On the graph provided, plot the temperature-versus-time data from the first trial in order to estimate the initial and final temperatures of the aqueous mixture.

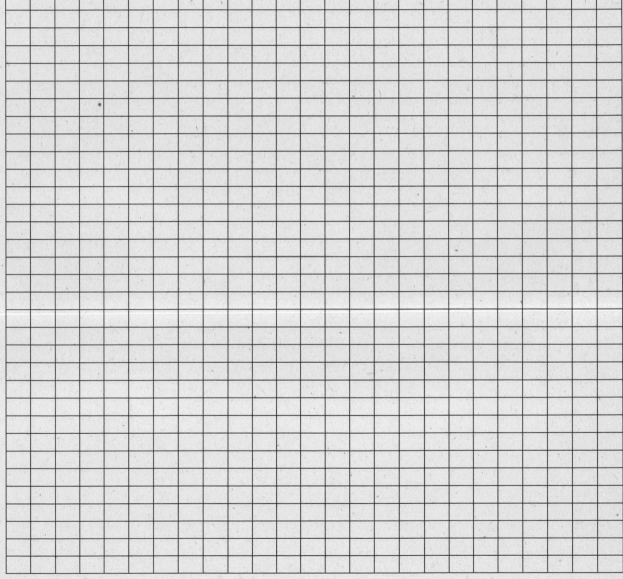

Name: _____   Date: _____

Lab Instructor: _____   Lab Section: _____

18. On the graph provided, plot the temperature-versus-time data from the second trial in order to estimate the initial and final temperatures of the aqueous mixture.

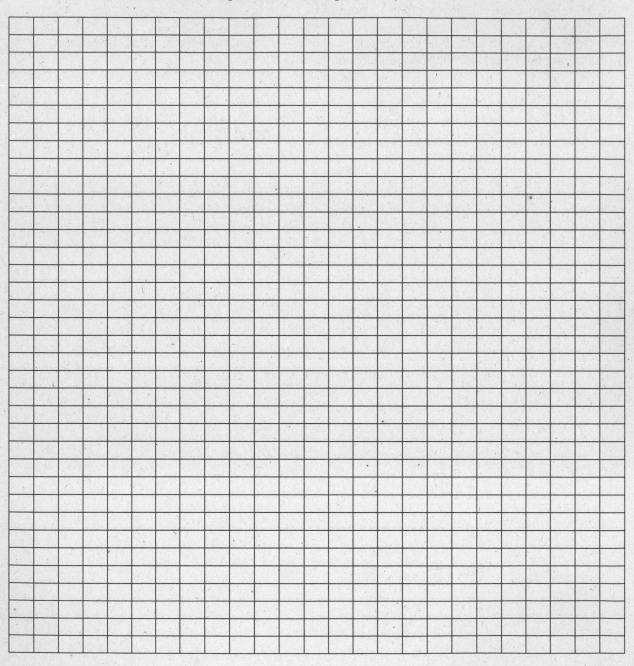

Name: _____   Date: _____

Lab Instructor: _____   Lab Section: _____

19. Calculate the enthalpy change for this neutralization reaction using the data from the first trial:

20. Calculate the enthalpy change for this neutralization reaction using the data from the second trial:

21. Calculate the enthalpy change per mole for this neutralization reaction, ΔH_{neut}, using the data from the first trial:

22. Calculate the enthalpy change per mole for this neutralization reaction, ΔH_{neut}, using the data from the second trial:

23. The average of the two ΔH_{neut} values: _____

Name: _____   Date: _____

Lab Instructor: _____   Lab Section: _____

Part III: Dissolution

24. Mass of lidded Styrofoam cup and stirbar: _____

25. Trial #1. Indicate in the data table below when mixing occurred.

Elapsed Time (s)	T (°C)	Elapsed Time (s)	T (°C)	Elapsed Time (s)	T (°C)	Elapsed Time (s)	T (°C)

Name: _____   Date: _____

Lab Instructor: _____   Lab Section: _____

26. Mass of lidded Styrofoam cup, stirbar, and aqueous mixture: _____

27. Mass of aqueous mixture (subtract 24 from 26): _____

28. Mass of lidded Styrofoam cup and stirbar: _____

29. Trial #2. Indicate in the data table below when mixing occurred.

Elapsed Time (s)	T (°C)	Elapsed Time (s)	T (°C)	Elapsed Time (s)	T (°C)	Elapsed Time (s)	T (°C)

Name: _____   Date: _____

Lab Instructor: _____   Lab Section: _____

30. Mass of lidded Styrofoam cup, stirbar, and aqueous mixture: _____

31. Mass of aqueous mixture (subtract 28 from 30): _____

32. On the graph provided, plot the temperature-versus-time data from the first trial in order to estimate the initial and final temperatures of the aqueous mixture.

Name: _____   Date: _____

Lab Instructor: _____   Lab Section: _____

33. On the graph provided, plot the temperature-versus-time data from the second trial in order to estimate the initial and final temperatures of the aqueous mixture.

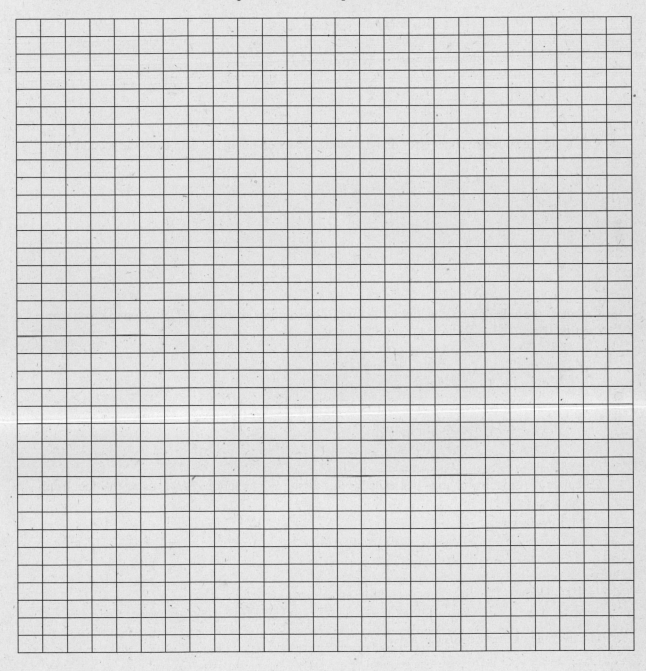

Name: _____   Date: _____

Lab Instructor: _____   Lab Section: _____

34. Calculate the enthalpy change for this dissolution reaction using the data from the first trial:

35. Calculate the enthalpy change for this dissolution reaction using the data from the second trial:

36. Calculate the enthalpy change per mole for this dissolution reaction, ΔH_{diss}, using the data from the first trial:

37. Calculate the enthalpy change per mole for this dissolution reaction, ΔH_{diss}, using the data from the second trial:

38. The average of the two ΔH_{diss} values: _____

Name: _____   Date: _____

Lab Instructor: _____   Lab Section: _____

39. Calculation of ΔH of formation of solid $(NH_4)_2SO_4$ using the known heats of formation of aqueous NH_3 [-81.2 kJ/mol for 1.5 M $NH_3(aq)$] and aqueous H_2SO_4 (calculated in the prelab assignment) and the measured heats of reaction:

40. Calculation of the percent error in the heat of formation determination. The accepted value for the heat of formation of solid $(NH_4)_2SO_4$ is -1180.85 kJ/mol.

Name: _____   Date: _____

Lab Instructor: _____   Lab Section: _____

EXPERIMENT 17

Enthalpy of Formation of an Ammonium Salt

POSTLABORATORY QUESTIONS

1. Explain why the straight line on the left has a rising slope, whereas the straight line on the right has a falling slope. What can you deduce about the initial temperature of the starting materials in this example?

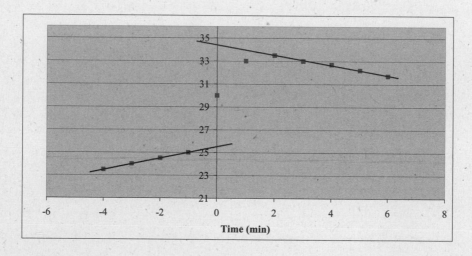

2. Sketch a sample curve for an endothermic reaction where the starting materials begin at a temperature that is initially the same as room temperature.

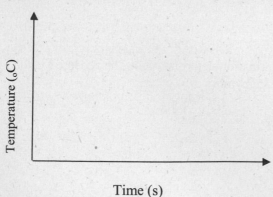

Time (s)

3. What are some sources of error in this experiment?

EXPERIMENT 18

Hot- and Coldpacks

OBJECTIVE

To design either a hot- or coldpack that undergoes a 3°C temperature change in the most cost-effective manner possible. In the first week of the experiment the heat capacity of a calorimeter and the heats of solution of several ionic salts will be determined by constant-pressure calorimetry. During the second week of the experiment the hot- or coldpack will be designed and its efficacy tested.

INTRODUCTION

The importance of energy changes in chemical systems hardly can be overstated. From the food we eat to fuel our bodies to the combustion of gasoline used to power automobiles, energy changes of chemical systems are a near omnipresent, and in some cases essential, component of our lives. Of particular importance in chemistry are those chemical processes for which the energy change involves an absorption or release of heat. Heat is the energy transferred as the result of a difference in temperature. The heat transferred during a chemical process is given the symbol q. For chemical processes occurring at constant pressure, the heat is equivalent to the enthalpy change (ΔH):

$$q = \Delta H \tag{1}$$

The branch of chemistry that deals with such energy changes is called *thermochemistry*.

In thermochemistry, specific definitions are given to the terms *system* and *surroundings*. The system is composed of those items whose energy change is of interest to the experimenter—usually the components of a chemical reaction, although the definition certainly can be broadened. Everything else makes up the surroundings. The experimenter defines what is included in the system on the basis of his or her interest and convenience. A logical consequence of this binary classification is that any energy lost (or gained) by the system is gained (or lost) by the surroundings.

In most cases, only a small part of the surroundings engages in heat transfer with the system; items far removed from the system interact negligibly with it. The art of measuring the heat or enthalpy change associated with a given chemical system is accomplished by constructing a well-defined surroundings that retain (or donate) the heat lost (or gained) by the system. The device used to serve as the well-defined surroundings is called a *calorimeter*. A calorimeter is an insulated container equipped with a stirrer, thermometer, and loosely fitting lid (to maintain the contents at a constant atmospheric pressure). The heat lost (or gained) by the system is transferred to (or from) the surroundings, the calorimeter, and its contents, i.e.,

$$-q_{system} = q_{surroundings} \tag{2}$$

the negative sign in Equation (2) implies a loss of heat. A measurement of the temperature change of the calorimeter and its contents allows for calculation of the heat transferred, an amount equivalent to the enthalpy change.

The temperature change (ΔT) of an object is proportional to the heat transferred. Each object has a particular capacity for absorbing heat, its own heat capacity (C). The heat capacity of an object is the constant of proportionality in the defining relationship between the temperature change of an object and the heat absorbed or released by it:

$$q = C \cdot \Delta T \tag{3}$$

The heat capacity is the amount of heat necessary to change the temperature of an object by 1°C. A related property is the specific heat capacity (s), the quantity of heat required to change the temperature of 1 g of a substance by 1°C:

$$q = m \cdot s \cdot \Delta T \tag{4}$$

In this experiment the heat change on dissolution of an ionic salt in water is to be determined. The generic reaction for this process is

$$M_yX_z(s) \xrightarrow{\text{H}_2\text{O}} yM^{z+}(aq) + zX^{y-}(aq) \tag{5}$$

The enthalpy change accompanying this process is termed the *heat of solution* (ΔH_{soln}). In the calorimetry experiments under investigation in this experiment it is most convenient to define the dissolution reaction as the system. The surroundings then are composed of the calorimeter and the aqueous mixture. Given these definitions of the system and surroundings, the following relationships hold:

$$-q_{dissolution} = q_{calorimeter} + q_{mixture} \tag{6}$$

$$-\Delta H_{soln} = C_{calorimeter} \cdot \Delta T_{calorimeter} + m_{mixture} \cdot s_{mixture} \cdot \Delta T_{mixture} \tag{7}$$

An electronic balance allows for the mass of the aqueous mixture to be measured easily in the laboratory. Measuring the temperature changes of the calorimeter and the aqueous mixture, which are equivalent on account of their thermal contact, is accomplished easily by observing the initial and final thermometer values. Therefore, if the heat capacity of the calorimeter and the specific heat capacity of the aqueous mixture (usually assumed to be equivalent to that of water, 4.184 J/g · °C) are known, the heat of solution can be simply calculated.

Your group's purpose in this experiment is to design the most cost-effective hot- or coldpack (your lab instructor will inform you which) that undergoes a 3°C temperature change. Your group will perform this experiment over two weeks. In the first week of the experiment your group will determine the heat capacity of a calorimeter and the heat of solution of one of the seven ionic salts listed in the Equipment and Reagents section of this experiment. Other student groups in the class will determine the heats of solution of the other six ionic salts. Your instructor will collect all the student-determined heat of solution values and relay them to you, thus providing you with heats of solution for all seven ionic salts. During the second week of the experiment your group will use the heats of solution and the Materials Price List (Table 18.1) to determine the most cost-effective manner of constructing a hot- or coldpack. The efficacy of your hot- or coldpack will be tested by another group in your laboratory section, just as your group will evaluate another's hot- or coldpack. The criteria used in the testing are given at the end of this experiment.

Table 18.1: Materials Price List

Item	Cost
Tap water	$0.001/mL
Distilled water	$0.02/mL
Snack-sized baggie	$0.04 each
Rubber band	$0.01 each
Twist tie	$0.01 each
Ammonium chloride (NH_4Cl)	$0.03/g
Ammonium nitrate (NH_4NO_3)	$0.03/g
Lithium chloride (LiCl)	$0.06/g
Potassium acetate (KCH_3CO_2)	$0.03/g
Potassium chloride (KCl)	$0.02/g
Potassium iodide (KI)	$0.10/g
Sodium acetate ($NaCH_3CO_2$)	$0.02/g

ADDITIONAL READING

Read the sections in the "Laboratory Techniques" chapter at the beginning of this Lab Manual on cleaning glassware, handling chemicals, heating liquids and solutions, and weighing prior to performing this experiment.

 Safety Precautions:

- Protective eyewear approved by your institution must be worn at all times while you are in the laboratory.

- If any of the reagents used in this experiment should come in contact with your skin, rinse with water for 15 minutes.

EXPERIMENT

Week 1

Part I: Determination of the Heat Capacity of the Calorimeter

First, it is necessary for your group to determine the heat capacity of your calorimeter. Place approximately 50 mL of distilled water in a beaker. Determine and record the mass of this filled beaker. Pour this water into a dry calorimeter. Weigh the now empty beaker to determine the mass of water transferred to the calorimeter by difference. Place the lid on the calorimeter, and measure the temperature of this water every minute until it remains constant. Record the temperature values.

Place about 50 mL of distilled water in a second beaker. Determine and record the mass of this filled beaker. Place this beaker on a hotplate, and heat it slowly to increase its temperature by approximately 5–6°C. Turn off the hotplate, wait for the temperature to stabilize, and record the temperature of this warm water.

Immediately after measuring the temperature, pour the warm water into the calorimeter. Place the lid on the calorimeter, and swirl it gently. Insert the thermometer through the lid, and record the equilibrium temperature, the final temperature at which the water stabilizes. Weigh the beaker that earlier contained the warm water to determine the mass of warm water transferred to the calorimeter by difference (see Figure 18.1).

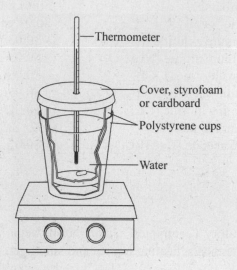

Figure 18.1
Coffee cup calorimeter.

Pour the warm water out of the calorimeter, fill it completely with room-temperature water, and allow it to stand for at least 5 minutes. (It is necessary for the water to stand in the calorimeter for at least 5 minutes to ensure that the calorimeter, which has absorbed a small amount of heat, has returned to room temperature.)

Calculate the heat capacity of your calorimeter.

Repeat this determination of the heat capacity of the calorimeter until at least three close values are achieved. *Note:* It is necessary to dry the calorimeter thoroughly before each use.

Part II: Determination of the Heat of Solution for a Salt

Design and perform an experiment that allows your group to determine the molar heat of solution for one of the salts listed in the Available Equipment and Reagents section. Your instructor will assign your group a salt in class. *Use no more than 5 g of any salt per trial.* Repeat the molar heat of solution determination until at least three close values are achieved. Submit your calculated values of the molar heat of solution to your instructor prior to leaving lab.

Week 2

Part III: Construction of a Hot- or Coldpack

Your group is to design and build the most cost-effective hot- or coldpack (your lab instructor will inform you which) that undergoes a 3°C temperature change. To do this, use the molar heats of solution for the seven salts provided by your instructor. (These are the values determined by the class during the first week of this experiment.) Also consider the Materials Price List (Table 18.1). Choose which salt and other materials lead to the most cost-effective hot- or coldpack design.

Part IV: Testing Another Group's Hot- or Coldpack Design

After you have designed, constructed, and tested (and redesigned?) your hot- or coldpack submit it to another group for testing. The testing requires three trials. Provide the group that will perform the testing of your hot- or coldpack with everything they need to test your hot- or coldpack three times.

Just as another group will be testing your hot- or coldpack design, so too will your group test the hot- or coldpack design of another. Perform the testing by completing the Hot- or Coldpack Evaluation Sheet.

Equipment and Reagents

To perform this experiment, you will have access to all the equipment in your lab drawer, a calorimeter (two Styrofoam cups) with lid, a thermometer, a hotplate, the electronic balances, snack-sized Ziploc baggies, rubber bands, and twist ties.

The following salts also will be available: NH_4Cl, NH_4NO_3, CH_3CO_2K, KCl, KI, $LiCl$, and CH_3CO_2Na.

Waste Disposal. Dispose of all chemical waste as directed by your instructor.

Name: _____   Date: _____

Lab Instructor: _____   Lab Section: _____

EXPERIMENT 18

Hot- and Coldpacks

PRELABORATORY QUESTIONS

1. In Part I of the experiment, determination of the heat capacity of the calorimeter, there are three components that participate in heat transfer: the hot water, the cold water, and the calorimeter. Which of these components will gain heat? Which will lose heat? Write an equation that relates these three heats.

2. Part I of the procedure was performed by a student, and the following data were recorded:

   Mass of 100 mL beaker and hot water = 95.614 g
   Mass of 100 mL beaker = 50.551 g
   Mass of 50 mL beaker and cold water = 70.093 g
   Mass of 50 mL beaker = 31.248 g
   Initial temperature of cold water = 24.2°C
   Initial temperature of hot water = 45.8°C
   Final temperature of calorimeter and contents = 35.2°C

   Use these data to calculate the heat capacity of the calorimeter.

Name: _____   Date: _____

Lab Instructor: _____   Lab Section: _____

3. Write balanced total ionic equations for the aqueous dissolution reaction of each of the seven ionic salts listed in the Equipment and Reagents section.

4. Look up the Material Safety Data Sheet (MSDS; **http://msds.ehs.cornell.edu/msdssrch.asp**) of one of the seven ionic salts listed in the Equipment and Reagents section and list any hazards associated with it.

Name: _____   Date: _____

Lab Instructor: _____   Lab Section: _____

EXPERIMENT 18

Hot- and Coldpacks

RESULTS/OBSERVATIONS

Week 1

Part I: Determination of the Heat Capacity of the Calorimeter

1. Calorimeter heat capacity determination:

Trial #	Mass of H_2O and Beaker #1 (g)	Mass of Beaker #1 (g)	Initial Temp. of H_2O in Calorimeter (°C)	Mass of H_2O in Beaker #2 (g)	Temp. of H_2O in Beaker #2 (°C)	Equilibrium Temp. of Calorimeter and Its Contents (°C)	Mass of Beaker #2 (g)
1							
2							
3							
4							

2. Identity of ionic salt examined: _____

3. Write any other pertinent data in the space below. Clearly indicate both the property and the amount.

Name: _____   Date: _____

Lab Instructor: _____   Lab Section: _____

4. Calculation of the calorimeter heat capacity using the data collected in trial #_____ (choose a trial):

5. Calculated calorimeter heat capacity: _____ (Trial #1) _____ (Trial #2)

   _____ (Trial #3) _____ (Trial #4)

6. Average calculated calorimeter heat capacity: _____ (average of three trials)

Part II: Determination of the Heat of Solution for a Salt

7. Write any pertinent data in the space below. Clearly indicate both the quantity and the amount.

Name: _____   Date: _____

Lab Instructor: _____   Lab Section: _____

8. Calculation of the molar heat of solution of _____ (insert name of salt) using the data
    collected in trial # _____ (choose a trial):

9. Calculated molar heat of solution of _____ (insert name of salt): _____ (Trial #1)

    _____ (Trial #2)

    _____ (Trial #3)

    _____ (Trial #4)

10. Average calculated molar heat of solution of _____ (insert name of salt):

    _____ (average of three trials)

Week 2

Part III: Construction of a Hot- or Coldpack

11. Was your group assigned a _____ hot- or _____ coldpack (check one)?

12. Salt chosen for use in hot- or coldpack: _____

Name: _____   Date: _____

Lab Instructor: _____   Lab Section: _____

13. Write any pertinent data in the space below. Clearly indicate both the property and the amount.

14. Perform any pertinent calculations in the space below. Clearly indicate both the property and the result.

Name: _____   Date: _____

Lab Instructor: _____   Lab Section: _____

Part IV: Testing Another Group's Hot- or Coldpack Design

15. Write any pertinent data and calculations in the space below. Clearly indicate both the property and the amount.

Hot- or Coldpack Evaluation Sheet

Names of designers: _____   _____

                       _____   _____

Names of testers: _____   _____

                       _____   _____

Are you are testing a _____ hotpack or _____ coldpack (check one)?

Does the hot- or coldpack you are testing do each of the following?

	Yes	No
Is it easy to use?		
Is it safe?		
Does it consistently produce a 3°C temperature change?		
Is any variation in the temperature change within 20% of the desired value?		

Complete the data table on the following page before answering the questions below.

Average cost per °C change: _____

What is your overall opinion of the hot- or coldpack you are testing on a scale of 1 to 5?

1   2   3   4   5

Fails in   Average   Performs the desired function
all ways.   with perfect ease and safety.

> *Return this completed form to the group which designed this product after you have completed it. The design group will attach this form to their final report.*

Hot- or Coldpack Testing Evaluation Sheet

	Trial 1	Trial 2	Trial 3
Type of water used: distilled or tap?			
Volume of H_2O (mL)?			
Salt used?			
Mass of salt (g)?			
Cost and nature of additional items (if applicable)?			
Total cost of salt?			
Cost of baggie?			
Total cost per pack?			
Time to ΔT_{max} (s)?			
ΔT_{max} (°C)?			
Average ΔT_{max} (°C)?			
Cost per °C change?			
Average cost per °C change?			

(*Note:* The questions in italics are to be answered by the product designers. The others are to be answered by the product testers.)

Place any comments on the hot- or coldpack or the procedure used in testing it in this space:

Name: _____   Date: _____

Lab Instructor: _____   Lab Section: _____

EXPERIMENT 18

Hot- and Coldpacks

POSTLABORATORY QUESTIONS

1. Explain why the straight line on the left has a rising slope, whereas the straight line on the right has a falling slope. What can you deduce about the initial temperature of the starting materials in this example?

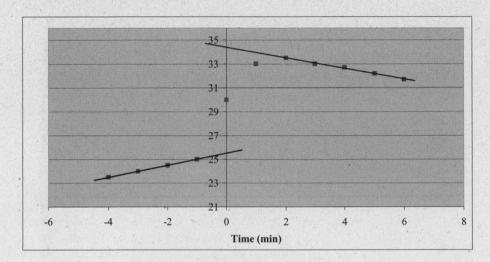

Name: _____   Date: _____

Lab Instructor: _____   Lab Section: _____

2. What are the desired characteristics of a salt used in making a cost-effective hot- or coldpack?

3. How is the constant pressure of the constant-pressure calorimetry measurements performed in this experiment maintained?

4. What are some sources of error in this experiment?

EXPERIMENT 19

Thermodynamics of Dissolution of Potassium Nitrate

OBJECTIVE

To determine values for the thermodynamic functions ΔH°, ΔS°, and ΔG° for the dissolution of potassium nitrate (KNO_3). The solubility of KNO_3 will be measured at a variety of temperatures. The K_{sp} at each temperature will be calculated from these data. A plot of $\ln K_{sp}$ versus inverse temperature will be used to extract the values of ΔH° and ΔS°. The values of ΔH° and ΔS° then can be used to calculate ΔG°.

INTRODUCTION

The usefulness of thermodynamic quantities to chemistry can hardly be overstated. Quantities such as the Gibbs free-energy change (ΔG), the enthalpy change (ΔH), and the entropy change (ΔS) can be used to predict what types of chemical and physical processes are possible and under what conditions. They can be used to calculate properties of the equilibrium state, such as the equilibrium constant (K). Calculation of the heat transferred during a chemical or physical process is possible via them. Obtaining the values of these quantities, be it by theory or experiment, is highly desirable. Conveniently, these thermodynamic quantities are related to one another in such a way that experimental measurement of a few allows for calculation of the rest. This experiment will exploit these relationships in determining K, ΔG°, ΔH°, and ΔS° for the dissolution of potassium nitrate (KNO_3):

$$KNO_3(s) \rightleftharpoons K^+(aq) + NO_3^-(aq) \tag{1}$$

The standard Gibbs free-energy change (ΔG°) associated with a chemical or physical process is related to the changes in the standard enthalpy (ΔH°) and entropy (ΔS°) by the expression

$$\Delta G^\circ = \Delta H^\circ - T\Delta S^\circ \tag{2}$$

where T is the temperature. The standard free-energy change for a reaction is also related to the equilibrium constant by the expression

$$\Delta G^\circ = -RT \ln K \tag{3}$$

where R is the universal gas constant. For the specific case of a dissolution reaction, $K = K_{sp}$. Combining equations (2) and (3) yields the relation

$$\ln K_{sp} = -\frac{\Delta H^\circ}{RT} + \frac{\Delta S^\circ}{R} \tag{4}$$

Equation (4) has the form of a straight line ($y = mx + b$):

$$\underbrace{\ln K_{sp}}_{y} = \underbrace{-\frac{\Delta H^{\circ}}{R}}_{m} \underbrace{\frac{1}{T}}_{x} + \underbrace{\frac{\Delta S^{\circ}}{R}}_{b} \tag{5}$$

If $\ln K_{sp}$ is plotted as a function of $1/T$, a straight line with a slope of $-\Delta H^{\circ}/R$ and a y-intercept of $\Delta S^{\circ}/R$ should result. The ΔH° can be determined by multiplying the measured slope by $-R$, and the ΔS° can be calculated by multiplying the measured y-intercept by R.

In this experiment the solubility of KNO_3 will be measured as a function of temperature. The K_{sp} at each temperature will be calculated from these data. A plot of $\ln K_{sp}$ versus inverse temperature will be used to extract the values of ΔH° and ΔS°. The values of ΔH° and ΔS° then can be used to calculate ΔG°.

ADDITIONAL READING

Read the parts of the "Laboratory Techniques" chapter at the beginning of this Lab Manual on cleaning glassware, handling chemicals, heating liquids and solutions, burets, and weighing prior to performing this experiment.

 Safety Precautions:

■ Protective eyewear approved by your institution must be worn at all times while you are in the laboratory.

■ If any of the reagents used in this experiment should come in contact with your skin rinse with water for 15 minutes.

PROCEDURE

Obtain a buret, rinse it twice with distilled water, and then fill it with distilled water and attach it to a ring stand with a buret clamp.

Assemble a hot-water bath by placing a 400 mL beaker about three-quarters full of water on a hotplate. Add a few boiling chips to the water. Heat the water, but do not allow it to boil.

For sample 1, weigh out the amount of KNO_3 specified in Table 19.1 using the electronic balance. It is not necessary to weigh out the exact amount listed in the table. However, it is necessary to know the precise mass of solid KNO_3 in the sample, so record the mass to the nearest milligram.

Place the solid KNO_3 in a 16 mm. testtube. Add 5.00 mL of distilled water to the testtube from the buret. As you add the water, make sure that it washes any crystals adhering to the sides of the testtube to the bottom. As was the case with the mass, it is not necessary to dispense exactly 5.00 mL, but it is necessary to know the precise volume of water added to the sample. Record the initial and

Table 19.1

Sample	Mass of KNO_3 (g)
1	2.50
2	4.00
3	5.25
4	6.50
5	8.00

final buret readings. Calculate the volume of water dispensed by difference. Stir the solution gently and carefully with a glass stirring rod until the solid is evenly suspended in the water. Note what happens to the temperature of the solution as it is stirred. Record your observation.

Clamp the testtube to a ring stand above the hot-water bath, and insert the testtube into the hot-water bath. Put a second clamp on the ring stand above the clamp holding the testtube. Put the thermometer through a split one-hole rubber stopper, and use this second clamp to support the thermometer. Immerse the thermometer in the testtube. Stir the solution in the testtube occasionally with the thermometer until the solid has dissolved completely.

Once the crystals have dissolved completely, remove the testtube from the hot-water bath. Leave the thermometer in the testtube. Let the solution cool, stirring occasionally. When crystals begin to form (do not let the solution become opaque with crystals but just slightly cloudy), dip the testtube into the hot-water bath, and then remove it and stir the solution. Repeat this process until the crystals dissolve completely. Record the temperature at which all the crystals just dissolve. If the solid dissolves too quickly to determine the precise temperature at which dissolution is complete, cool the solution again until crystals re-form, and repeat the dipping process. It might be necessary to repeat this process several times until a precise determination of the temperature of dissolution is achieved.

Repeat this procedure for each of the other samples listed in Table 19.1.

Waste Disposal. All chemical waste should be disposed of as directed by your instructor.

Name: _____ Date: _____

Lab Instructor: _____ Lab Section: _____

EXPERIMENT 19

Thermodynamics of Dissolution of Potassium Nitrate

PRELABORATORY QUESTIONS

1. Use the standard thermodynamic functions below to calculate ΔH°, ΔS°, and ΔG° for the reaction to be examined in this experiment:

$$KNO_3(s) \rightleftharpoons K^+(aq) + NO_3^-(aq) \qquad (1)$$

Substance	ΔH°_f (kJ/mol)	ΔG°_f (kJ/mol)	S° (J/mol · K)
K(s)	0	0	64.2
K$^+$(aq)	−252.4	−283.3	103
KNO$_3$(s)	−494.6	−381.8	133.1
NO$_3^-$(aq)	−207.4	−111.3	146.4

2. What is the equilibrium expression for Reaction (1)?

3. More KNO$_3$ dissolves in hot water than in cold. Does K_{sp} of KNO$_3$ increase, decrease, or stay the same with increasing temperature? Explain.

4. If K_{sp} increases with increasing temperature, is the reaction endothermic or exothermic? Explain.

Name: _____   Date: _____

Lab Instructor: _____   Lab Section: _____

EXPERIMENT 19

Thermodynamics of Dissolution of Potassium Nitrate

RESULTS/OBSERVATIONS

1. Mass of KNO_3 solid samples:

Sample	Mass of KNO_3 (g)
1	
2	
3	
4	
5	

2. Volumes of distilled water added to each solid sample:

Sample	Initial Buret Volume (mL)	Final Buret Volume (mL)	Volume Dispensed (mL)
1			
2			
3			
4			
5			

3. Observation of temperature as each sample was stirred:

Sample	Observation
1	
2	
3	
4	
5	

Name: _____   Date: _____

Lab Instructor: _____   Lab Section: _____

4. Temperature measurements:

Sample	T (°C) Dissolution	T (K) Dissolution
1		
2		
3		
4		
5		

5. Show the $[K^+]$ calculation for sample 1. Assume the volume of the solution to be equivalent to the volume of solvent (H_2O) used.

6. Show the $[NO_3^-]$ calculation for sample 1. Assume the volume of the solution to be equivalent to the volume of solvent (H_2O) used.

7. Show the K_{sp} of KNO_3 calculation using the data for sample 1.

Name: _____   Date: _____

Lab Instructor: _____   Lab Section: _____

8. Table of results:

Sample	$[K^+]$ (M)	$[NO_3^-]$ (M)	K_{sp}	$\ln K_{sp}$	T (K)	$1/T$ (K^{-1})
1						
2						
3						
4						
5						

9. On the graph provided, plot $\ln K_{sp}$ as a function of $1/T$.

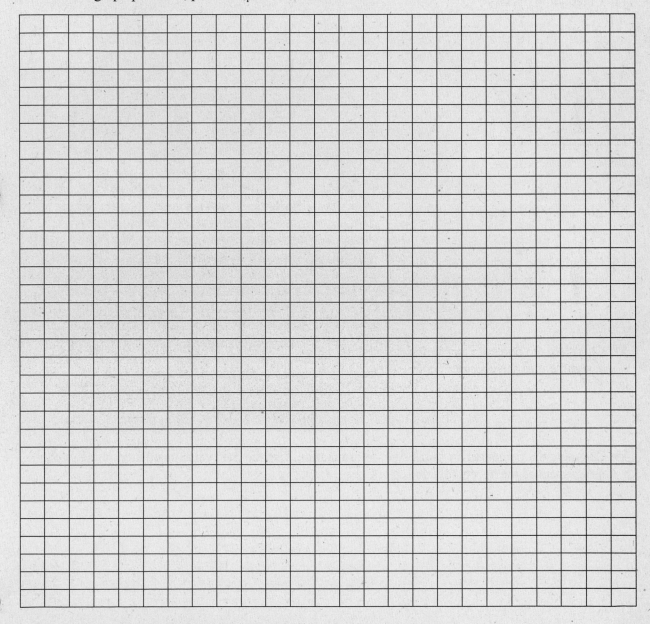

Name: _____   Date: _____

Lab Instructor: _____   Lab Section: _____

10. Calculation of $\Delta H°$:

11. Calculation of $\Delta S°$:

12. Calculation of $\Delta G°$:

13. Calculate the percent error in the $\Delta H°$ determination. Use the $\Delta H°$ value calculated in the prelab assignment as the accepted value.

14. Calculate the percent error in the $\Delta S°$ determination. Use the $\Delta S°$ value calculated in the prelab assignment as the accepted value.

15. Calculate the percent error in the $\Delta G°$ determination. Use the $\Delta G°$ value calculated in the prelab assignment as the accepted value.

Name: _____   Date: _____

Lab Instructor: _____   Lab Section: _____

EXPERIMENT 19

Thermodynamics of Dissolution of Potassium Nitrate

POSTLABORATORY QUESTIONS

1. How does the assumption that solution volumes are equal to the volume of water added affect the determination of ΔS°?

2. According to the experimentally determined value of ΔH°, is the dissolution of KNO_3 exothermic or endothermic? Explain.

Name: _____   Date: _____

Lab Instructor: _____   Lab Section: _____

3. According to the experimentally determined value of $\Delta S°$, does the dissolution of KNO_3 lead to more or less disorder? Explain.

4. According to the experimentally determined value of $\Delta G°$, is the dissolution of KNO_3 spontaneous? Explain.

EXPERIMENT 20

Physical Properties of Pure Substances and Solutions

OBJECTIVE

To observe, and in some cases quantify, a number of physical properties of pure substances and solutions: boiling, freezing, sublimation, triple-point behavior, and freezing-point depression.

INTRODUCTION

The properties of a substance are those characteristics used to describe or identify it. Properties can be classified as either physical or chemical. Physical properties such as color, odor, amount, temperature, viscosity, and boiling point can be determined without altering the chemical composition of the sample: Boiling water results in the conversion of liquid water to water vapor, a change in the physical state of the substance, yet at the molecular level both the liquid and the gas are composed of H_2O molecules. Chemical properties, conversely, do change the chemical composition of a sample: The combustibility of molecular hydrogen (H_2) results in its reaction with oxygen (O_2) to form a new substance, water.

   This experiment is concerned with physical properties of substances. More specifically, changes in the physical properties of substances will be observed in this experiment: The vaporization of liquid nitrogen (N_2), the freezing of a t-butanol solution, and the sublimation of carbon dioxide are a few of the physical changes that will be observed. The conditions at which each change occurs can be used to characterize the substance being observed. For example, the temperature-pressure conditions under which carbon dioxide sublimes are characteristic of carbon dioxide. The difference in the freezing points of pure t-butanol and a solution of t-butanol are characteristic of the amount of solute present in the solution. The amount of heat necessary to vaporize liquid nitrogen is characteristic of liquid nitrogen. Understanding the relationship between these external conditions and the physical properties of pure substances and solutions is the goal of this experiment.

   There are three parts to this experiment. A separate introduction and procedure is provided for each. Depending on the length of your lab period, you will have either one or two weeks to complete this experiment.

ADDITIONAL READING

Read the sections in the "Laboratory Techniques" chapter at the beginning of this Lab Manual on cleaning glassware, handling chemicals, weighing solids, heating liquids and solutions, and vacuum filtration prior to performing this experiment.

Part A: Phase Behavior of Water, Carbon Dioxide, and *t*-Butanol

Vapor pressure is a manifestation of the tendency of a liquid or solid to become gaseous. The vaporization of a liquid is called *evaporation*. The vaporization of a solid is called *sublimation*. For any pure substance there are certain conditions of temperature and pressure at which the tendency of the substance to change state in one direction (e.g., liquid to gas) is exactly balanced by its tendency to change state in the opposite direction (e.g., gas to liquid). Under these special conditions a state of equilibrium exists between the two phases. For a pure substance, such as water, there is a set of pressure-temperature points at each of which there is equilibrium between the gaseous and liquid phases. Since the liquid and gaseous phases coexist at these points, they are called *liquid-gas coexistence points*. The lines connecting coexistence points are called *coexistence curves* (see Figure 20.1). Each of the points along a liquid-gas coexistence curve is a boiling point; that is, the liquid will boil when the prevailing pressure is less than the equilibrium vapor pressure at that temperature. The particular boiling-point temperature that corresponds to an equilibrium vapor pressure of exactly 1 atmosphere is called the *normal boiling point*. For water this temperature is exactly 100.00°C (373.15 K). For prevailing pressures less than 1 atmosphere the substance will boil at temperatures less than the normal boiling point. The boiling of water at conditions other than those of the normal boiling point will be one of the changes examined in this part of the experiment.

Just as there are points at which equilibrium exists between the liquid and gaseous states, there are pressure-temperature coexistence points at which there is equilibrium between the solid and liquid phases; these are called *freezing points* or *melting points*. The normal freezing point of a substance occurs at a prevailing pressure of exactly 1 atmosphere.

The point at which all three phases, solid, liquid, and gas, are in equilibrium is called the *triple-point*. Fascinatingly enough, at the triple-point one should be able to see evidence of simultaneous boiling and freezing of the substance. For water, the triple-point occurs at 273.16 K and 0.00603 atm (by comparison, the freezing point of water at a prevailing pressure of 1.000 atm is 273.15 K.) In this experiment you will have an opportunity to observe 2-methyl-2-propanol, commonly referred to as *t*-butanol, at its triple-point.

The triple-point pressure for *t*-butanol, like that of water, occurs at a pressure much less than 1 atmosphere. This is not always the case. There are many substances that have triple-point pressures above 1 atmosphere. An example is carbon dioxide (CO_2), the solid form of which is called *dry ice* because, under ordinary pressure conditions, it sublimes rather than melting. The triple-point of CO_2 occurs at a pressure of 5.11 atmospheres and a temperature of −56.4°C. The normal sublimation point of carbon dioxide is −78.5°C. The physical state of carbon dioxide as a function of temperature and pressure is summarized in it phase diagram, shown in Figure 20.1. The melting, boiling, and sublimation of CO_2 will be observed in this experiment.

   Safety Precautions:

- Protective eyewear approved by your institution must be worn at all times while you are in the laboratory.

- Handle dry ice with tongs or forceps, not your bare hands. It is extremely cold, and handling it can cause injury to the skin.

- *t*-Butanol is flammable. Make sure that there are no open flames in the laboratory. In the event of a fire, alert your instructor. A fire in a small container can be smothered by covering the vessel with a watchglass. Use a fire extinguisher to eliminate a larger fire. If the fire is burning over too large an area to be extinguished easily, evacuate the area and activate the fire alarm. If your clothing should catch on fire, use the safety shower or a fire blanket to extinguish the flames.

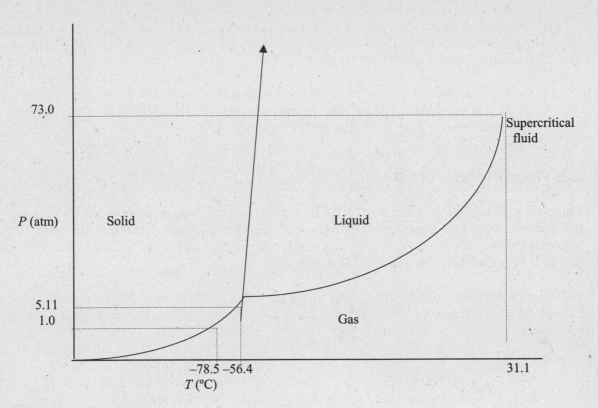

Figure 20.1
Phase diagram of carbon dioxide.

PROCEDURE

Part A: Phase Behavior of Water, Carbon Dioxide, and *t*-Butanol

A-1: Phase Behavior of Water

Obtain a no. 1 rubber stopper. Place 20 mL of distilled water in a 125 mL Florence flask, add two or three boiling chips, and place the flask on a hotplate. Heat the water to boiling. Note that before the liquid starts to boil, you should observe the formation of gas bubbles leaving the water; this does not, however, constitute boiling. Since dissolved gases, such as oxygen and nitrogen, become less soluble when the temperature rises, they form small bubbles and leave the water before the water starts to boil. Boiling, by contrast, is the vigorous formation of bubbles that occurs when the vapor pressure of a liquid exceeds the pressure of the gas or gases above the liquid. Bubbles form throughout the liquid owing to the rapid expansion that results when small portions of liquid are converted to gas. In the case of a container of water open to the atmosphere, boiling will occur when the vapor pressure of the liquid water exceeds the atmospheric pressure.

After the water has been boiling for at least 1 minute, use your crucible tongs to put the rubber stopper into the top of the flask as you shut off the hotplate. Grasp the neck of the flask with your testtube clamp, and remove the flask from the hotplate. Place the flask on the benchtop, and quickly push the stopper securely into the top of the flask.

Using the testtube clamp, hold the flask upside down, and run a small stream of cold water from the water tap over it for 5 to 10 seconds. What happens? Record your observations. Try these related experiments:

(a) See what happens if you place your (cool) hand on top of the hot inverted flask (make sure that it isn't so hot that you would burn your hand, of course). Alternatively, place a bit of crushed ice on top of the inverted flask.

(b) Set the flask aside and allow it to cool to room temperature, or cool it with crushed ice to an even lower temperature. Then, with the flask inverted, hold the neck of the flask tightly in your (hot) hand and see if you can get the water to boil.

A-2: Phase Behavior of CO_2

This part will be done as a demonstration by the instructor. Your instructor will go through this portion of the experiment several times, each time for a different group of students. Students should observe carefully what the instructor does, and record all their observations, just as if they had done the experiment themselves.

Obtain a Beral (plastic) pipet, and snip off the tip with a pair of scissors or a razor blade. Pour 20 mL of cold tap water into a *plastic* 100 mL beaker, and place the 100 mL beaker inside a *plastic* 600 mL beaker.

C AUTION: Handle dry ice with tongs or forceps, not your bare hands. It is extremely cold, and handling it can cause injury to the skin.

Using the forceps or a scoopula, place about four or five of the small chunks of dry ice in the pipet, enough to fill the bulb between one-fifth and one-quarter full. Then, after getting the solid CO_2 chunks down into the bottom, fold the plastic tube over twice, and seal it with a screw clamp. It is important to get a good seal, so be sure to place the folded tube in the center of the screw clamp and to tighten it firmly (see Figure 20.2).

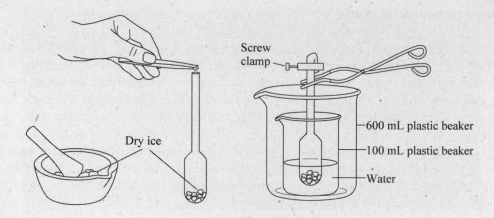

Figure 20.2
The mortar and pestle are for grinding the dry ice into small pieces, if necessary. The foreceps should be used to place the dry ice in the bulb of a Beral pipet. The plastic tube of the Beral pipet then is folded over twice and sealed with a screw clamp. The sealed pipet bulb then is submerged in a 100 mL plastic beaker partially filled with water. This 100 mL beaker then is placed within a 600 mL plastic beaker.

Pick up the pipet with crucible tongs, submerge it somewhat in the cold water (to keep the tube from frosting up) in the *plastic* 100 mL beaker that is inside the *plastic* 600 mL beaker, and observe what happens. **(Note the necessity for using plastic beakers. Glass beakers could break when the CO_2 pressure causes the pipet bulb to burst.)** As the CO_2 warms up, it should begin to liquefy, and you should see the triple-point, the point at which both solid and liquid are in equilibrium with the vapor. Shaking it a bit helps to see what is happening. After about a minute, the pressure should build up enough so that all the solid CO_2 will melt. Stand back from the apparatus a few feet because shortly after that, the pressure should increase rapidly, causing the pipet to rupture. To avoid getting sprayed with water, you might want to place a sheet of paper over the top of the beaker at this point. (If you had insufficient CO_2 inside the pipet to produce enough pressure to rupture the pipet, or you didn't seal the pipet well enough to prevent leaking, wait at least 2 minutes, release the screw clamp, put some more pieces of solid CO_2 in the pipet and repeat. If you continue to have problems, remove the screw clamp, and with the tip of the pipet bent over, use a pair of pliers to seal it off.) Note what happens, recording all your observations, including a description of the material remaining inside the ruptured pipet. When you have completed this part of the experiment, dispose of the pipet in the wastebasket.

A-3: Phase Behavior of t-Butanol

i: *Triple-point of* t-*Butanol*

In the hoods you will find several setups for measuring the triple-point of *t*-butanol (see Figure 20.3). The setup consists of a testtube with a sidearm and a rubber stopper drilled to fit a thermometer probe. Take a digital thermometer to one of the setups. Lubricate the hole in the stopper and the tip of your thermometer probe with glycerine, and using a twisting motion, slide the probe through the hole in the stopper. The thermometer probe should be inserted far enough to reach nearly but not entirely to the bottom of the sidearm testtube. Wash off the glycerine with water when you are finished, and dry the thermometer probe.

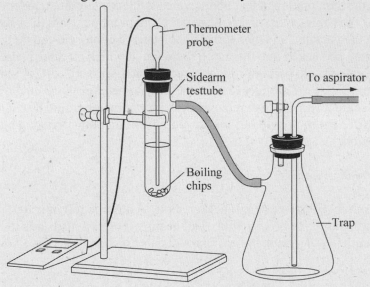

Figure 20.3
Setup for measuring the triple-point of *t*-butanol.

WARNING: *t*-Butanol is flammable. Make sure that there are no open flames in the laboratory.

t-Butanol freezes at a temperature near room temperature, so the reagent bottles will be heated just before lab starts to a high enough temperature to keep the *t*-butanol molten throughout the lab period. When you have gotten to this part of the experiment and are ready to use the *t*-butanol, see your lab instructor, who will provide it to you. Place at least five or six medium-sized boiling chips and enough *t*-butanol in the special sidearm testtube (about 3 cm) to cover the tip of the thermometer probe (the sensitive part). Recap the bottle of *t*-butanol, and return it to your instructor.

If you are using a setup already containing *t*-butanol in a sidearm testtube, melt the *t*-butanol, if necessary, by running hot water over the side of the testtube, and add five or six fresh boiling chips. (It is crucial to have *fresh* boiling chips. It isn't necessary to remove all the previously used chips, but if there get to be too many, decant the alcohol, but not the chips, into the sink, and wash it down the drain with hot water. Wash the boiling chips with warm water, decant the water, and throw the chips into the waste basket. Get a fresh sample of *t*-butanol.)

Clamp the testtube to the ring stand. Insert the stopper tightly into the testtube.

Turn on the water aspirator. The water should flow as fast as possible. Touch your finger to the end of the vacuum tube attached to the sidearm of the trap to make sure that there is good suction.

Turn off the water aspirator, and attach the vacuum tubing to the sidearm of the testtube. Turn on the water aspirator again. The temperature should begin to drop almost immediately. After a short period of time the liquid should start to boil. As soon as you observe boiling, reduce the water flow going through the aspirator slightly, not enough to cause the boiling to stop but enough to lengthen the time that it will take before the liquid also begins to freeze. Eventually, you should observe simultaneous boiling and freezing. Push the "HOLD" button on the digital thermometer when you observe simultaneous boiling and freezing. Record the temperature at that point, which is the triple-point temperature.

After you observe the triple-point, first remove the vacuum tubing from the sidearm of the testtube to relieve the vacuum, and then turn off the water aspirator (if you do this in reverse order, water may be sucked into the trap.) The point at which all three phases are present simultaneously, the triple-point of *t*-butanol, occurs at a pressure of 42.4 mm Hg. Record the triple-point pressure. Carefully remove the thermometer probe from the stopper. Do not dismantle the setup any more than removing the thermometer probe. It will be used by other students. A different setup will be used to get the boiling point, freezing point, and freezing-point depression.

ii: *Boiling Point of* t-*Butanol at Laboratory Pressure*

Prepare a hot-water bath by filling a 600 mL beaker about two-thirds full of water, placing it on a hot plate, adding two or three boiling chips, and heating it until it just starts to boil. Two students can share each hot-water bath, but each should carry out his or her own boiling-point determination.

WARNING: *t*-Butanol is flammable. Make sure that there are no open flames in the laboratory.

While you are waiting for the water to boil, clean and dry your largest (25 cmm) testtube, stand it in a beaker, and weigh it on the balance. Record the mass. Get the bottle of *t*-butanol from your instructor. Using your ruler, measure up from the bottom of the testtube about 4 cm; then pour *t*-butanol into the testtube until it is 4 cm deep. Reweigh the beaker and testtube in order to get the weight of *t*-butanol. Record these data. Calculate the mass of *t*-butanol by difference. You will need to know the mass of *t*-butanol later in Part B of the experiment.

Obtain a freezing-point apparatus and a split rubber stopper containing a wire stirrer. Insert your thermometer into the hole not containing the stirrer (notice that the stopper has been split to facilitate inserting the thermometer). Adjust the thermometer so that the bottom of the tip is inside the testtube and submerged beneath the surface of the liquid. Clamp the testtube to the ring stand.

Once the water in the hot-water bath is boiling turn off the hot plate. Using beaker tongs (*not* crucible tongs), place the beaker of hot water on the base of the ring stand under the testtube. Loosen the clamp, and immerse the testtube in the hot-water bath long enough to raise the temperature of the liquid to the boiling point. Record this boiling point. Your instructor will either show you how to read the barometer or give you a reading of the current barometric pressure in the laboratory. Record the barometric pressure.

iii: Freezing point of t-Butanol at Laboratory Pressure

Use beaker tongs to remove the hot-water bath from the ring stand. Then fill your 400 mL beaker half full of cold water, and by adding crushed ice if necessary, adjust the temperature to about 18°C. Place the cold-water bath on the ring stand.

Immerse the testtube containing the *t*-butanol in the cold-water bath, and stir continuously with the stirrer (to minimize supercooling) until the liquid freezes. Take the freezing point to be the point at which solid first appears. Record the freezing point. Check the temperature of the cold-water bath periodically to see that it remains at approximately 18°C, adding small chunks of ice if necessary. *Save the* t-*butanol and setup for use in Part B.* (If you will not be doing Part B until the next lab period, melt the mixture in the testtube in hot water before attempting to remove the thermometer. Then disassemble the apparatus and dispose of the solution in the sink, rinsing it down the drain with plenty of hot water.)

Waste Disposal (Part A). All chemical waste should be disposed of as directed by your instructor.

Part B: Colligative Properties—Freezing-point Depression

Colligative properties of solutions depend to a good approximation on the number of solute particles present in solution, regardless of their chemical identity; a mole of one kind of solute particles, no matter whether the particles are ions or molecules, has nearly the same effect on these colligative properties as a mole of any other solute particles. Examples of colligative properties include vapor-pressure lowering, boiling-point elevation, freezing-point depression, and osmotic pressure.

In this part of the experiment freezing-point depression will be used to determine the molecular weight of a solute. For a given solvent, the amount of lowering of the freezing point temperature (ΔT_f) is proportional to the concentration of dissolved particles, that is,

$$\Delta T_f = k_f \cdot m \tag{1}$$

The constant k_f is the freezing-point depression constant, or cryoscopic constant, for the particular solvent, and m is the molality of particles dissolved in the solvent. In freezing from a solution, most solvents crystallize out as crystals of pure solvent; thus, as the solvent disappears from the solution, the solution becomes more and more concentrated in solute, causing the freezing point to drop further and further. The freezing point of the solvent to use in your calculations is the point at which freezing just begins. The freezing-point depression constant of t-butanol is 8.3 kg · °C/mol. You will determine the freezing point of t-butanol in a solution of an unknown molecular solute (i.e., the solute does not dissociate into ions) dissolved in t-butanol and use that information, together with the freezing point of pure t-butanol, to determine the apparent molecular weight of the solute.

      Safety Precautions:

- Protective eyewear approved by your institution must be worn at all times while you are in the laboratory.

- t-Butanol is flammable. Make sure that there are no open flames in the laboratory. In the event of a fire, alert your instructor. A fire in a small container can be smothered by covering the vessel with a watchglass. Use a fire extinguisher to eliminate a larger fire. If the fire is burning over too large an area to be extinguished easily, evacuate the area and activate the fire alarm. If your clothing should catch on fire, use the safety shower or a fire blanket to extinguish the flames.

PROCEDURE

Part B: Colligative Properties—Freezing-Point Depression

NOTE: **If you have saved the *t*-butanol and setup from Part A, you may skip the italicized procedural steps below.**

WARNING: *t*-Butanol is flammable. Make sure that there are no open flames in the laboratory.

Clean and dry your largest (25 mm) testtube, stand it in a beaker, and weigh it on the balance. Record the mass. Get the bottle of t-butanol *from your instructor. Using your ruler, measure up from the bottom of the tube about 4 cm; then pour* t-butanol *into the testtube until it is 4 cm deep. Reweigh the beaker and testtube in order to get the weight of* t-butanol. *Record these data.*

Using the same apparatus and procedure as you used to get the freezing point in Part A, make a new reading of the freezing point of the pure t-butanol.

Reweigh the beaker and testtube of t-butanol to again determine the mass of t-butanol as was done in Part A-3-ii. Record the mass. Has the mass of t-butanol changed? Why might it be expected to?

Stand a clean, dry 10 mm testtube in a beaker, and obtain a sample of unknown solute in the testtube from your instructor. Weigh the beaker and testtube with the unknown in it on the balance. Record the mass.

If necessary, run hot water over the side of the testtube containing the frozen *t*-butanol until it liquefies. Then, without spilling the liquid, remove the thermometer-stirrer assembly, pour the unknown solute in with the *t*-butanol, and replace the thermometer and stirrer. Weigh the empty 10 mm testtube in the same beaker you used previously. Record the mass. Determine the mass of unknown solute added to the *t*-butanol by subtraction.

Place some crushed ice and water in a large beaker. Get an estimate of the freezing point of the solvent in the solution by setting the testtube in the crushed ice-water mixture until it begins to freeze.

Run hot water over the outside of the testtube until the temperature of the solution is *above 30°C,* and then clamp the testube to the ring stand. Immerse the testtube in a cold-water bath held at 5 to 10°C below the approximate freezing point of the mixture (by adding crushed ice to cold water). Stirring constantly, watch for freezing. Take the freezing point to be the temperature where you *first* see crystals.

Warm up the contents of the testtube with hot water, and repeat the freezing point determination.

Waste Disposal (Part B). All chemical waste should be disposed of as directed by your instructor.

Part C: Molar Heat of Vaporization of Nitrogen

(Suggestion to instructor: Do this first, at the beginning of the lab period, even though it is listed as Part C.)

This experiment will be carried out by a few members of the class under the supervision of the instructor. If you are not an active participant in this exercise you are still expected to watch what is done and record the data.

The normal boiling point of liquid nitrogen is –196°C. When a sample of liquid nitrogen is poured into heated liquid water the nitrogen quickly boils away. Since changing state from liquid to gas requires energy, a considerable amount of heat is required to boil away the nitrogen. This energy comes mostly from the hot water. As the nitrogen boils away, the temperature of the water drops. The amount of energy transferred from the water to the liquid nitrogen can be determined by considering the mass of water, the specific heat capacity of water (4.184 J/g · °C), and the drop in temperature of the water. If it is assumed that all the heat lost by the water is used to boil the nitrogen away, that is, vaporize it, the heat of vaporization per gram of the sample of liquid nitrogen can be calculated from the expression

$$\Delta H_{vap} = \frac{s_{H_2O} \cdot m_{H_2O} \cdot \Delta T_{H_2O}}{m_{N_2}} \tag{2}$$

To convert the heat of vaporization per gram to the heat of vaporization per mole, simply multiply the value of ΔH_{vap} per gram by the molar mass of N_2.

Safety Precautions:

■ Protective eyewear approved by your institution must be worn at all times while you are in the laboratory.

■ Only the lab instructor may handle liquid nitrogen. It is extremely cold, and handling it can cause injury to the skin.

PROCEDURE (FOR LAB INSTRUCTOR)

Part C: Molar Heat of Vaporization of Nitrogen

The instructor will need to recruit three students to assist with this part of the experiment:

(1) One to record the mass of hot water and to measure the temperature of the hot water just before adding the nitrogen.

(2) One to record masses of nitrogen on the blackboard.

(3) One to read off the time at 10-second intervals.

Label one 6 oz Styrofoam cup with the letter *W*, and label another cup with the letter *N*. Measure the masses of the two cups on the electronic balance, and list the masses on the blackboard.

Place 100 mL of distilled water in a 200 mL beaker, immerse a thermometer in the water, and heat the water on a hotplate until the temperature slightly exceeds 75°C.

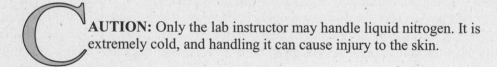

CAUTION: Only the lab instructor may handle liquid nitrogen. It is extremely cold, and handling it can cause injury to the skin.

Obtain a Styrofoam cup filled with liquid nitrogen.

Pour enough hot water into the preweighed cup labeled *W* to fill it one-third full, and then weigh the cup and contents on the electronic balance. Read the mass, and ask a student to write the mass on the blackboard. Return the cup to the benchtop, and immerse a thermometer in it.

Set the other preweighed cup, the one labeled *N*, on the electronic balance, and fill it slightly less than half-full of liquid nitrogen. This cup will lose an appreciable mass of nitrogen every second, so in order to get a reasonably accurate mass measurement at the time of mixing, it will be necessary to make a series of mass measurements at regular time intervals and then to prepare a graph of mass versus time. On the blackboard, list two columns. The first column should be a list of numbers—0, 10, 20, and 30—and will represent 10-second time intervals, i.e., 0 will represent the initial time reading, 10 will represent a time reading 10 seconds later, etc. The second column will represent the mass (cup + nitrogen) corresponding to each time reading. The plan will be to take four measurements of the mass, and then, while keeping track of the time, to carry the liquid nitrogen to the container of hot water and add the liquid nitrogen to the hot water precisely 10, 20, or 30 seconds after the fourth mass measurement. From a plot of mass on the y-axis and time on the x-axis, you will be able to extrapolate and estimate the mass at the time of mixing.

Ask a student to read off the time at 10-second intervals (i.e., "Start, 10, 20, 30, 40, 50, etc."). Note the mass of the nitrogen cup at each time, and record it on the blackboard. After having taken four mass measurements quickly go to the benchtop, note the temperature of the hot water, and then, *precisely* 10, 20, or 30 seconds after taking the fourth mass reading, add the liquid nitrogen to the hot water in the first cup. Write the initial temperature of the hot water on the blackboard.

As soon as the nitrogen has boiled away, measure the final temperature of the water, and record the final temperature on the blackboard. Pour the water down the drain.

Heat up the remaining water in the 200 mL beaker to 75°C, and then repeat the experiment with the other half of the liquid nitrogen.

All students should record the data (masses of empty cups, masses of hot-water cups used in each trial, masses and times for the nitrogen cups, temperature of the hot water in each trial, and final temperature for each trial).

Waste Disposal (Part C). All chemical waste should be disposed of as directed by your instructor.

Name: _____   Date: _____

Lab Instructor: _____   Lab Section: _____

EXPERIMENT 20

Physical Properties of Pure Substances and Solutions

PRELABORATORY QUESTIONS

1. Draw a qualitatively correct phase diagram for water. Label the triple-point and the critical point with the temperature-pressure combinations that describe them. Label the normal melting point and boiling point with the temperature-pressure combination that describes it.

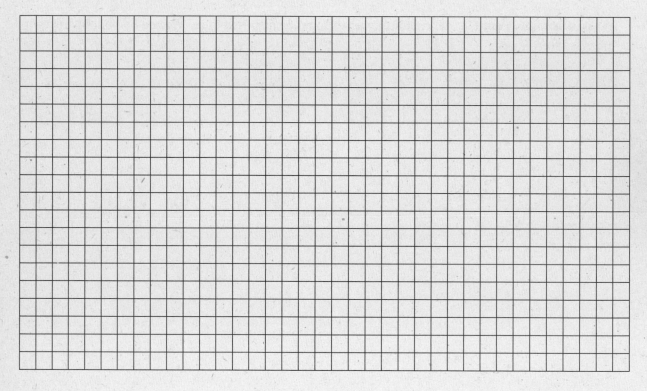

What is the unusual feature of the phase diagram of water?

2. Explain the difference between molarity and molality.

Name: _____   Date: _____

Lab Instructor: _____   Lab Section: _____

3. Look up the MSDS of *t*-butanol (**http://msds.ehs.cornell.edu/msdssrch.asp**). What are the hazards of *t*-butanol? What first-aid measures should be taken in the event of an accident?

4. As most solutions begin to freeze, the solvent freezes out as pure crystals of solid solvent. As a pure substance freezes, the temperature remains constant, but as a solvent in a solution freezes out, the temperature continues to drop gradually. If the temperature of the freezing mixture keeps dropping, how do you determine the freezing point of the solvent in the solution?

5. Why does the temperature continue to drop gradually as solvent in a solution freezes out?

Name: _____   Date: _____

Lab Instructor: _____   Lab Section: _____

EXPERIMENT 20

Physical Properties of Pure Substances and Solutions

RESULTS/OBSERVATIONS

Part A: Phase Behavior of Water, Carbon Dioxide, and *t*-Butanol

A-1: Phase Behavior of Water

1. What happened when a small stream of cold water was run over the bottom of the Florence flask?

2. What happened when your (cool) hand or a bit of crushed ice was placed on top of the inverted flask?

3. What happened when the flask was cooled to room temperature and, with the flask inverted, was held tightly in your (hot) hand?

4. Explain your observations in question 1 above using the phase diagram of water.

5. Explain your observations in question 2 above using the phase diagram of water.

6. Explain your observations in question 3 above using the phase diagram of water.

Name: _____   Date: _____

Lab Instructor: _____   Lab Section: _____

A-2: Phase Behavior of CO_2

7. Observations of CO_2.

8. Explanation of CO_2 observations using the phase diagram of CO_2.

A-3: Phase Behavior of t-Butanol

A-3-i: Triple-point of t-Butanol

9. Measured triple-point temperature of *t*-butanol: _____

10. Triple-point pressure of *t*-butanol: _____

A-3-ii: Boiling Point of t-Butanol at Laboratory Pressure

11. Mass of empty testtube and beaker: _____

12. Mass of testtube, beaker, and *t*-butanol: _____

13. Mass of *t*-butanol (subtract 11 from 12): _____

14. Boiling point of *t*-butanol at barometric pressure: _____

15. Barometric pressure: _____

A-3-iii: Freezing point of t-Butanol at Laboratory Pressure

16. Freezing point of *t*-butanol: _____

Name: _____ Date: _____

Lab Instructor: _____ Lab Section: _____

17. Use the values determined in Part A-3 to construct a phase diagram for *t*-butanol. Label the *x*- and *y*-axes; identify the solid, liquid, and gaseous regions; and label the triple-point and the boiling and freezing points at laboratory pressure.

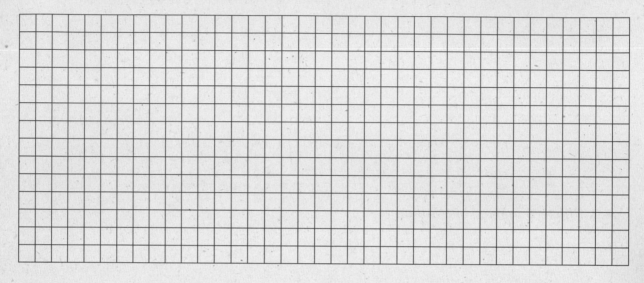

Part B: Colligative Properties—Freezing-Point Depression

NOTE: If you have saved the *t*-butanol and setup from Part A, you will not enter data in the italicized steps below.

18. *Mass of empty testtube and beaker:* _____

19. *Mass of testtube, beaker, and* t-*butanol:* _____

20. *Mass of* t-butanol *(subtract 18 from 19):* _____

21. *Freezing point of* t-butanol: _____

22. Mass of testtube, beaker, and *t*-butanol: _____

23. Mass of *t*-butanol (by subtraction): _____

24. Mass of beaker, testtube, and unknown solute: _____

25. Mass of beaker and testtube: _____

26. Mass of unknown solute (subtract 25 from 24): _____

Name: _____   Date: _____

Lab Instructor: _____   Lab Section: _____

27. Freezing point of the solution: _____ (trial 1), _____ (trial 2)

28. Calculate the molecular weight of the unknown solute.

Part C: Molar Heat of Vaporization of Nitrogen

29. Mass of the cup labeled W: _____

30. Mass of the cup labeled N: _____

31. Mass of cup labeled W filled one-third full with hot water: _____ (trial 1)

32. Mass of hot water in cup labeled W (subtract 29 from 31): _____ (trial 1)

33. Initial temperature of the hot water in the cup labeled W: _____ (trial 1)

34. Trial 1 data:

Time (s)	Mass of Cup N and $N_2(l)$	Mass of $N_2(l)$
0		
10		
20		
30		
40		
50		

35. Final temperature of the hot water in the cup labeled W: _____ (trial 1)

36. Mass of cup labeled W filled one-third full with hot water: _____ (trial 2)

37. Mass of hot water in cup labeled W (subtract 29 from 36): _____ (trial 2)

Name: _____   Date: _____

Lab Instructor: _____   Lab Section: _____

38. Initial temperature of the hot water in the cup labeled W: _____ (trial 2)

39. Trial 2 data:

Time (s)	Mass of Cup N and $N_2(l)$	Mass of $N_2(l)$
0		
10		
20		
30		
40		
50		

40. Final temperature of the hot water in the cup labeled W: _____ (trial 2)

41. Plot the mass of $N_2(l)$ as a function of time for the data in trial 1 to determine the mass of $N_2(l)$ vaporized.

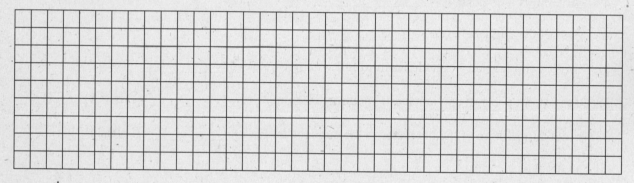

42. Plot the mass of $N_2(l)$ as a function of time for the data in trial 2 to determine the mass of $N_2(l)$ vaporized.

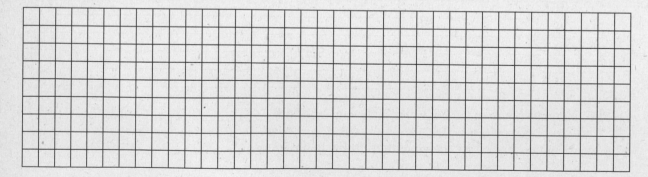

Name: _____   Date: _____

Lab Instructor: _____   Lab Section: _____

43. Calculation of the molar heat of vaporization of $N_2(l)$ for the data in trial 1.

44. Calculation of the molar heat of vaporization of $N_2(l)$ for the data in trial 2.

45. Average value of the molar heat of vaporization of $N_2(l)$: _____

46. Calculate the percent error in the molar heat of vaporization of $N_2(l)$ determination. The

    accepted value is 7.11 kJ/mol. _____

Name: _____   Date: _____

Lab Instructor: _____   Lab Section: _____

EXPERIMENT 20

Physical Properties of Pure Substances and Solutions

POSTLABORATORY QUESTIONS

1. Refer to Figure 20.1 when answering the following.

    (a) What physical state is CO_2 in at $-110°C$ and 2 atm?

    (b) If the temperature is increased to 0°C while keeping the pressure constant at 2 atm, what happens?

    (c) If the pressure is now increased to 70 atm while keeping the temperature constant at 0°C, what happens?

    (d) If the temperature increased to 100°C while keeping the pressure constant at 70 atm, what happens?

2. If 1.10 g of an unknown compound reduces the freezing point of 75.22 g benzene from 5.53 to 4.99°C, what is the molar mass of the compound? (The k_f of benzene is 5.12 mol · °C/kg.)

3. What temperature change would 300.0 g of water undergo in vaporizing 10.0 g of N_2 if the procedure in Part C were followed?

Molecular Weight via Freezing-Point Depression

OBJECTIVE

To determine the molecular weight of an unknown compound via freezing-point depression.

INTRODUCTION

Unsurprisingly, the physical properties of a solution differ from those of the pure solvent. What is surprising is that certain physical properties of a solution depend very much less on the chemical identity of the solute than on its amount. Physical properties that display this behavior are known as *colligative properties*. To a first approximation, colligative properties are independent of the chemical identity of the solute, the magnitude of each colligative property depending only on the number of solute particles present in solution. The four colligative properties are vapor-pressure lowering, boiling-point elevation, freezing-point depression, and osmotic pressure.

The practical applications of these colligative properties are widespread, and in many cases familiar. Freezing-point depression is the phenomenon that explains why salt scattered over roads in the winter leads to the melting of ice. It also explains why a mixture of water and antifreeze in a car radiator doesn't solidify when winter temperatures fall to 0°C or below. It is boiling-point elevation that prevents the water–antifreeze solution in a car radiator from boiling when it absorbs heat from a hot car engine. Osmotic pressure allows trees to transport sap to their uppermost branches and leaves.

In this experiment you will make use of freezing-point depression to determine the molecular weight of an unknown compound. For a given solvent, the amount of lowering of the freezing-point temperature (ΔT_f) is proportional to the concentration of dissolved particles, that is,

$$T_{f(\text{solvent})} - T_{f(\text{solution})} = \Delta T_f = k_f \cdot m \tag{1}$$

The constant k_f is the freezing-point depression constant, or cryoscopic constant, specific to the solvent, and m is the molality of particles dissolved in the solvent. The freezing-point depression constants for a selected number of solvents are listed in Table 21.1.

Table 21.1: Freezing-Point Depression Constants of Liquids

Solvent	k_f (°C/molal)
Water	1.86
t-Butanol	8.3
p-Xylene	4.3
Cyclohexane	20.2

Molality is a measure of concentration defined by the relation

$$\text{Molality } (m) = \frac{\text{moles of solute}}{\text{mass of solvent (kg)}} \qquad (2)$$

Since colligative properties depend on the number of solute particles present, it is essential to know the particle form each solute takes when dissolved in solution. For example, a 0.1 m aqueous glucose ($C_6H_{12}O_6$) solution,

$$C_6H_{12}O_6(s) \xrightarrow{\;H_2O\;} C_6H_{12}O_6(aq) \qquad (3)$$

exhibits a freezing-point depression nearly half that of a 0.1 m aqueous sodium chloride (NaCl) solution,

$$NaCl(s) \xrightarrow{\;H_2O\;} Na^+(aq) + Cl^-(aq) \qquad (4)$$

because NaCl dissociates in aqueous solution to form ions, providing twice as many moles of solute particles as does glucose.

ADDITIONAL READING

Read the sections in the Laboratory Techniques chapter at the beginning of this Lab Manual on cleaning glassware, handling chemicals, and weighing prior to performing this experiment.

 Safety Precautions:

- Protective eyewear approved by your institution must be worn at all times while you are in the laboratory.

- Cyclohexane, *t*-butanol, and *p*-xylene are flammable liquids. No open flames should be used during this experiment.

- If any of the chemicals used in this experiment spill on your skin, rinse it with water for 15 minutes

EXPERIMENT

You are to design and implement a procedure that allows for determination of the molecular weight of an unknown compound by way of freezing-point depression. Your unknown will be a molecular, not an ionic, compound; i.e., it doesn't dissociate in solution. Your lab instructor will inform you as to whether your unknown is a polar or nonpolar compound.

Design your experimental procedure around the following restrictions:

1. Use no more than 20 g of solvent per experimental trial.

2. Use no more than 2 g of your unknown compound per experimental trial.

3. To ensure accurate experimental results you should stir your liquid continuously while determining its freezing point.

4. Take the freezing point of the solution to be the temperature at which solid particles just begin to appear.

5. Repeat your measurements at least three times to ensure sufficient experimental precision.

EQUIPMENT AND REAGENTS

To perform this experiment, you will have access to all the equipment in your lab drawer, a split two-hole rubber stopper containing a wire stirrer, a hotplate, *t*-butanol, *p*-xylene, cyclohexane, distilled water, rock salt, and ice.

Waste Disposal. Dispose of all chemical waste as directed by your instructor.

Name: _____   Date: _____

Lab Instructor: _____   Lab Section: _____

EXPERIMENT 21

Molecular Weight via Freezing-Point Depression

PRELABORATORY QUESTIONS

1. Classify each of the four solvents available for use in this experiment—water, *t*-butanol, *p*-xylene, and cyclohexane—as either polar or nonpolar. The Lewis structures of *t*-butanol, *p*-xylene, and cyclohexane are shown below.

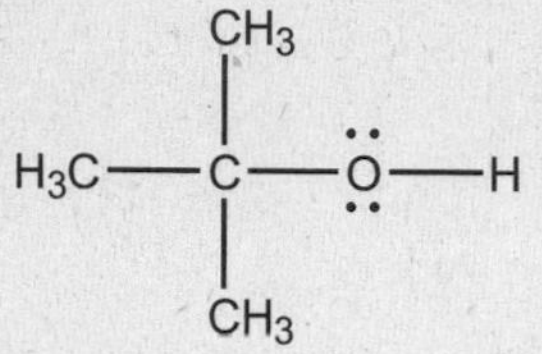

t-Butanol 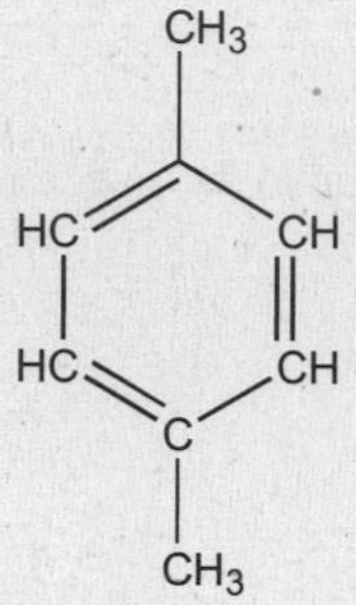   *p*-Xylene 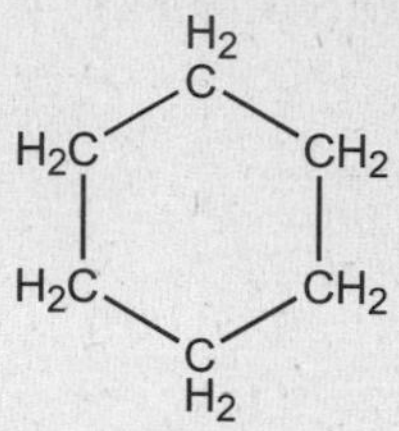   Cyclohexane

2. Using the rule of thumb "like dissolves like," which solvent(s) could be used in this experiment with a polar unknown compound? Which solvent(s) could be used with a nonpolar unknown compound?

3. The larger the measured value of ΔT_f, the more precise the determined molar mass. Given this fact, is it advantageous for a solvent in a freezing-point depression experiment to have a freezing-point constant of a large or small magnitude? Explain your answer.

Name: _____   Date: _____

Lab Instructor: _____   Lab Section: _____

EXPERIMENT 21

Molecular Weight via Freezing-Point Depression

RESULTS/OBSERVATIONS

1. Unknown compound identification number: _____

2. Solvent used: _____

3. Write any other pertinent data in the space below. Clearly indicate both the property and the amount.

4. Calculation of molecular weight using the data collected in trial _____ (choose a trial):

5. Molecular weight of unknown compound: _____ (trial 1), _____ (trial 2),

    _____ (trial 3), _____ (trial 4)

6. Molecular weight of unknown compound: _____ (average of three trials)

Name: _____   Date: _____

Lab Instructor: _____   Lab Section: _____

EXPERIMENT 21

Molecular Weight via Freezing-Point Depression

POSTLABORATORY QUESTIONS

1. Calculate the freezing point of a 0.850 m aqueous solution of glucose ($C_6H_{12}O_6$).

2. Which will display a greater ΔT_f, a 1.0 m aqueous solution of NaCl or a 0.75 m aqueous solution of $CaCl_2$? Explain.

3. Which will display a greater ΔT_f, a 1.0 m aqueous solution of ethylene glycol or a 1.0 m solution of ethylene glycol in t-butanol? Explain.

4. Why is it important to always take the freezing point of the solution to be the temperature at which solid particles just begin to appear?

Kinetics of
Ferroin Reactions

OBJECTIVE

The kinetics of one of three ferroin reactions will be investigated. For the reaction investigated the rate will be measured for a number of trials, each varying in the initial reactant concentrations and temperature. These data will be used to determine the reaction orders with respect to each reactant using both the integrated-rate-law method and the initial-rates method and the energy of activation.

INTRODUCTION

Chemical kinetics is the area of chemistry concerned with the rates at which chemical reactions occur. The range over which chemical reaction rates vary is enormous. Some chemical reactions, such as explosions, occur in a fraction of a second. The reactions responsible for the ripening of fruits and vegetables occur over a period of days. Those responsible for the human aging process have a time scale of years.

The rate at which a specific chemical reaction occurs is affected by the reactant concentrations and temperature. The manner in which this rate depends on these two variables can be expressed in the form of a rate law. For the generalized reaction

$$a \, A + b \, B \rightarrow \text{products} \tag{1}$$

the rate law has the form

$$\text{Rate} = k[A]^{\alpha}[B]^{\beta} \tag{2}$$

The exponents α and β are the orders of reaction in species A and B, respectively. The orders of reaction are a measure of how sensitive the rate of reaction is to the concentration of the reactant species. In general, they are unrelated to the stoichiometric coefficients a and b in the balanced chemical equation. For most reactions, the orders of reaction are small, positive integer values, but while less common, it is possible for the orders of reaction to be negative, zero, or even fractional. The constant of proportionality, k, is termed the *rate constant*. The rate constant is a function of temperature. The temperature dependence of the rate constant is given by the Arrhenius equation,

$$k = Ae^{-E_a/RT} \tag{3}$$

In this equation, A is known as the *frequency factor* or *pre-exponential factor;* it is related to the number of collisions per second between reactants that occur with the correct geometry for reaction. E_a is the activation energy, the minimum energy required for two colliding reactants to react. R is the universal gas constant (8.3.14 J/mol · K). T is the temperature, expressed in units of Kelvin. As the value of E_a increases, the rate constant decreases. Increases in temperature increase the value of the rate constant.

In this experiment you will study the kinetics of one of three possible ferroin reactions. Ferroin is the common name for tris(1,10-phenanthroline)iron(II) ion, an octahedral complex composed of three 1,10-phenanthroline molecules bonded to a central Fe^{2+}. Since the molecule 1,10-phenanthroline is often referred to by the abbreviation *phen,* ferroin will be represented as $Fe(phen)_3^{2+}$ in this experiment. The structure of phen is an interconnected set of three rings, two of which contain nitrogen atoms (see Figure 22.1).

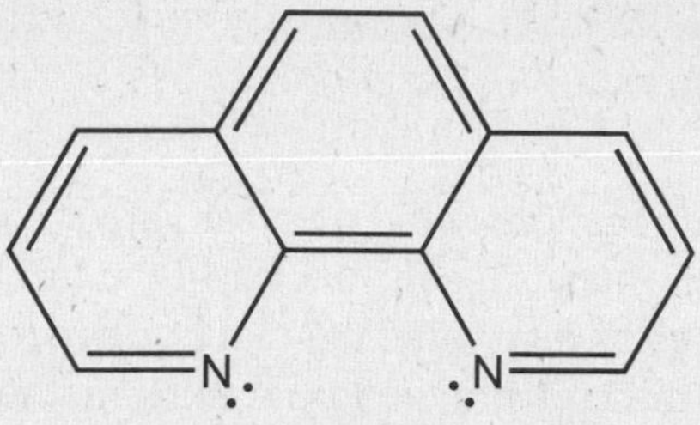

1,10-phenanthroline (phen)

Figure 22.1

The lone pairs on these nitrogen atoms are donated to the central Fe^{2+} in the complex ion, forming coordinate covalent (dative) bonds between the iron and each nitrogen atom. When drawing chemical structures, chemists often represent phen as two N atoms connected by a curved line (see Figure 22.2).

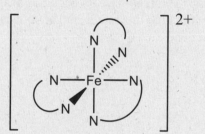

tris (1,10-phenanthroline) iron(II) (ferroin)

Figure 22.2

The three ferroin reactions examined in this experiment are

$$Fe(phen)_3^{2+}(aq) + 3\ Cu^{2+}(aq) \rightarrow Fe^{2+}(aq) + 3\ Cu(phen)^{2+}(aq) \tag{4}$$

$$Fe(phen)_3^{2+}(aq) + P_2O_7^{4-}(aq) \rightarrow [Fe(phen)_2(P_2O_7)]^{2-}(aq) + phen(aq) \tag{5}$$

$$Fe(phen)_3^{2+}(aq) + 6\ H_2O(l) + 3\ H^+(aq) \rightarrow Fe(OH_2)_6^{2+}(aq) + 3\ (phen)H^+(aq) \tag{6}$$

Your lab section will be divided into a number of groups, and each group will be assigned to study one of three possible ferroin reactions. The generic rate law for these ferroin reactions is

$$\text{Rate} = k\left[\text{Fe(phen)}_3^{2+}\right]^{\alpha}[X]^{\beta} \tag{7}$$

where X represents the second reactant (Cu^{2+}, $P_2O_7^{-4}$, or H^+).

In each case the reaction rate will be determined by measuring the decrease in concentration of ferroin with time during the initial stages of the reaction. The concentration of ferroin will be monitored spectroscopically. Ferroin is an intensely colored material that absorbs light strongly at 510 nm. None of the other reactants or products absorbs light of this wavelength. The amount of light absorbed, the absorbance, at 510 nm is directly proportional to the concentration of ferroin. By measuring absorbance at 510 nm as a function of elapsed time during the reaction, the concentration of ferroin as a function of reaction time can be monitored.

You will carry out a series of experiments to find α, the order of the reaction in ferroin, and β, the order of the reaction in the other reactant (Cu^{2+}, $P_2O_7^{4-}$, or H^+). To simplify matters, the same initial concentration of ferroin will be used in each trial; however, two different concentrations of X will be used. By sharing data with another student, you also will obtain data at a second temperature.

To obtain α, the order of reaction in ferroin, the expressions for the integrated forms of the rate laws will be used. For a general first-order reaction

$$a\,\text{A} \rightarrow \text{products} \tag{8}$$

the integrated form of the rate law is

$$\ln[A]_t = -kt + \ln[A]_0 \tag{9}$$

where k is the rate constant, t is time, $[A]_0$ is the initial concentration of A, and $[A]_t$ is the concentration of A at time t. For a general second-order reaction the integrated rate law has the form

$$\frac{1}{[A]_t} = kt + \frac{1}{[A]_0} \tag{10}$$

Both these equations have the familiar form of a straight line ($y = mx + b$):

$$\underbrace{\ln[A]_t}_{y} = \underbrace{-k}_{m}\underbrace{t}_{x} + \underbrace{\ln[A]_0}_{b} \tag{11}$$

and

$$\underbrace{\frac{1}{[A]_t}}_{y} = \underbrace{k}_{m}\underbrace{t}_{x} + \underbrace{\frac{1}{[A]_0}}_{b} \tag{12}$$

If a plot of ln(ferroin absorbance) as a function of t yields a straight line, the reaction is first order with respect to ferroin (i.e., $\alpha = 1$), the slope of the line is equal to $-k$, and the y-intercept is equal to ln(ferroin absorbance)$_0$. However, if a plot of 1/(ferroin absorbance) as a function of time yields a straight line, the reaction is second order in ferroin (i.e., $\alpha = 2$), the slope of the line is equal to k, and the y-intercept is equal to 1/(ferroin absorbance)$_0$. Since it is not known beforehand whether the reactions are first or second order in ferroin, both plots will have to be constructed for each trial. In total, you will generate eight graphs: two plots for each of your two experimental trials and two plots for each of the two trials performed by another student at a different temperature.

Because it is colorless, the concentration of X cannot be monitored directly in this experiment, so determination of β, the order of reaction in X, must be accomplished by indirect means. The approach used to determine β will be based on concentration-dependent changes in the ferroin half-life ($t_{1/2}$), the time it takes for half the original amount of ferroin to react. The average rate of reaction during the first half-life is $1/t_{1/2}$. Thus the faster the rate, the shorter the half-life. The order of the reaction in X will be determined by analyzing what happens to the average rate during the first half-life when the concentration of X is doubled while holding the ferroin concentration constant. If you find that doubling the concentration of X

(a) has *no effect* on the average reaction rate, then the order of reaction in X is zero;

(b) approximately *doubles* the average rate, then the order of reaction in X is one; or

(c) approximately *quadruples* the average rate, then the order of reaction in X is two.

(Why are the doubling and quadrupling in b and c only approximate? Note that the average rate of reaction during the first half-life is not the same as the initial rate of reaction. Since the concentrations of both reactants decrease as the reaction proceeds, the reaction rate also slows during the course of the reaction. Because of this, the average rate of reaction during the first half-life is *lower* than the initial rate. If the order of the reaction in X were one, a doubling of the concentration of X would lead to an exact doubling of the initial rate but only an approximate doubling of the average rate.)

By measuring the reaction rate at two temperatures it is possible to determine the activation energy required for the reaction of ferroin and X. Performing the reaction at two different temperatures, T_1 and T_2, leads to the determination of two different rate constants, k_1 and k_2. Each of these rate constants obeys the Arrhenius equation [Equation (3)]. The ratio of the rate constants,

$$\frac{k_1}{k_2} = \frac{Ae^{-E_a/RT_1}}{Ae^{-E_a/RT_2}} \tag{13}$$

can be simplified to eliminate the frequency factor

$$\frac{k_1}{k_2} = e^{E_a/RT_2 - E_a/RT_1} \tag{14}$$

Taking the logarithm of both sides yields:

$$\ln\left(\frac{k_1}{k_2}\right) = \frac{E_a}{R}\left(\frac{1}{T_2} - \frac{1}{T_1}\right) \tag{15}$$

which can be solved for E_a.

In this experiment you will study the kinetics of one of three possible ferroin reactions. For the reaction you are assigned you will determine the reaction orders of both reactants and E_a.

ADDITIONAL READING

Read the sections in the Laboratory Techniques chapter at the beginning of this Lab Manual on cleaning glassware, handling chemicals, heating liquids and solutions, spectrometers, pipets, and burets prior to performing this experiment.

 Safety Precautions:

■ Protective eyewear approved by your institution must be worn at all times while you are in the laboratory.

■ If any of the reagents used in this experiment should spill on your skin, rinse the affected area with water for 15 minutes.

PROCEDURE

The experiment will be performed by teams of two (or possibly three) students. One student should mix the reagents while the other times the reaction. Your team will be assigned to Group A, B, or C. Each group will consist of several teams of students. The teams in Group A will study the reaction between ferroin and Cu^{2+}, the teams in Group B will study the reaction between ferroin and $P_2O_7^{4-}$, and the teams in Group C will study the reaction between ferroin and H^+. Each team will carry out the reaction *at one assigned temperature,* either room temperature or 10ºC above room temperature. At the end of the experiment, you will exchange data with a team in your group that worked at the other temperature.

Each team will carry out a total of three reactions as shown below. In the first reaction, a specific amount of distilled water and either Cu^{2+}, $P_2O_7^{4-}$, or H^+ will be allowed to react with a specific amount of ferroin. In the second reaction, the same amount of ferroin will be used, but the concentration of Cu^{2+}, $P_2O_7^{4-}$, or H^+ will be doubled. The third experiment will check for artifacts by testing whether or not ferroin reacts in the absence of a reagent. If the absorbance remains relatively constant during the third reaction, you may assume that the reaction with distilled water alone does not occur at an appreciable rate and can be ignored safely.

Group	Reaction	Mixture before Adding Ferroin
A	1	1.0 mL 0.02 M Cu(NO$_3$)$_2$ + 3.5 mL distilled H$_2$O
A	2	2.0 mL 0.02 M Cu(NO$_3$)$_2$ + 2.5 mL distilled H$_2$O
A	3	4.5 mL distilled H$_2$O
B	1	1.0 mL 0.10 M Na$_4$P$_2$O$_7$ + 3.5 mL distilled H$_2$O
B	2	2.0 mL 0.10 M Na$_4$P$_2$O$_7$ + 2.5 mL distilled H$_2$O
B	3	4.5 mL distilled H$_2$O
C	1	1.0 mL 1.0 M HNO$_3$ + 3.5 mL distilled H$_2$O
C	2	2.0 mL 1.0 M HNO$_3$ + 2.5 mL distilled H$_2$O
C	3	4.5 mL distilled H$_2$O

The absorbance of the solution will be measured with a spectrometer. Because the reactions proceed considerably faster at the higher temperature, the majority of the spectrometers will be assigned to teams of students running their reactions at 10°C above room temperature. Since the reactions are much faster at the higher temperature, *teams assigned to perform the reactions at 10°C above room temperature must run the entire set of reactions twice.* The reactions carried out at room temperature require about an hour and a half, so only one run will be possible within the lab period.

Turn the spectrometer on. Allow at least 10 minutes for it to warm up.

Obtain a buret, rinse it with distilled water, mount it on a ring stand, and fill it with distilled water. Use the buret to measure out water as called for in the directions below.

If your team has been assigned to work at the higher temperature, assemble a warm-water bath using a 600 mL beaker and a hotplate. Gently heat the water. Monitor the temperature with a thermometer. Adjust the temperature-control knob on the hotplate to maintain a temperature about 10°C above room temperature.

Obtain four clean, dry cuvets, and number each with a wax pencil near the top. To avoid fingerprints on the optical surfaces of the cuvets (which would cause erroneous measurements), only touch them near the very top.

Label a small clean testtube with an F (for ferroin). Put about 3 mL of 5.0×10^{-4} M ferroin in the tube. If you are performing the reaction at the elevated temperature, place this testtube in the warm-water bath.

In a testtube, obtain about 5 mL of the reactant solution [0.02 M $Cu(NO_3)_2$, 0.10 M $Na_4P_2O_7$, or 1.0 M HNO_3] assigned to your group.

Use a measuring pipet to carefully add the appropriate amount of reactant solution for Reaction 1 to cuvet 1. Use the buret to add the appropriate amount of distilled water for Reaction 1 to this same cuvet. Prepare cuvets 1 and 2 for Reactions 2 and 3 by adding the appropriate amounts of both distilled water and reactant solution. Add 5 mL of distilled water to the fourth cuvet. This cuvet will be used to calibrate the zero absorbance point on the spectrometer. If you are performing the reaction at the elevated temperature, place these cuvets in the warm-water bath.

Once the warm-water bath has stabilized at a temperature about 10°C above room temperature record the temperature of the bath. Allow the solutions about 10 minutes to equilibrate to the temperature of the bath. If you are working at room temperature use a thermometer to determine the temperature of the ferroin solution directly; record this temperature.

After allowing the spectrometer 10 minutes to warm up, set the wavelength to 510 nm. To standardize the instrument, use the blank, which is the distilled water in cuvet 4. Holding the cuvet by the lip, wipe the outside of the cuvet with a clean Kimwipe to remove any water, fingerprints, or stains that might reduce the intensity of light transmitted to the detector. Place the cuvet in the sample compartment, pushing it fully down. Align the mark on the cuvet with that on the sample holder, and close the sample compartment. Adjust the calibration dial so that the spectrometer reads zero absorbance.

Make sure that you can see the clock in your laboratory or have a watch with a second hand. Also get a small measuring pipet and a piece of parafilm.

Make sure that the spectrometer is free before performing the following three steps. (Since you will be sharing the spectrometer with others, you should coordinate your measurements with the other teams.)

Using the pipet, transfer 0.50 mL of the 5.0×10^{-4} M ferroin solution to cuvet 1, immediately recording the time of day. Cover the testtube with a small piece of parafilm, and mix the contents by inverting the covered testtube two times. (**Note:** Too much parafilm will prevent the tube from being inserted into the spectrometer correctly. If the parafilm causes the tube to sit at an angle in the spectrometer, your measurements will be incorrect.) As quickly as possible, wipe off the tube with a Kimwipe, place the tube in the spectrophotometer, and record the time and absorbance. If applicable, immediately return the testtube to the warm-water bath.

Noting the time of addition, use the measuring pipet to add 0.50 mL of ferroin to cuvet 2. Cover the tube with parafilm, mix the solution, and measure the absorbance. If applicable, quickly return the cuvet to the warm-water bath.

Noting the time of the addition, add 0.50 mL of the ferroin solution to cuvet 3. Cover the tube with parafilm, mix the solution, and measure the absorbance. If applicable, quickly return the cuvet to the warm water bath.

Continue to take readings of the absorbance and time for each of your three solutions at evenly spaced intervals until the absorbance of tubes 1 and 2 drops to less than one-half of their initial values. (In other words, take data for at least one half-life.) If you are operating at the elevated temperature, your readings should be taken no more than 3 minutes apart. (If possible, a shorter measurement time would be better.) If you are operating at room temperature, your readings should be taken at 10-minute intervals.

If you are working at the elevated temperature perform another run.

Important: If you ran your reaction at room temperature, obtain a set of data from others in your group who did the same reaction but at the elevated temperature. Similarly, if you ran your reactions at the elevated temperature, get some data from a group who did their reactions at room temperature. You must have this information to calculate the activation energy.

Waste Disposal. Dispose of all chemical waste as directed by your instructor.

Name: _____   Date: _____

Lab Instructor: _____   Lab Section: _____

EXPERIMENT 22

Kinetics of Ferroin Reactions

PRELABORATORY QUESTIONS

1. Figure 22.3 summarizes the relationship between the color of light, its wavelength, and its
    complement.

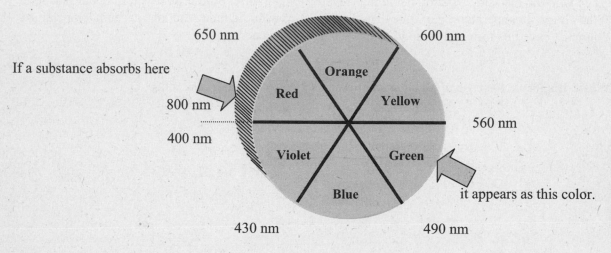

Figure 26.1

Figure 22.3
A color wheel summarizing the relationship between the color of light,
its wavelength, and its complement.

(a) If ferroin absorbs strongly at 510 nm, what color do you expect a pure solution of ferroin
     to be?

(b) If you place a blue solution in a visible spectrometer, at what wavelength (in nanometers)
     would you expect to observe the maximum absorbance?

(c) During the course of the reaction, do you expect the color of the ferroin solution to become
     darker or lighter? Explain.

Name: _____   Date: _____

Lab Instructor: _____   Lab Section: _____

2. Explain, with reference to the appropriate equation, why the average initial rate of reaction will quadruple when the concentration of X is doubled if the reaction is second order in X.

3. Why is it important to keep the concentration of ferroin constant when the concentration of X is doubled?

4. Explain, with reference to equation(s) and graph(s), how you could determine k, the rate constant for the reaction, if you were assigned a reaction that is second order in ferroin and zeroth order in X.

5. Explain why chemical reaction rates typically increase with increasing reactant concentration.

6. Explain why chemical reaction rates typically increase with increasing temperature.

Name: _____   Date: _____

Lab Instructor: _____   Lab Section: _____

EXPERIMENT 22

Kinetics of Ferroin Reactions

RESULTS/OBSERVATIONS

1. Identity of assigned reactant solution (Cu^{2+}, $P_2O_7^{4-}$, or H^+): _____

2. Measured temperature at which reaction occurred: _____

3. Time and absorbance measurements for Reaction 1:

Time of Day	Elapsed Time (s)	Absorbance	ln(Absorbance)	1/Absorbance

Name: _____   Date: _____

Lab Instructor: _____   Lab Section: _____

4. Time and absorbance measurements for Reaction 2:

Time of Day	Elapsed Time (s)	Absorbance	ln(Absorbance)	1/Absorbance

5. Time and absorbance measurements for Reaction 3:

Time of Day	Elapsed Time (s)	Absorbance

Name: _____   Date: _____

Lab Instructor: _____   Lab Section: _____

6.   Time and absorbance measurements for Reaction 1 (if a second run was completed):

Time of Day	Elapsed Time (s)	Absorbance	ln(Absorbance)	1/Absorbance

7.   Time and absorbance measurements for Reaction 2 (if a second run was completed):

Time of Day	Elapsed Time (s)	Absorbance	ln(Absorbance)	1/Absorbance

Name: _____   Date: _____

Lab Instructor: _____   Lab Section: _____

8. Time and absorbance measurements for Reaction 3 (if a second run was completed):

Time of Day	Elapsed Time (s)	Absorbance

9. Time and absorbance measurements for Reaction 1 from another team in the same group.

Temperature at which reaction occurred: _____

Time of Day	Elapsed Time (s)	Absorbance	ln(Absorbance)	1/Absorbance

Name: _____   Date: _____

Lab Instructor: _____   Lab Section: _____

10. Time and absorbance measurements for Reaction 2 from another team in the same group.

Time of Day	Elapsed Time (s)	Absorbance	ln(Absorbance)	1/Absorbance

11. Time and absorbance measurements for Reaction 3 from another team in the same group.

Time of Day	Elapsed Time (s)	Absorbance

Name: _____   Date: _____

Lab Instructor: _____   Lab Section: _____

12. Time and absorbance measurements for Reaction 1 from another team in the same group (if a second run was completed). Temperature at which reaction occurred: _____

Time of Day	Elapsed Time (s)	Absorbance	ln(Absorbance)	1/Absorbance

13. Time and absorbance measurements for Reaction 2 from another team in the same group (if a second run was completed).

Time of Day	Elapsed Time (s)	Absorbance	ln(Absorbance)	1/Absorbance

Name: _____   Date: _____

Lab Instructor: _____   Lab Section: _____

14. Time and absorbance measurements for Reaction 3 from another team in the same group (if a second run was completed).

Time of Day	Elapsed Time (s)	Absorbance

Name: _____   Date: _____

Lab Instructor: _____   Lab Section: _____

15. Plot of ln(absorbance) as a function of t for Reaction 1 near room temperature,

_____ (insert the actual temperature value).

272 Experiment 22: Kinetics of Ferroin Reactions

Name: _____ Date: _____

Lab Instructor: _____ Lab Section: _____

16. Plot of ln(absorbance) as a function of t for Reaction 1 at an elevated temperature of
_____ (insert the actual temperature value). Plot both sets of data on this graph. Use
different symbols (e.g., circles and squares) for the two different runs.

Name: _____   Date: _____

Lab Instructor: _____   Lab Section: _____

17. Plot of 1/absorbance as a function of *t* for Reaction 1 near room temperature, _____
    (insert the actual temperature value).

Name: _____   Date: _____

Lab Instructor: _____   Lab Section: _____

18. Plot of 1/absorbance as a function of t for Reaction 1 at an elevated temperature of
_____ (insert the actual temperature value). Plot both sets of data on this graph. Use
different symbols (e.g., circles and squares) for the two different runs.

Name: _____   Date: _____

Lab Instructor: _____   Lab Section: _____

19. Plot of ln(absorbance) as a function of t for Reaction 2 near room temperature, _____
(insert the actual temperature value).

Name: _____   Date: _____

Lab Instructor: _____   Lab Section: _____

20. Plot of ln(absorbance) as a function of t for Reaction 2 at an elevated temperature of
_____ (insert the actual temperature value). Plot both sets of data on this graph. Use
different symbols (e.g., circles and squares) for the two different runs.

Name: _____   Date: _____

Lab Instructor: _____   Lab Section: _____

21. Plot of 1/absorbance as a function of *t* for Reaction 2 near room temperature, _____ (insert the actual temperature value).

Name: _____ Date: _____

Lab Instructor: _____ Lab Section: _____

22. Plot of 1/absorbance as a function of t for Reaction 2 at an elevated temperature of
_____ (insert the actual temperature value). Plot both sets of data on this graph. Use
different symbols (e.g., circles and squares) for the two different runs.

Name: _____   Date: _____

Lab Instructor: _____   Lab Section: _____

23. Based on these graphs, is the reaction between ferroin and _____ (insert your assigned reactant) first or second order in ferroin? Explain.

24. Use the data collected by both your team and the other to determine the ferroin half-life ($t_{1/2}$) and average rate ($1/t_{1/2}$) under the various reaction conditions.

	Reaction 1 at Room Temperature	Reaction 2 at Room Temperature	Reaction 1 at an Elevated Temperature	Reaction 2 at an Elevated Temperature	Reaction 1 at an Elevated Temperature (2nd run)	Reaction 2 at an Elevated Temperature (2nd run)
$t_{1/2}$						
$1/t_{1/2}$						

25. Based on the ferroin half-life values, what is the reaction order in _____ (insert your assigned reactant)? Explain.

26. Calculation of the rate constant at room temperature _____ (insert the actual temperature) using the data from Reaction 1:

Name: _____   Date: _____

Lab Instructor: _____   Lab Section: _____

27. Determined values of the rate constant at room temperature: _____ , _____

    _____ , _____

28. Average value of the rate constant at room temperature: _____

29. Calculation of the rate constant at an elevated temperature _____ (insert the actual temperature) using the data from Reaction 1:

30. Determined values of the rate constant at an elevated temperature: _____ , _____

    _____ , _____

31. Average value of the rate constant at an elevated temperature: _____

32. Calculation of the activation energy (E_a) using the average values of the rate constant at the two temperatures:

33. Based on your data, does ferroin react to any appreciable extent with distilled water? Explain.

Name: _____   Date: _____

Lab Instructor: _____   Lab Section: _____

EXPERIMENT 22

Kinetics of Ferroin Reactions

POSTLABORATORY QUESTIONS

1. Why was Reaction 3 (ferroin reaction with water) performed?

2. Why must both reactant solutions be at the same temperature prior to mixing?

3. Why was the total volume of reagents after mixing kept constant in Reactions 1–3?

4. If 1.0×10^{-3} M ferroin was used in place of 5.0×10^{-4} M ferroin in Reactions 1–3, how would the following be affected—increase, decrease, or remain the same?

   (a) The reaction rate.

   (b) The value of α.

Name: _____   Date: _____

Lab Instructor: _____   Lab Section: _____

(c) The value of β.

(d) The value of k.

(e) The value of E_a.

Chemical Kinetics:
Iodine Clock Reaction

OBJECTIVE

To determine the complete concentration and temperature dependence of the reaction rate for the reaction between peroxydisulfate ion and iodide ion in aqueous solution.

INTRODUCTION

Peroxydisulfate ion reacts with iodide ion in aqueous solution to give iodine and sulfate ion:

$$2\,I^- + S_2O_8^{2-} \longrightarrow I_2 + 2\,SO_4^{2-} \tag{1}$$

In this experiment you will determine the rate at which Reaction (1) occurs. The rate of a chemical reaction depends on reactant concentrations and temperature. The manner in which this rate depends on these variables can be expressed in the form of a rate law,

$$\text{Rate} = k\left[I^-\right]^x\left[S_2O_8^{2-}\right]^y \tag{2}$$

where x and y are the orders of reaction in iodide and peroxydisulfate ions, respectively, and k is the specific rate constant; the specific rate constant is a function of temperature.

The initial rate of this reaction will be determined by measuring the time required to generate a certain amount of iodine according to Reaction (1). The same amount of iodine will be produced in every trial.

Two other reactions will be used to signal when this constant amount of iodine has been produced. The first of these two reactions involves thiosulfate ion ($S_2O_3^{2-}$), a certain identical amount of which should be added to the reaction mixture in each trial. Thiosulfate reacts with iodine as fast as iodine is produced, converting it back to iodide ion:

$$2\,S_2O_3^{2-} + I_2 \rightarrow S_4O_6^{2-} + 2\,I^- \tag{3}$$

Because Reaction (3) is so fast relative to Reaction (1) iodine will not have a chance to build up in solution until all the thiosulfate has been consumed. Any buildup of iodine in the solution indicates that the thiosulfate has been used up, and this, of course, means that the constant amount of iodine has been produced.

To visually detect the presence of excess iodine in the solution you will add some starch to one of the solutions before mixing the iodide and peroxydisulfate. As soon as iodine begins to build up in solution, it will react to form a dark-blue complex with starch:

$$I_2 + \text{starch} \rightarrow I_2\text{-starch complex} \tag{4}$$
$$\text{(Dark blue)}$$

The time period to the appearance of the dark-blue color, measured from the time of mixing the peroxydisulfate with the iodide in the presence of starch and thiosulfate, is the time it takes to produce a certain constant amount of iodine. The initial rate is inversely proportional to the time period; the faster the rate, the shorter the time period.

Rates of reaction are temperature-dependent because by increasing the temperature of your reaction you are giving each of the reactants, on average, a bit more energy. Since the reactants need to have a specific amount of energy to react, the specific rate constant k (and thus the overall reaction rate) usually increases with increasing temperature. This behavior is described by the Arrhenius equation,

$$k = Ae^{-E_a/RT} \tag{5}$$

In this equation A is known as the *frequency factor* or *pre-exponential factor;* theoretically, A is related to the number of collisions per second between iodide and peroxydisulfate ions that occur with the correct geometry for reaction. E_a is the activation energy, the minimum energy required for two colliding reactants to react. R is the universal gas constant (8.3.14 J/mol · K). T is the temperature.

Combining the Arrhenius equation with the generalized rate law for the reaction of iodide with peroxydisulfate ions shows the complete concentration and temperature dependence of the reaction rate:

$$\text{Rate} = Ae^{-E_a/RT}\left[I^-\right]^x\left[S_2O_8^{2-}\right]^y \tag{6}$$

Your purpose in this experiment is to determine the complete concentration and temperature dependence of the reaction rate for the reaction between peroxydisulfate ion and iodide ion in aqueous solution.

ADDITIONAL READING

Read the sections in the Laboratory Techniques chapter at the beginning of this Lab Manual on cleaning glassware, handling chemicals, weighing, and heating liquids and solutions prior to performing this experiment.

 Safety Precautions:

■ Protective eyewear approved by your institution must be worn at all times while you are in the laboratory.

■ If any of the reagents used in this experiment should spill on your skin, rinse the affected area with water for 15 minutes.

EXPERIMENT

You are to design and perform experiments in order to determine the complete concentration and temperature dependence of the reaction rate for the reaction between peroxydisulfate ion and iodide ion in aqueous solution. You will have two lab sessions to accomplish this.

Design your experiment around the following restrictions:

1. Because these reactions are very sensitive to the presence of impurities, it is *essential* to use scrupulously clean glassware.

2. As an additional precaution, a drop of 0.1 M EDTA (ethylenediaminetetraacetate) should be added to each reaction mixture to remove impurities that otherwise might catalyze the reaction.

3. Each reaction mixture must contain the same amount of $Na_2S_2O_3$. Otherwise, the amount of I_2 consumed by Reaction (3) will vary from one trial to another, resulting in misleading data.

4. In each reaction mixture the amount of $Na_2S_2O_3$ should be at least 10 times less than the amount of $(NH_4)_2S_2O_8$ or an average as opposed to the initial rate is measured.

5. About 0.2 g (about half as much as you could pile on a dime) of soluble starch should be added to each reaction mixture.

6. If necessary, use distilled water as a solvent. *Do not* use tap water. Tap water contains impurities that might catalyze the reaction, distorting the experimental results.

7. Repeat each of your measurements at least three times to ensure sufficient experimental precision.

Equipment and Reagents

To perform this experiment, you will have access to all the equipment in your lab drawer and the following:

0.16 M KI(aq)
0.12 M $(NH_4)_2S_2O_8$(aq)
0.0055 M $Na_2S_2O_3$(aq)
0.10 M EDTA(aq)
Distilled water
Solid soluble starch indicator
Hotplate
Ice

Waste Disposal. All chemical waste should be disposed of as directed by your instructor.

Name: _____   Date: _____

Lab Instructor: _____   Lab Section: _____

EXPERIMENT 23

Chemical Kinetics: Iodine Clock Reaction

PRELABORATORY QUESTIONS

1. Combining which two of the five chemicals to be used in this experiment starts the reaction?

2. What signals the time to produce a certain constant amount of iodine in this experiment?

3. Will the time to produce a certain constant amount of I_2 increase or decrease if the amount of $Na_2S_2O_3$ used initially is increased, all other reagent concentrations being kept constant?

4. What would happen if the number of moles of $S_2O_3^{2-}$ present initially in an experimental trial were more than twice the number of moles of $S_2O_8^{2-}$ present initially and greater than the moles of I^- present initially?

Name: _____   Date: _____

Lab Instructor: _____   Lab Section: _____

5. Consider a trial of this experiment in which 40. mL of 0.16 M KI, 20. mL of 0.0055 M $Na_2S_2O_3$,
    10.0 mL of 0.12 M $(NH_4)_2S_2O_8$, 30. mL of distilled water, a drop of 0.1 M EDTA, and 0.2 g of
    soluble starch compose the initial reaction mixture. (Assume that the drop of EDTA and the 0.2
    g of soluble starch have a negligible effect on the solution volume.)

    (a) Calculate the initial concentration of I^- (the concentration *after* mixing but *before* reacting).

    (b) Calculate the initial concentration of $S_2O_8^{2-}$ (the concentration *after* mixing but *before*
        reacting).

    (c) Calculate the moles of I_2 produced by Reaction (1) when the timing is stopped?

    (d) Determine the initial rate of reaction for this trial if timing was stopped after 5.0 minutes.
        The initial rate of reaction should be expressed in moles per liter of I_2 per second.

Name: _____   Date: _____

Lab Instructor: _____   Lab Section: _____

EXPERIMENT 23

Chemical Kinetics: Iodine Clock Reaction

RESULTS/OBSERVATIONS

1. Write any pertinent data in the space below. Clearly indicate both the property and the amount.

Name: _____   Date: _____

Lab Instructor: _____   Lab Section: _____

2. Calculation of the orders of reaction in iodide and peroxydisulfate ions:

3. Reaction order in iodide: _____. Reaction order in peroxydisulfate: _____.

Name: _____   Date: _____

Lab Instructor: _____   Lab Section: _____

4. Calculation of the energy of activation (E_a) and the frequency factor (A):

5. The value of E_a: _____. The value of A: _____.

Name: _____   Date: _____

Lab Instructor: _____   Lab Section: _____

EXPERIMENT 23

Chemical Kinetics: Iodine Clock Reaction

POSTLABORATORY QUESTIONS

1. Consider the generalized reaction

$$A \rightarrow products$$

for which the following data were obtained:

Trial	[A] (M)	Rate (M/s)	Temperature (K)
1	0.50	2.0×10^{-5}	300.
2	0.25	5.0×10^{-6}	300.
3	0.50	3.0×10^{-3}	330.

(a) Determine the rate law for this reaction.

(b) Calculate the value of the rate constant at 300. K.

(c) Calculate the rate of the reaction at 300. K if the concentration of A is 1.0×10^{-3} M.

Name: _____   Date: _____

Lab Instructor: _____   Lab Section: _____

    (d) Calculate the value of the rate constant at 330. K.

    (e) Determine the values of E_a and A.

EXPERIMENT 24

Spectroscopic Determination of an Equilibrium Constant

OBJECTIVE

To determine the equilibrium constant for a reaction. The concentration of one of the reaction species will be monitored spectroscopically. The other concentrations will be determined stoichiometrically. The equilibrium constant will be determined from a number of experimental trials, each trial having different initial reactant concentrations.

INTRODUCTION

All chemical reactions occur so as to approach a state of equilibrium. Once chemical equilibrium is attained the concentrations of all reactants and products remain constant with time. The equilibrium state of a chemical reaction is characterized by an equilibrium expression. For the general reaction

$$a\,A + b\,B \rightleftharpoons c\,C + d\,D \tag{1}$$

the equilibrium expression has the form

$$K = \frac{[C]^c\,[D]^d}{[A]^a\,[B]^b} \tag{2}$$

The reactant and product concentrations in the equilibrium expression are those at equilibrium, each raised to a power equal to its stoichiometric coefficient in the balanced chemical equation. K is a temperature-dependent constant called the *equilibrium constant.* By convention, the units are omitted when reporting equilibrium constants.

In this experiment the value of the equilibrium constant for the reaction between iron(III) (Fe^{3+}) and thiocyanate ion (SCN^-),

$$Fe^{3+}(aq) + SCN^-(aq) \rightleftharpoons \underset{\text{(Orange)}}{FeSCN^{2+}(aq)} \tag{3}$$

will be determined. The equilibrium expression for this reaction is

$$K = \frac{\left[FeSCN^{2+}\right]}{\left[Fe^{3+}\right]\left[SCN^-\right]} \tag{4}$$

To calculate the value of K it is necessary to determine the concentration of each of the species in the solution at equilibrium. In this experiment, the equilibrium $FeSCN^{2+}$ concentration will be measured spectrophotometrically, taking advantage of the fact that $FeSCN^{2+}$ is the only highly colored species in the solution. From the determined $FeSCN^{2+}$ equilibrium concentration and the known initial concentrations of Fe^{3+} and SCN^-, the stoichiometry of the reaction can be used to calculate the equilibrium concentrations of Fe^{3+} and SCN^-.

To spectrophotometrically determine the equilibrium concentration of $FeSCN^{2+}$ the absorbance of a reaction mixture at equilibrium will be measured. The absorbance (A) is related to the amount of the light-absorbing species in a sample according to the relation

$$A = \varepsilon dc \tag{5}$$

Equation (5) is called the *Beer-Lambert law,* or simply *Beer's law.* The absorbance is unitless. The concentration, c, is usually given in units of moles per liter (M). The distance light travels through the sample, the so-called pathlength d, is commonly expressed in centimeters. The quantity ε is called the *molar absorptivity* or *extinction coefficient* and has units of $M^{-1} \cdot cm^{-1}$; the molar absorptivity is the characteristic of the substance that expresses how much light is absorbed at a particular wavelength by a particular substance.

There are two approaches to calculating the $FeSCN^{2+}$ equilibrium concentration from its absorbance. One is to use Equation (5) directly, which requires knowledge of the values of ε and d. The second depends on having a standard solution of $FeSCN^{2+}$, a solution in which the concentration of $FeSCN^{2+}$ is known. This second approach will be used in this experiment.

A standard solution of $FeSCN^{2+}$ with a known concentration c_1 will have a measured absorbance A_1 that satisfies Beer's law:

$$\frac{A_1}{c_1} = \varepsilon d \tag{6}$$

A second solution of $FeSCN^{2+}$ with an unknown concentration c_2 and a measured absorbance A_2 also will satisfy Beer's law:

$$\frac{A_2}{c_2} = \varepsilon d \tag{7}$$

If the absorbances of these two samples were measured at the same wavelength using the same cuvet, Equations (6) and (7) can be equated:

$$\frac{A_1}{c_1} = \frac{A_2}{c_2} \tag{8}$$

Thus the known concentration and measured absorbance of the standard $FeSCN^{2+}$ solution can be used to calculate the unknown $FeSCN^{2+}$ concentration in a second solution once its absorbance is measured.

A standard solution of $FeSCN^{2+}$ can be prepared by starting with a small known concentration of SCN^- and adding such a large excess of Fe^{3+} that essentially all the SCN^- is converted to $FeSCN^{2+}$. Under these conditions, it can be assumed that the final number of moles of $FeSCN^{2+}$ is equal to the initial number of moles of SCN^-.

Your purpose in this experiment is to determine the equilibrium constant for Reaction (3). You will do this a number times, each trial involving different initial Fe^{3+} and SCN^- concentrations. For each trial the absorbance of $FeSCN^{2+}$ will be measured, as it will be for a standard solution of $FeSCN^{2+}$. Equation (8) then will be used to calculate the equilibrium concentrations of $FeSCN^{2+}$. Using the $FeSCN^{2+}$ equilibrium concentration and the known initial concentrations of Fe^{3+} and SCN^-, the stoichiometry of the reaction will be used to calculate the equilibrium concentrations of Fe^{3+} and SCN^- for each trial. These values then will be substituted into the equilibrium expression to calculate K.

ADDITIONAL READING

Read the sections in the Laboratory Techniques chapter at the beginning of this Lab Manual on cleaning glassware, handling chemicals, and spectrometers prior to performing this experiment.

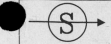

Safety Precautions:

- Protective eyewear approved by your institution must be worn at all times while you are in the laboratory.

- If any of the reagents used in this experiment should spill on your skin, rinse the affected area with water for 15 minutes.

PROCEDURE

Turn the spectrometer on. Allow at least 10 minutes for it to warm up. Thoroughly clean six 16 mm testtubes. Rinse these testtubes thoroughly with distilled water and flame dry them. Label the testtubes 1 through 6. To each of the 16 mm testtubes add 5.0 mL of 2.0×10^{-4} M KSCN.

To testtube 1 add 5.0 mL of 0.20 M Fe(NO$_3$)$_3$ and stir. The resulting solution is your standard solution of FeSCN^{2+}.

For the other testtubes, do the following: Measure 10.0 mL of 0.20 M Fe(NO$_3$)$_3$ in a 25 mL graduated cylinder, fill to 25.0 mL with distilled water, and stir thoroughly to mix. Measure 5.0 mL of the resulting 0.080 M Fe^{3+} into testtube 2. Discard all but 10.0 mL of the 0.080 M Fe^{3+}, refill the graduated cylinder to 25.0 mL with distilled water, and stir thoroughly. Add 5.0 mL of the resulting 0.032 M Fe^{3+} to testtube 3. Again discard all but 10.0 mL of the contents of the graduated cylinder, refill to 25.0 mL with distilled water, stir, and add 5.0 mL to testtube 4. Repeat with testtubes 5 and 6.

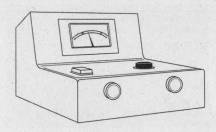

Figure 24.1
The Spec 20.

After allowing the spectrometer (see Figure 24.1) 10 minutes to warm up, set the wavelength to 455 nm. All absorbance measurements in this experiment should be made at 455 nm. To standardize the instrument, use a blank, which is distilled water in a cuvet. Holding the cuvet by the lip, wipe the outside of the cuvet with a clean Kimwipe to remove any fingerprints or stains that might reduce the intensity of light transmitted to the detector. Place the cuvet in the sample compartment, pushing it fully down. Align the mark on the cuvet with that on the sample holder, and close the sample compartment. Adjust the calibration dial that so the spectrometer reads zero absorbance.

Remove the cuvet from the spectrometer, and empty it into a waste beaker. Rinse the cuvet several times with the solution in testtube 1. Fill the cuvet three-quarters full with the solution from testtube 1. Holding the cuvet by the lip, wipe the outside of the cuvet clean with a Kimwipe, place the cuvet in the sample compartment, align the mark on the cuvet with that on the sample holder, and close the sample compartment. Read and record the absorbance value.

Repeat the absorbance measurements for the solutions in testtubes 2 through 6.

Waste Disposal. Dispose of all chemical waste as directed by your instructor.

Name: _____   Date: _____

Lab Instructor: _____   Lab Section: _____

EXPERIMENT 24

Spectroscopic Determination of an Equilibrium Constant

PRELABORATORY QUESTIONS

1. Calculate the initial concentrations of Fe^{3+} and SCN^- for the six solutions to be prepared in this experiment (the solutions prepared in testtubes 1–6). These are the concentrations just after the two solutions are mixed in the large testtube (thereby diluting each other) but before any $FeNCS^{2+}$ has formed. Assume that the volumes are additive and that the $Fe(NO_3)_3$ and KSCN are each completely dissociated. Insert your answers into the table below.

Testtube	Initial $[Fe^{3+}]$	Initial $[SCN^-]$
1		
2		
3		
4		
5		
6		

Name: _____   Date: _____

Lab Instructor: _____   Lab Section: _____

2. If the equilibrium $[FeSCN^{2+}]$ in testtube 2 was determined to be 9.5×10^{-5} M, calculate the equilibrium $[Fe^{3+}]$. (The equilibrium $[Fe^{3+}]$ is equal to the initial $[Fe^{3+}]$ less the equilibrium $[FeSCN^{2+}]$. This follows from the stoichiometry of the reaction.)

3. If the equilibrium $[FeSCN^{2+}]$ in testtube 2 was determined to be 9.5×10^{-5} M, calculate the equilibrium $[SCN^-]$.

4. For the solution in testtube 1 why can you assume that the equilibrium concentration of $FeSCN^{2+}$ is equal to the initial concentration of SCN^- in the testtube?

5. Why must the testtubes be completely dry before use?

6. Why is it important to rinse the cuvet several times with the solution to be analyzed before filling it to measure the absorbance?

7. Why must all the absorbance measurements be made at the same wavelength?

Name: _____   Date: _____

Lab Instructor: _____   Lab Section: _____

EXPERIMENT 24

Spectroscopic Determination of an Equilibrium Constant

RESULTS/OBSERVATIONS

1. Absorbance measurements:

Testtube	Absorbance
1	
2	
3	
4	
5	
6	

2. Calculation of the $[FeSCN^{2+}]$ for the solution in testtube 1:

3. Calculation of the equilibrium $[FeSCN^{2+}]$ for the solution in testtube 2:

4. Calculation of the equilibrium $[Fe^{3+}]$ for the solution in testtube 2:

Name: _____   Date: _____

Lab Instructor: _____   Lab Section: _____

5. Calculation of the equilibrium $[SCN^-]$ for the solution in testtube 2:

6. Calculation of the equilibrium constant (K) for the solution in testtube 2:

7. The equilibrium concentrations of $[FeSCN^{2+}]$, $[Fe^{3+}]$, and $[SCN^-]$, and K:

Testtube	$[FeSCN^{2+}]$	$[Fe^{3+}]$	$[SCN^-]$	K
1				
2				
3				
4				
5				
6				

8. The average value of K from the five determinations above: _____

Name: _____   Date: _____

Lab Instructor: _____   Lab Section: _____

EXPERIMENT 24

Spectroscopic Determination of an Equilibrium Constant

POSTLABORATORY QUESTIONS

1. How reasonable was your assumption that all the SCN^- in testtube 1 was converted to $FeSCN^{2+}$? (Using your average value of K, show how to calculate the percentage of SCN^- converted to $FeSCN^{2+}$.)

2. In this experiment, the reaction

$$Fe^{3+}(aq) + SCN^-(aq) \rightleftharpoons FeSCN^{2+}(aq)$$

was studied. Competing somewhat with this reaction are the related reactions

$$FeSCN^{2+}(aq) + SCN^-(aq) \rightleftharpoons Fe(SCN)_2^+(aq)$$

and

$$Fe(SCN)_2^+(aq) + SCN^-(aq) \rightleftharpoons Fe(SCN)_3(aq)$$

To ignore these competing reactions how must the equilibrium constants of these latter two reactions compare with that of the reaction being studied?

Chemical Equilibrium and Le Châtelier's Principle

OBJECTIVE

To observe how the equilibrium distribution of products and reactants in a number of chemical reactions is affected by a variety of factors, such as temperature and changes in initial product and reactant concentrations. In each instance, the observations are to be explained, be it quantitatively or qualitatively, using the principles of chemical equilibrium.

INTRODUCTION

Chemical equilibrium is that state characterized by a constancy of reactant and product concentrations with time. All chemical reactions occur so as to approach a state of chemical equilibrium. For the general reaction

$$a\,A + b\,B \rightleftharpoons c\,C + d\,D \tag{1}$$

the equilibrium concentrations of reactants and products satisfy the expression

$$K = \frac{[C]^c[D]^d}{[A]^a[B]^b} \tag{2}$$

Equation (2) is termed the *equilibrium expression,* and K is termed the *equilibrium constant.*

The value of the equilibrium constant for a given reaction depends only on temperature. Even though the individual equilibrium concentrations might change, when substituted into the equilibrium expression and evaluated, the resulting K is the same. For example, consider the hypothetical reaction

$$2\,X + Z \rightleftharpoons X_2Z \qquad K = 4.0 \text{ (at 25°C)} \tag{3}$$

The equilibrium expression for this reaction is

$$K = \frac{[X_2Z]}{[X]^2[Z]} \tag{4}$$

The equilibrium mixture at 25°C could contain 0.10 M X, 0.25 M Z, and 0.010 M X$_2$Z,

$$\frac{[0.010]}{[0.10]^2[0.25]} = 4.0 \tag{5}$$

or 0.66 M X, 0.20 M Z, and 0.35 M X$_2$Z,

$$\frac{[0.35]}{[0.66]^2[0.20]} = 4.0 \tag{6}$$

or one of countless other reactant and product concentration mixtures. By convention, the units are omitted when reporting equilibrium constants.

The magnitude of K indicates whether the equilibrium mixture is composed primarily of reactants or products. A reaction with a large K will have an equilibrium mixture composed primarily of products. A small K indicates a reaction that will have an equilibrium mixture composed primarily of reactants. If K for a reaction is an intermediate value, on the order of 1, the equilibrium mixture will contain comparable amounts of both reactants and products.

To determine whether a reaction is at equilibrium or, if it is not, how the reactant and product concentrations will change in order to proceed toward equilibrium, the reaction quotient (Q) is determined. The reactant quotient is defined similarly to the equilibrium constant, except the reactant and product concentrations are not necessarily those at equilibrium. For Reaction (1) the reaction quotient has the form

$$Q = \frac{[C]^c[D]^d}{[A]^a[B]^b} \tag{7}$$

If $Q = K$, the reaction is at equilibrium. If $Q < K$, the reaction will produce products and consume reactants; it will shift to the right. If $Q > K$, the reaction will consume products and produce reactants; it will shift to the left.

Often chemists would like to maximize the amount of a particular species produced in a chemical reaction. The equilibrium constant itself cannot be changed (except by changing the temperature), but the reaction conditions can be changed to increase the amount of a species produced. The manner in which changes in reaction conditions affect the composition of the equilibrium mixture can be understood qualitatively in terms of Le Châtelier's principle: *When a change is imposed on a system at equilibrium, the system will adjust, if possible, by shifting in the direction that will minimize the effect of the change.* For example, consider the general reaction

$$A + Z \rightleftharpoons 2\,AZ \tag{8}$$

which is characterized by the equilibrium expression

$$K = \frac{[AZ]^2}{[A][Z]} \tag{9}$$

How can the amount of AZ produced by Reaction (8) be maximized? By imposing a change that shifts the reaction to the right. There are three ways this can be done (without changing the temperature). One way would be to add more reactant A to the equilibrium mixture. According to Le Châtelier's principle, the reaction then will shift to decrease the amount of A in the mixture. Thus the reaction will shift to the right, consuming some A and Z and producing more AZ. A second way would be to add more Z to the mixture, which would have an effect analogous to adding A. The third way would be to remove some AZ from the reaction mixture. The reaction then would shift to the right to produce more AZ.

Le Châtelier's principle also can be understood in terms of the reaction quotient. Consider Reaction (8) again. Initially, the reaction is at equilibrium, and $Q = K$. The addition of more A will cause Q,

$$Q = \frac{[AZ]^2}{[A][Z]} \tag{10}$$

to momentarily decrease. Since $Q < K$, the reaction will shift to the right in an effort to reestablish equilibrium, consequently producing more AZ. Similar explanations can be used to show the effect of adding more Z or the removal of AZ.

Le Châtelier's principle even can be used to predict the direction the reaction will shift when the temperature is varied. To do this, consider heat to be either a reactant, something that is consumed during the process of the reaction (endothermic), or a product, something that is produced during the process of the reaction (exothermic). For example, assume Reaction (8) to be exothermic:

$$A + Z \rightleftharpoons 2\ AZ + heat \tag{11}$$

Raising the temperature is equivalent to adding heat to the equilibrium mixture. According to Le Châtelier's principle, the reaction will shift to decrease the heat. The only way the reaction can decrease the heat is by shifting to the left, consuming some AZ and producing A and Z. Note that unlike changes in concentration, changes in temperature alter the value of the equilibrium constant. Thus adding heat to Reaction (11) decreases the value of K.

This experiment is composed of four parts, A, B, C, and D. Each part examines aspects of chemical equilibrium. The specific background information and procedure necessary for understanding and completing each part of the experiment follow. You will have two weeks to complete the experiment. Your purpose in each part is to explain your observations, be it quantitatively or qualitatively, using the principles of chemical equilibrium.

ADDITIONAL READING

Read the sections in the Laboratory Techniques chapter at the beginning of this Lab Manual on cleaning glassware, handling chemicals, weighing solids, heating liquids and solutions, centrifugation, decantation, and burets prior to performing this experiment.

Part A: Supersaturation and the Effect of Temperature on Equilibrium

Solubility is defined as the maximum amount of solute that can be dissolved in a certain amount of a specific solvent at equilibrium. The solubility of a substance depends on the solvent and the temperature. For example, the solubility of NaCl in water is 35.7 g/100 mL H_2O at 0°C and 39.12 g/100 mL H_2O at 100°C.

A solution containing the maximum equilibrium amount of solute that can be dissolved in a certain amount of a specific solvent is said to be *saturated*. If more solute is added to a saturated solution it will remain undissolved. A solution that contains less than the maximum equilibrium amount of solute that can be dissolved in a certain amount of a specific solvent is termed *unsaturated*. If more solute is added to an unsaturated solution it eventually will dissolve. Under the proper conditions, it is possible to prepare a solution that contains more than the maximum equilibrium amount of solute that can be dissolved in a certain amount of a specific solvent. Such a solution is termed *supersaturated*. Supersaturation is a nonequilibrium condition. Eventually, the supersaturated solution will come to equilibrium, precipitating the excess solid as it does so.

In this experiment, you will investigate the solubility behavior of sodium acetate trihydrate ($NaC_2H_3O_2 \cdot 3H_2O$) dissolved in water. This salt dissolves in water according to the reaction

$$NaC_2H_3O_2 \cdot 3\,H_2O(s) \xrightleftharpoons{\quad H_2O \quad} Na^+(aq) + C_2H_3O_2^-(aq) + 3\,H_2O(l) \tag{12}$$

Sodium acetate trihydrate is much more soluble in hot water than in cold water. A saturated solution of $NaC_2H_3O_2 \cdot 3H_2O$ prepared at a high temperature and cooled slowly to room temperature usually will result in the formation of a supersaturated solution.

Why don't supersaturated solutions always achieve equilibrium quickly, precipitating out their excess solute in the process? For a solid to precipitate from solution numerous solute particles need to come together and orient themselves in a manner characteristic of the solid. Because of the large number of particles involved, this is an unlikely event, causing precipitation to be delayed significantly.

Often it is possible to accelerate the process of crystallization by adding a seed crystal to the solution. The seed crystal provides a surface for the solute particles to adsorb onto, increasing the likelihood of many solute particles coming together and forming a crystal.

In this part of the experiment you will prepare a supersaturated solution of $NaC_2H_3O_2 \cdot 3H_2O$. You then will test the efficacy of two different seed crystals, $NaC_2H_3O_2 \cdot 3H_2O$ and $CoCl_2 \cdot 6H_2O$.

   Safety Precautions:

- Protective eyewear approved by your institution must be worn at all times while you are in the laboratory.

PROCEDURE

Part A: Supersaturation and the Effect of Temperature on Equilibrium

In this part of the experiment you will prepare a supersaturated solution of $NaC_2H_3O_2 \cdot 3H_2O$. To prevent crystallization of the supersaturated solution until after the seed crystal is added it is essential that the testtube used to prepare the solution be scrupulously clean and scratch-free. Moreover, the initial addition of $NaC_2H_3O_2 \cdot 3H_2O$ to the testtube cannot leave behind any crystals on the upper portion of the testtube; otherwise, they might later fall into the supersaturated solution and induce crystallization. To prevent this latter occurrence, tear a sheet of notebook paper into quarters, then tear one of the quarters the long way into eighths, and, finally, roll two of the eighths into tight cylinders. Insert the cylinders into two of your smallest (13 × 100 mm) clean testtubes so that they reach nearly to the bottom.

Weigh out 1.5 g of $NaC_2H_3O_2 \cdot 3H_2O$. Carefully pour the solid into the paper cylinder in the first testtube. Do this above the sink, and then rinse any spilled solid down the drain. Gently tap the paper cylinder to dislodge any adhering crystals, and carefully slide the paper cylinder out of the testtube. Repeat with the second testtube. Rinse the pieces of notebook paper in the sink to remove any remaining crystals, wringing the pieces out and placing them in the waste basket. Obtain some distilled water in a small beaker. Use a dropper to add 8 drops of distilled water to the solid $NaC_2H_3O_2 \cdot 3H_2O$ in each testtube. If necessary, use the drops of water to wash any adhering crystals of $NaC_2H_3O_2 \cdot 3H_2O$ to the bottoms of the testtubes. Place the testtubes in a testtube rack.

Fill a 100 mL beaker three-quarters full of water, and add two or three boiling chips to make a hot-water bath. Begin heating the beaker of water on a hotplate. Clamp one of the testtubes containing the sodium acetate solution to a ring stand. Adjust the clamp to lower the testtube into the hot-water bath. Continue heating the water until the solid dissolves entirely; then turn off the hotplate.

After allowing the water to cool a bit, adjust the clamp to raise the testtube out of the hot water. Carefully remove the tube from the clamp. Holding the top of the testtube securely with one hand, rap the testtube briskly a few times with your finger to mix the hot solution. (This mixing process is necessary because the solution is probably more concentrated at the bottom than at the top.) After mixing, stand the testtube in a testtube rack. Repeat for the second testtube. If crystals reappear in either testtube, dip the testtube into the hot-water bath until the solid dissolves again.

Fill a 200 mL beaker half full with cold water. Stand the two testtubes in the cold-water bath for about 3 minutes to bring the solutions to room temperature. If no crystals are present, the preparation of the supersaturated solution is complete. If crystals are present, reheat, mix, and cool the solution again. Record the appearance of the solutions in the testtubes.

Obtain two or three crystals of $NaC_2H_3O_2 \cdot 3H_2O$ on a small piece of notebook paper and two or three crystals of $CoCl_2 \cdot 6H_2O$ on a different sheet of paper. Holding one of the testtubes in your hand, drop a crystal of $CoCl_2 \cdot 6H_2O$ into the testtube. Quickly add one of the $NaC_2H_3O_2 \cdot 3H_2O$ crystals to the other tube. Observe what happens in each case. One crystallization should occur considerably faster than the other. Record your observations. After crystallization occurs, feel the bottom of each testtube. Record your observations. Explain your various observations using balanced chemical equations where applicable.

Waste Disposal (Part A). Dispose of all chemical waste as directed by your instructor.

Part B: Effect of Concentration on Equilibrium and the Calculation of *K*

Antimony trichloride ($SbCl_3$) undergoes hydrolysis (reaction with water) in aqueous solutions to form a white precipitate of antimony oxychloride (SbOCl) according to the reaction:

$$SbCl_3(aq) + H_2O(l) \rightleftharpoons SbOCl(s) + 2\ HCl(aq) \tag{13}$$

The equilibrium constant for this reaction has the form

$$K = \frac{[HCl]^2}{[SbCl_3]} \tag{14}$$

where, as is customary, the concentrations of the pure liquid, H_2O, and the pure solid, SbOCl, are not included when writing the equilibrium expression (their concentrations are constants that are incorporated into the value of the equilibrium constant).

The equilibrium constant for this reaction can be calculated if the equilibrium concentrations of HCl and $SbCl_3$ are known. Direct measurement of these concentrations is difficult. However, they can be determined indirectly.

The indirect method involves preparing a nonequilibrium solution containing known amounts of both $SbCl_3$ and HCl. This starting solution will have such a high concentration of HCl that the reaction cannot come to equilibrium. Water will be added to this starting solution, decreasing the concentrations of both $SbCl_3$ and HCl. When the $SbCl_3$ and HCl concentrations have been decreased sufficiently, the reaction will achieve equilibrium, an occurrence denoted by the formation of the white solid, SbOCl. If the addition of water is stopped at exactly the point where the reaction comes to equilibrium, it is reasonable to assume that only tiny amounts of $SbCl_3$ and HCl have reacted. Using this assumption, one can calculate the equilibrium concentrations of $SbCl_3$ and HCl from the initial amounts of these chemicals and the total volume of the solution at equilibrium.

    Safety Precautions:

- Protective eyewear approved by your institution must be worn at all times while you are in the laboratory.

- Chemical burns can result when the acidic $SbCl_3$ solution or 6 *M* HCl solution comes in contact with skin. If you spill either solution on your skin, immediately wash the affected area with water. Continue washing with water for 15 minutes. Have a classmate notify your instructor.

PROCEDURE

Part B: Effect of Concentration on Equilibrium and the Calculation of K

B-1: Dilution of an Acidic Antimony Trichloride Solution

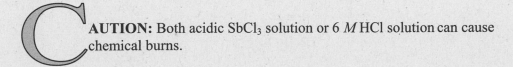

CAUTION: Both acidic $SbCl_3$ solution or 6 *M* HCl solution can cause chemical burns.

Using the buret your instructor has prepared for dispensing the $SbCl_3$ solution, measure 5.0 mL of the acidic antimony trichloride solution (0.5 *M* $SbCl_3$ in 6 *M* HCl) into a 25 × 200 mm testtube. Fill a buret with distilled water. Carefully add 8.0 mL of distilled water from the buret to the testtube, and then stir the mixture with a clean glass stirring rod. Record your observations of the solution. Continue to add distilled water in 1 mL increments, stirring the mixture in the testtube and recording your observations, until a total of 15 mL of water has been added. Explain your observations using balanced chemical equations where applicable.

B-2: Addition of Hydrochloric Acid to Antimony Oxychloride

Using the buret your instructor has set up for dispensing the HCl solution, add 5 mL of 6 *M* HCl to the testtube from Part B-1. Mix thoroughly. Record your observations. Explain your observations using balanced chemical equations where applicable.

B-3: Determination of the Equilibrium Constant

Carefully repeat the experiment of Part B-1 so as to obtain an accurate measurement of the equilibrium point. Starting with a fresh 5 mL sample of the acidic antimony trichloride solution, very slowly and carefully add distilled water from the buret until a slight milkiness appears in the solution, a milkiness that persists even after mixing. From your observations of Part B-1, you should be able to estimate approximately how much water will be required for this. When you have added almost enough water to cause the change, start adding the water one drop at a time. Record the volume of water added. Calculate the value of the equilibrium constant.

Waste Disposal (Part B). Dispose of all chemical waste as directed by your instructor.

Part C: Effects of Concentration, Solvent, and Temperature on Equilibrium

Cobalt(II) can exist as either a solid ionic salt or a complex ion in solution. Typical salts of cobalt(II) have the general formula $CoX_2 \cdot 6H_2O$, where X can represent nitrate (NO_3^-), acetate ($CH_3CO_2^-$), or chloride (Cl^-) ions. In aqueous solutions cobalt(II) usually is bonded to six water molecules to form the complex ion hexaaquocobalt(II) ($[Co(H_2O)_6]^{2+}$), which has an octahedral geometry (see Figure 25.1) and a pink color.

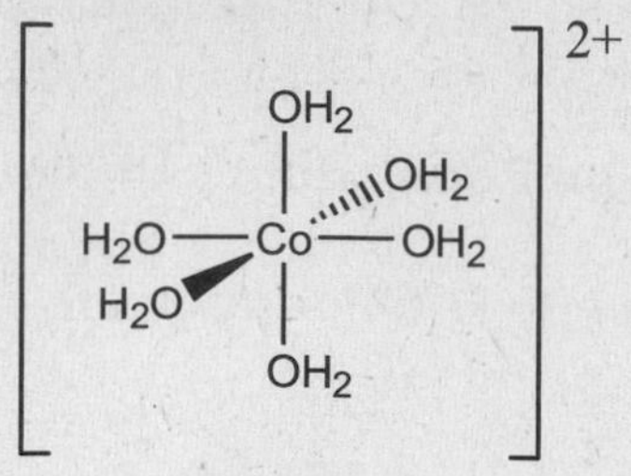

$$[Co(H_2O)_6]^{2+}$$

Figure 25.1
The octahedral geometry of the complex ion hexaaquocobalt(II).

Chloride ion can react with $[Co(H_2O)_6]^{2+}$ to form $[CoCl_4]^{2-}$ according to the reaction:

$$[Co(H_2O)_6]^{2+} + 4\ Cl^- \rightleftharpoons [CoCl_4]^{2-} + 6\ H_2O \qquad (15)$$
$$\text{(Pink)} \qquad\qquad\qquad \text{(Blue)}$$

In this reaction the coordination number of the cobalt changes from 6 to 4, and the structure of the ion changes from octahedral to tetrahedral (see Figure 25.2).

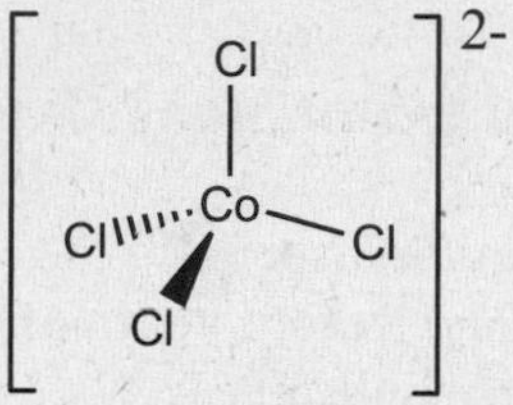

$$[CoCl_4]^{2-}$$

Figure 25.2
The tetrahedral geometry of the complex ion tetrachlorocobalt(II).

More important, $[CoCl_4]^{2-}$ has a deep-blue color, which means that the relative equilibrium concentrations of $[Co(H_2O)_6]^{2+}$ (a pink substance) and $[CoCl_4]^{2-}$ (a blue substance) can be tracked by eye. Interestingly, because the blue tetrahedral anion is approximately 100 times more deeply colored than the pink octahedral cation, only a small concentration of $[CoCl_4]^{2-}$ is necessary to turn the reaction mixture blue.

In this part of the experiment you will investigate the effects of changing chloride concentration, solvent, and temperature on the equilibrium of Reaction (15).

 Safety Precautions:

■ Protective eyewear approved by your institution must be worn at all times while you are in the laboratory.

■ Concentrated hydrochloric acid (HCl) is very corrosive and highly irritating to the lungs. Avoid breathing the vapors. Gloves must be worn while handling concentrated HCl. If you spill HCl on your skin, immediately wash with water for 15 minutes, and alert your instructor. If you get HCl on your clothing, the clothing should be removed immediately, and the area of contact should be rinsed thoroughly with water. Notify your instructor in the event of an HCl spill.

PROCEDURE

Part C: Effects of Concentration, Solvent, and Temperature on Equilibrium

C-1: The Effect of Solvent on Equilibrium

Using the tip of your spatula, obtain two or three crystals of each of the two available cobalt(II) salts, $CoCl_2 \cdot 6H_2O$ and $Co(NO_3)_2 \cdot 6H_2O$. Record the color of each solid. Place the $CoCl_2 \cdot 6H_2O$ crystals in a labeled, dry 16 mm testtube, and add 5 mL of water to dissolve the solid. Repeat this for the $Co(NO_3)_2 \cdot 6H_2O$ crystals. Record the appearance of each solution.

Repeat this procedure using absolute ethyl alcohol as the solvent and then again using acetone as the solvent.

Explain your observations using balanced chemical equations where applicable.

C-2: The Effect of Chloride Concentration on Equilibrium

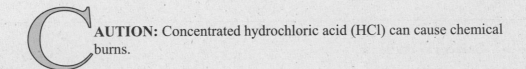

CAUTION: Concentrated hydrochloric acid (HCl) can cause chemical burns.

Obtain approximately 20 mL of concentrated (12 M) HCl in your 50 mL beaker, covering the beaker afterward with a watch glass. Place 5.0 mL of the aqueous 0.4 M $Co(NO_3)_2$ solution in a 25 × 200 mm testtube. Record the color of the solution. Add 2.0 mL of concentrated HCl to the $Co(NO_3)_2$ solution, swirl to mix thoroughly, and record the solution color. Repeat this three more times.

Next, add six successive 2.0 mL increments of distilled water to the $Co(NO_3)_2$–HCl solution, mixing thoroughly after each addition. Record your observations after each dilution.

The first experiment will be repeated, only now the concentrated HCl will be added to a solution of $Co(NO_3)_2$ dissolved in ethyl alcohol. Place 5.0 mL of 0.4 M $Co(NO_3)_2$ dissolved in ethyl alcohol in a 25 × 200 mm testtube. Add concentrated HCl to the solution dropwise until a distinct blue color is observed. Record the number of drops required.

Explain your observations using balanced chemical equations where applicable.

C-3: The Effect of Temperature on Equilibrium

Fill a 400 mL beaker about half full of water, add a magnetic stirbar, and bring the water to a boil on a stirrer-hotplate. Obtain 10.0 mL of 0.4 M $CoCl_2$ in a 30 mL beaker. Add 6.0 mL of concentrated HCl, and mix thoroughly with a stirring rod. The resulting solution should be violet (i.e., between the original pink-red and bright blue). If it is not violet, adjust the color by carefully adding distilled water or concentrated HCl, depending on the color of the solution, a drop at a time. Divide the violet solution equally into three 16 mm testtubes. Place one testtube in some ice in a 200 mL beaker, heat a second gently to 80 to 90°C in the hot-water bath, and allow the third to remain at room temperature. Record the colors of each solution. Allow the solutions in testtubes 1 and 2 to return to room temperature. Record your observations.

Explain your observations using balanced chemical equations where applicable.

Waste Disposal (Part C). Dispose of chemical waste as directed by your instructor.

Part D: The Effect of Solubility on Completeness of Reaction

A cation can be precipitated from a solution of one of its soluble salts by the addition of an anion that forms an insoluble salt. For example, silver nitrate ($AgNO_3$) is a soluble salt, as is sodium chloride (NaCl). But addition of NaCl to a solution of $AgNO_3$ forms the insoluble salt AgCl:

$$Ag^+(aq) + Cl^-(aq) \rightleftharpoons AgCl(s) \qquad (16)$$

Of course, *insoluble* is a relative term. Reaction (16), like any chemical reaction, is subject to equilibrium conditions, equilibrium conditions that in this case strongly favor the formation of a precipitate. Still, at equilibrium some small amounts of Ag^+ and Cl^- are present in solution. If the supernatant (the liquid above the solid) is filtered from the precipitate in Reaction (16), the Ag^+ ions that remain in solution can be precipitated using an anion that forms a salt of even lesser solubility. Similarly, if solid AgCl is placed in water some small amount of the solid will dissolve to form Ag^+ and Cl^-.

In this part of the experiment the relative solubilities of some lead salts will be investigated.

    Safety Precautions:

- Protective eyewear approved by your institution must be worn at all times while you are in the laboratory.

- Lead compounds are toxic. Avoid contact with skin, and wash your hands thoroughly when you are finished.

PROCEDURE

Part D: The Effect of Solubility on Completeness of Reaction

In a small testtube mix 2 mL of 1 M $Pb(NO_3)_2$ and 4 mL of 1 M NaCl. Record your observations.

C**AUTION:** Lead compounds are toxic. Avoid contact with skin, and wash your hands thoroughly when you are finished.

Label a large beaker "Waste Lead Products." Centrifuge the mixture and discard the supernatant in the waste beaker. Wash the precipitate with 5 mL of distilled water and centrifuge. Discard the supernatant in the waste beaker. Add 3 mL of distilled water to the precipitate, and mix with a glass stirring rod for 3 minutes. Centrifuge, and then pour the supernatant into a clean small testtube. Discard the precipitate in the waste beaker. Record your observations.

Add 2 mL of 1 M Na_2CO_3 to the previously isolated supernatant and mix thoroughly with a glass stirring rod. Record your observations. Centrifuge the mixture for 5 minutes, and then pour the supernatant into a clean small testtube. Label the testtube containing the precipitate with a 1, and set it aside for later use.

Add 2 mL of 0.1 M Na$_2$S to the supernatant isolated in the preceding step, and mix with your stirring rod. Record your observations. Centrifuge the mixture for 5 minutes. There might be only a small amount of precipitate coating the bottom of the testtube. Discard the supernatant liquid in the waste beaker. Add four or five drops of 6 M HNO$_3$ to the solid. Record your observations.

To the solid in testtube 1 add four or five drops of 6 M HNO$_3$, and stir the mixture with a glass stirring rod. Record your observations.

Explain your observations using balanced chemical equations where applicable.

On the basis of your observations, arrange the following compounds in order of decreasing solubility: PbCO$_3$, PbCl$_2$, Pb(NO$_3$)$_2$, and PbS.

Waste Disposal (Part D). Dispose of all chemical waste as directed by your instructor.

Name: _____   Date: _____

Lab Instructor: _____   Lab Section: _____

EXPERIMENT 25

Chemical Equilibrium and Le Châtelier's Principle

PRELABORATORY QUESTIONS

1. Consider the hypothetical reaction

$$A_2(aq) + XZ(aq) \rightleftharpoons AX(aq) + AZ(aq) \quad K = 1.0 \times 10^{-4} \text{ (at 25°C)}$$

(a) At 25°C, which are favored for this reaction at equilibrium, reactants or products?

(b) At 25°C, for the following reactant and product concentrations, $[A_2] = 0.020 \ M$, $[XZ] = 1.5 \times 10^{-4} \ M$, $[AX] = 0.55 \ M$, and $[AZ] = 1.0 \ M$, is the reaction at equilibrium? If the reaction is not at equilibrium, in which direction will it shift to achieve equilibrium, left or right?

2. Consider the hypothetical endothermic reaction:

$$AX(aq) + Z_2(s) \rightleftharpoons AXZ_2(aq) \quad K = 7.8 \text{ (at 25°C)}$$

(a) What is the equilibrium expression for this reaction?

(b) If each of the following changes is made, will the reaction shift to the left, the right, or remain unchanged?

    i. Add AX.

    ii. Add Z_2.

    iii. Remove AXZ_2.

    iv. Increase the temperature.

Name: _____   Date: _____

Lab Instructor: _____   Lab Section: _____

EXPERIMENT 25

Chemical Equilibrium and Le Châtelier's Principle

RESULTS/OBSERVATIONS

Part A: Supersaturation and the Effect of Temperature on Equilibrium

1. Appearance of the two supersaturated solutions: _____

2. Observation of addition of a $CoCl_2 \cdot 6H_2O$ seed crystal: _____

3. Observation of addition of a $NaC_2H_3O_2 \cdot 3H_2O$ seed crystal: _____

4. Compare the two rates of crystallization: _____

5. Observation on feeling the bottom of the testtube to which a $CoCl_2 \cdot 6H_2O$ seed crystal was

    added: _____

6. Observation on feeling the bottom of the testtube to which a $NaC_2H_3O_2 \cdot 3H_2O$ seed crystal

    was added: _____

7. Explanation of observations using balanced chemical equations:

Name: _____ Date: _____

Lab Instructor: _____ Lab Section: _____

Part B: Effect of Concentration on Equilibrium and the Calculation of K

B-1: Dilution of an Acidic Antimony Trichloride Solution

8.

mL of H_2O Added	Observations
8	
9	
10	
11	
12	
13	
14	
15	

9. Explanation of observations using balanced chemical equations:

Name: _____   Date: _____

Lab Instructor: _____   Lab Section: _____

B-2: Addition of Hydrochloric Acid to Antimony Oxychloride

10. Observations after addition of 5 mL of 6 *M* HCl to the testtube from Part B-1:

11. Explanation of observations using balanced chemical equations:

B-3: Determination of the Equilibrium Constant

12. The initial volume of water in the buret: _____

13. The final volume of water in the buret: _____

14. Volume of water added to achieve equilibrium: _____

15. Calculation of the equilibrium constant:

Name: _____   Date: _____

Lab Instructor: _____   Lab Section: _____

Part C: Effect of Concentration, Solvent, and Temperature on Equilibrium

C-1: The Effect of Solvent on Equilibrium

16. Color of $CoCl_2 \cdot 6H_2O$ crystals: _____

17. Color of $Co(NO_3)_2 \cdot 6H_2O$ crystals: _____

18. Appearance of the solution made by dissolving $CoCl_2 \cdot 6H_2O$ crystals in water: _____

19. Appearance of the solution made by dissolving $Co(NO_3)_2 \cdot 6H_2O$ crystals in water: _____

20. Appearance of the solution made by dissolving $CoCl_2 \cdot 6H_2O$ crystals in ethyl alcohol:
_____

21. Appearance of the solution made by dissolving $Co(NO_3)_2 \cdot 6H_2O$ crystals in ethyl alcohol:
_____

22. Appearance of the solution made by dissolving $CoCl_2 \cdot 6H_2O$ crystals in acetone:
_____

23. Appearance of the solution made by dissolving $Co(NO_3)_2 \cdot 6H_2O$ crystals in acetone:
_____

24. Explanation of observations using balanced chemical equations:

Name: _____   Date: _____

Lab Instructor: _____   Lab Section: _____

C-2: The Effect of Chloride Concentration on Equilibrium

25. Initial color of the aqueous $Co(NO_3)_2$ solution: _____

26. Color of the $Co(NO_3)_2$ solution after the first 2.0 mL addition of concentrated HCl: _____

27. Color after the second 2.0 mL addition of concentrated HCl: _____

28. Color after the third 2.0 mL addition of concentrated. HCl: _____

29. Color after the fourth 2.0 mL addition of concentrated HCl: _____

30. Color after the first 2.0 mL addition of H_2O: _____

31. Color after the second 2.0 mL addition of H_2O: _____

32. Color after the third 2.0 mL addition of H_2O: _____

33. Color after the fourth 2.0 mL addition of H_2O: _____

34. Color after the fifth 2.0 mL addition of H_2O: _____

35. Color after the sixth 2.0 mL addition of H_2O: _____

36. Initial color of the $Co(NO_3)_2$ dissolved in ethyl alcohol solution: _____

37. Number of drops of concentrated HCl necessary to change the solution color to blue: _____

38. Explanation of observations using balanced chemical equations:

Name: _____ Date: _____

Lab Instructor: _____ Lab Section: _____

C-3: The Effect of Temperature on Equilibrium

39. Initial color of the solution in the three testtubes: _____

40. Color of the solution in testtube 1 when heated to 80–90°C: _____

41. Color of the solution in testtube 2 when cooled in the ice bath: _____

42. Color of the solution in testtube 1 when it returned to room temperature: _____

43. Color of the solution in testtube 2 when it returned to room temperature: _____

44. Explanation of observations using balanced chemical equations:

Name: _____   Date: _____

Lab Instructor: _____   Lab Section: _____

Part D: The Effect of Solubility on Completeness of Reaction

45. Observation on mixing 2 mL of 1 M Pb(NO$_3$)$_2$ and 4 mL of 1 M NaCl: _____

46. Observation of the supernatant solution isolated: _____

47. Observation on adding 2 mL of 1 M Na$_2$CO$_3$ to the previously isolated supernatant:

_____

48. Observations on adding 2 mL of 0.1 M Na$_2$S to the supernatant isolated in the preceding

step: _____

49. Observations on adding four or five drops of 6 M HNO$_3$ to the solid: _____

50. Observations on adding four or five drops of 6 M HNO$_3$ to the solid in testtube 1:

_____

51. Explanation of observations using balanced chemical equations:

52. On the basis of your observations arrange the following compounds in order of decreasing
solubility: PbCO$_3$, PbCl$_2$, Pb(NO$_3$)$_2$, and PbS: _____ > _____ > _____ > _____

Name: _____   Date: _____

Lab Instructor: _____   Lab Section: _____

EXPERIMENT 25

Chemical Equilibrium and Le Châtelier's Principle

POSTLABORATORY QUESTIONS

1. Is an unsaturated solution an equilibrium or a nonequilibrium situation? Explain.

2. Explain how a supersaturated solution of a salt that undergoes an exothermic dissolution reaction would be prepared.

3. Why can't the initial solution in Part B-1 come to equilibrium until water is added?

Name: _____   Date: _____

Lab Instructor: _____   Lab Section: _____

4. Consider the reaction below:

$$Fe^{3+}(aq) + SCN^-(aq) \rightleftharpoons FeSCN^{2+}(aq)$$
    (Yellow)   (Red)

In each of the cases below, assume that the reaction equilibrium is such that the solution initially has an orange coloration; i.e., the equilibrium mixture contains appreciable amounts of both Fe^{3+} and $FeSCN^{2+}$.

(a) Enough water is added to double the initial volume. What color is the resulting solution? Explain.

(b) Some aqueous $AgNO_3$ is added to the solution. (Neglect any change in volume.) What color is the resulting solution? Explain. [**Hint:** $AgNO_3(aq) + SCN^-(aq) \rightarrow AgSCN(s) + NO_3^-(aq)$.]

(c) Solid NaOH is added to the solution. What color is the resulting solution? Explain. [**Hint:** $Fe^{3+}(aq) + 3 NaOH(s) \rightarrow Fe(OH)_3(s) + 3 Na^+(aq)$.]

(d) Solid $Fe(NO_3)_3$ is added to the solution. What color is the resulting solution? Explain.

Name: _____   Date: _____

Lab Instructor: _____   Lab Section: _____

5. Three calcium compounds have the following relative solubilities: $CaSO_4 > CaCO_3 > Ca_3(PO_4)_3$.

(a) A solution of Na_2SO_4 was added to a solution of $CaCl_2$. The resulting mixture was centrifuged, and the supernatant was discarded. The precipitate was washed with 5 mL of distilled water and centrifuged. The supernatant was discarded. A few milliliters of distilled water was added to the precipitate and mixed with a glass stirring rod. The mixture was centrifuged, and the supernatant was poured into a clean small testtube.

   A few milliliters of Na_3PO_4 solution was added to the previously isolated supernatant. Is a precipitate likely to form? If so, write a balanced chemical equation to describe its formation.

(b) A solution of Na_3PO_4 was added to a solution of $CaCl_2$. The resulting mixture was centrifuged, and the supernatant was discarded. The precipitate was washed with 5 mL of distilled water and centrifuged. The supernatant was discarded. A few milliliters of distilled water was added to the precipitate and mixed with a glass stirring rod. The mixture was centrifuged and the supernatant poured into a clean small testtube.

   A few milliliters of Na_2CO_3 solution was added to the previously isolated supernatant. Is a precipitate likely to form? If so, write a balanced chemical equation to describe its formation.

EXPERIMENT 26

pKₐ of an Unknown Acid-Base Indicator

pK_a of an Unknown
Acid-Base Indicator

OBJECTIVE

To measure, both qualitatively and quantitatively, the pK_a of an unknown acid-base indicator. The qualitative pK_a determination will be made visually with the help of a pH meter. The quantitative pK_a measurement will use a visible spectrophotometer and a pH meter.

INTRODUCTION

Acid-base indicators are substances whose color depends on the pH of the solution to which they are added. Chemists typically use acid-base indicators to mark the end-point of titrations. But because they are inexpensive and easy to use, acid-base indicators are also used frequently to make approximate pH determinations. For example, acid-base indicators are often the vital component of kits for testing soil and swimming pool pH.

An acid-base indicator is nothing more than a weak acid that appears one color in its protonated form (HIn) and another in its deprotonated form (In⁻):

$$HIn(aq) + H_2O(l) \rightleftharpoons H_3O^+(aq) + In^-(aq) \tag{1}$$

(Color 1) (Color 2)

When a small amount of indicator is added to a solution, it does not affect the pH of the solution. Because it is present in such relatively small amounts, the indicator equilibrium is controlled by the prevailing $[H_3O^+]$ in solution. In acidic solution, the $[H_3O^+]$ is relatively large. By Le Châtelier's principle, a large $[H_3O^+]$ will shift the Reaction (1) equilibrium to the left, causing the solution to appear color 1. In a basic solution, the $[H_3O^+]$ is relatively small, the Reaction (1) equilibrium is shifted to the right, and the solution appears color 2. For pH's where the [HIn] and [In⁻] are comparable, the solution will appear as a mixture of the two colors. Put another way, the indicator color changes when [HIn] ≈ [In⁻]. For example, the acid-base indicator bromothymol blue is yellow in its protonated form and blue in its deprotonated form. At low pH's where the HIn form predominates, the solution appears yellow. At high pH's where the In⁻ form is the primary species, the solution appears blue. When the [HIn] is comparable with that of [In⁻], the solution will appear green.

Different indicator solutions change color at different pH's. To understand why this is so, start with the equilibrium expression for Reaction (1):

$$K_a = \frac{\left[H_3O^+\right]\left[In^-\right]}{\left[HIn\right]} \tag{2}$$

Taking the negative logarithm of both sides and rearranging yields:

$$-\log K_a = -\log\left[H_3O^+\right] - \log\frac{\left[In^-\right]}{[HIn]} \tag{3}$$

$$pH = pK_a + \log\frac{\left[In^-\right]}{[HIn]} \tag{4}$$

i.e., a description of the indicator equilibrium in terms of the Henderson–Hasselbalch equation. At the color transition, $[HIn] \approx [In^-]$, so $pH \approx pK_a$. Since each indicator has its own pK_a value, each will change color at a different pH. For example, the indicator bromothymol blue has a pK_a of 7.1. It changes color over the pH range 6.1–8.1.

The color change of an indicator occurs over a range because the criterion $[HIn] \approx [In^-]$ applies to a range of concentrations. Any time appreciable amounts of both HIn and In^- are present in solution, both species will contribute to the observed color. As a rule of thumb, appreciable amounts of both species are present in solution when $0.10 < [In^-]/[HIn] < 10$.

The visible color of a solution can be quantified by measuring its absorbance using a visible spectrometer. For example, if a solution appears red in white light (normal room light), the solution must be selectively transmitting red light. This implies that the solution is absorbing the complementary color, green. On a visible spectrometer, this solution would be expected to have a maximum absorbance in the violet region of the visible spectrum, which corresponds to 490–560 nm. Figure 26.1 summarizes the relationship between the color of light, its wavelength, and its complement.

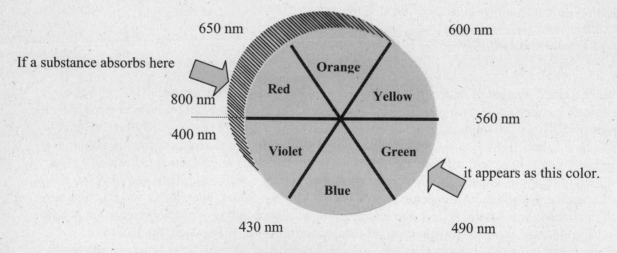

Figure 26.1
A color wheel summarizing the relationship between the color of light, its wavelength, and its complement.

Figure 26.2 shows the visible absorbance spectrum of a very acidic solution of bromothymol blue. Under very acidic conditions, a solution of bromothymol blue appears yellow. Since the sample appears yellow, it must be selectively transmitting yellow light and absorbing the complementary color, violet. As is shown in Figure 26.2, the very acidic solution of bromothymol blue has an absorbance maximum near 430 nm, a wavelength associated with violet light.

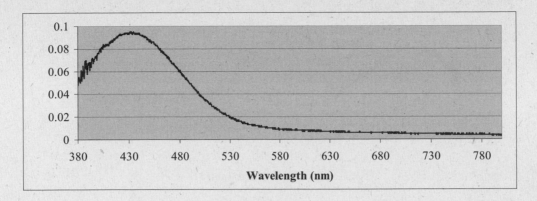

Figure 26.2
The visible absorbance spectrum of a very basic solution of bromothymol blue. Since the sample absorbs violet light (430 nm peak), it will transmit the complementary color, yellow.

Figure 26.3 shows the visible absorbance spectrum of a very basic solution of bromothymol blue. Under very basic conditions, a solution of bromothymol blue appears blue. The sample must be selectively transmitting blue light and absorbing the complementary color, orange. This solution has an absorbance maximum at about 615 nm, a wavelength associated with orange light.

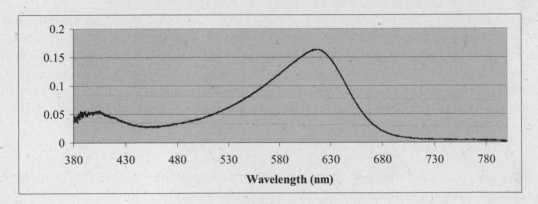

Figure 26.3
The visible absorbance spectrum of a very basic solution of bromothymol blue. Since the sample absorbs orange light (615 nm peak), it will transmit the complementary color, blue.

Figure 26.4 shows the visible absorbance spectrum of a nearly neutral solution of bromothymol blue. Under these pH conditions, a solution of bromothymol blue appears green. Figure 26.4 shows that the sample is absorbing both violet (430 nm peak) and orange (615 nm peak) light, so it must be transmitting both yellow and blue light, the combination of which is green light.

Notice in the three absorbance spectra that the peak heights at 430 and 615 nm change as the prevailing solution pH changes. Consider the 430 nm absorbance. The HIn species is responsible for this absorbance. Under very acidic conditions where [HIn] is relatively large, the absorbance at 430 nm is large (see Figure 26.2). For very basic conditions where [HIn] is relatively small, the absorbance at 430 nm is small (see Figure 26.3). At a pH near 7 where [HIn] is moderate, the absorbance at 430 nm is moderate (see Figure 26.4). Clearly, the absorbance at 430 nm is related to the amount of HIn present in solution. Similarly, the absorbance at 615 nm tracks the amount of In– in solution.

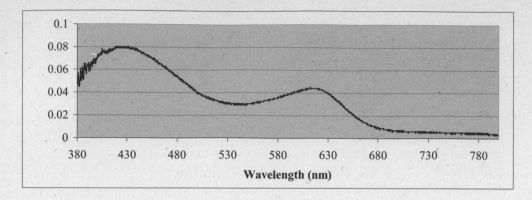

Figure 26.4
The visible absorbance spectrum of a nearly neutral solution of bromothymol blue. Since the
sample is absorbing both violet (430 nm peak) and orange (615 nm peak) light, it must be
transmitting both yellow and blue light, the combination of which is green light.

The relationship between absorbance and the amount of the light-absorbing species in a sample is

$$A = \varepsilon dc \tag{5}$$

Equation (5) is called the *Beer–Lambert law,* or simply *Beer's law.* The absorbance A is unitless. The
concentration c usually is given in units of moles per liter (M). The distance light travels through the
sample, the so-called pathlength d, is commonly expressed in centimeters. The quantity ε is called the
molar absorptivity or *extinction coefficient* and has units of $M^{-1} \cdot cm^{-1}$; the molar absorptivity is the
characteristic of the substance that expresses how much light is absorbed at a particular wavelength
by a particular substance. To emphasize this last point, Equation (5) is sometimes expressed as

$$A_\lambda = \varepsilon_\lambda dc \tag{6}$$

because the values of A and ε depend on the wavelength of light; if the wavelength is changed, both
A and ε change. Indeed, an absorbance spectrum is nothing more than a plot illustrating how A (or ε)
varies with wavelength.

Your purpose in this experiment is to first qualitatively and then quantitatively measure the pK_a
of an unknown acid-base indicator.

 Safety Precautions:

■ Protective eyewear approved by your institution must be worn at all times while you are in the
laboratory.

■ If either 0.1 M HCl or 0.1 M NaOH solutions should spill on your skin, rinse the affected area
with water for 15 minutes.

ADDITIONAL READING

Read the sections in the Laboratory Techniques chapter at the beginning of this Lab Manual on cleaning glassware, handling chemicals, pH measurements, and spectrometers prior to performing this experiment.

EXPERIMENT

You are to design and perform experiments in order to first qualitatively and then quantitatively measure the pK_a of an unknown acid-base indicator. You will have two lab sessions to accomplish this.

Design your experiment to accomplish the following:

1. Using only your eyes, the acid and base solutions, the pH meter, and any additional distilled water that you might require, estimate the pK_a of your unknown indicator.

2. Determine the wavelength of maximum absorbance (λ_{max}) of your unknown indicator in both its protonated and deprotonated forms (i.e., find the λ_{max} values of HIn and In⁻). In order to be confident that your procedure is valid, you are required to test it first on an acid-base indicator with known λ_{max} values. Methyl orange has a λ_{max} of 497 nm in its HIn form and a λ_{max} of 463 nm in its In⁻ form.

3. Accurately measure the pK_a of your unknown indicator solution using the visible spectrometer. For best accuracy, absorbance measurements should be made at or near λ_{max}. Before applying your procedure to your unknown indicator, you must test it on methyl orange (pK_a = 3.46).

Design your experiment around the following restrictions:

1. You will not be given the concentration of your unknown indicator.

2. The amount of indicator in solution must be small, or the indicator equilibrium will not be controlled by the prevailing [H_3O^+] in solution.

3. Beer's law works best when applied to dilute solutions. For best results, the measured maxima should have absorbances of less than 0.2.

Equipment and Reagents

To perform this experiment, you will have access to all the equipment in your lab drawer as well as the following items:

0.10 M HCl
0.10 M NaOH
Methyl orange (unknown concentration)
pH meter
Visible spectrophotometer

Waste Disposal. Dispose of all chemical waste as directed by your instructor.

Name: _____   Date: _____

Lab Instructor: _____   Lab Section: _____

EXPERIMENT 26

pK_a of an Unknown Acid-Base Indicator

PRELABORATORY QUESTIONS

1. In what wavelength range would you expect a green solution to have its maximum absorbance?

2. Consider Reaction (1). When you place the indicator in a very basic solution, will it appear color 1, color 2, or a combination of both colors? In a very acidic solution? What about when the pH of the solution is equal to the pK_a of the indicator?

3. If the measured absorbances of a solution of bromothymol blue are found to be the same at both 430 and 615 nm, can it be concluded that the [HIn] and [In⁻] are equivalent? Explain.

4. Suppose that you start with a 500 mL beaker of water to which is added two drops of an acid-base indicator. The indicator has a pK_a equal to x. By adding a few drops of concentrated acid (or base) you are able to shift the pH of the solution significantly. Fill in the following data for this solution. You should express your pH values in terms of the pK_a of the indicator (e.g., pH = 4 pK_a − π).

[HIn]	[In⁻]	pH
99.%	1.0%	
90.%	10.%	
50.%	50.%	
10.%	90.%	
1.0%	99.%	

Name: _____   Date: _____

Lab Instructor: _____   Lab Section: _____

5.   Use your answer to question 4 to plot [HIn], expressed as a percentage, as a function of pH. Assume the pK_a of the indicator to be 5.0. Repeat this plot, only now assume the pK_a of the indicator to be 9.0. How do these two plots differ in appearance?

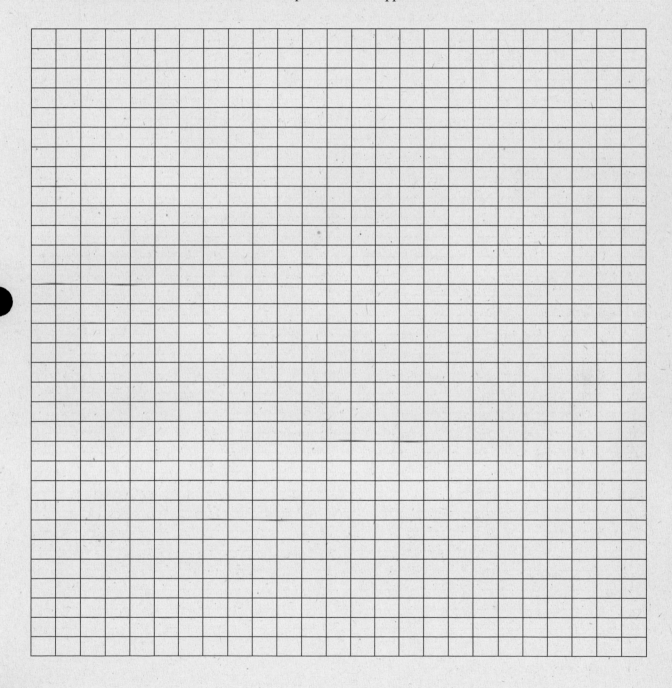

Name: _____   Date: _____

Lab Instructor: _____   Lab Section: _____

6. Use your answer to question 4 to plot [In⁻], expressed as a percentage, as a function of pH.
    Assume the pK_a of the indicator to be 3.0. Repeat this plot, only now assume the pK_a of the
    indicator to be 8.0. How do these two plots differ in appearance?

Name: _____   Date: _____

Lab Instructor: _____   Lab Section: _____

EXPERIMENT 26

pK_a of an Unknown Acid-Base Indicator

RESULTS/OBSERVATIONS

1. Estimated unknown indicator pK_a: _____

2. Experimentally determined λ_{max} of the HIn form of methyl orange: _____

3. Experimentally determined λ_{max} of the In$^-$ form of methyl orange: _____

4. λ_{max} of the HIn form of the unknown indicator: _____

5. λ_{max} of the In$^-$ form of the unknown indicator: _____

6. Experimentally determined pK_a value for methyl orange: _____

7. pK_a of the unknown indicator: _____

Name: _____   Date: _____

Lab Instructor: _____   Lab Section: _____

8. Write any other pertinent data in the space below. Clearly indicate both the property and the amount.

Name: _____   Date: _____

Lab Instructor: _____   Lab Section: _____

9. Perform any relevant calculations in the space below. Perform any plots on the provided graph paper.

Name: _____   Date: _____

Lab Instructor: _____   Lab Section: _____

Name: _____   Date: _____

Lab Instructor: _____   Lab Section: _____

Name: _____   Date: _____

Lab Instructor: _____   Lab Section: _____

EXPERIMENT 26

pK_a of an Unknown Acid-Base Indicator

POSTLABORATORY QUESTIONS

1. Explain why the color of an acid-base indicator changes as the prevailing pH changes.

2. Why do equal concentrations of HIn and In$^-$ not necessarily have the same absorbance, even if their absorbances are measured in cuvets of the same size at the same wavelength?

pH Titrations of
Strong and Weak Acids

OBJECTIVE

To perform both strong acid-strong base (HCl–NaOH) and weak acid-strong base (CH$_3$CO$_2$H–NaOH) titrations. The titrations will be monitored using both acid-base indicators and a pH meter. A sample of household vinegar will be analyzed for its acid content by titration with NaOH.

INTRODUCTION

A monoprotic acid (HA) dissociates in water to form hydronium ion (H$_3$O$^+$) and its conjugate base (A$^-$). Strong acids essentially dissociate completely in water:

$$HA(aq) + H_2O(l) \rightarrow H_3O^+(aq) + A^-(aq) \tag{1}$$

so the concentration of H$_3$O$^+$ is equal to the initial HA concentration. Weak acids only partially dissociate in water:

$$HA(aq) + H_2O(l) \rightleftharpoons H_3O^+(aq) + A^-(aq) \tag{2}$$

The concentration of H$_3$O$^+$ can be found by setting up an equilibrium expression, where the equilibrium constant is given the special symbol K_a:

$$K_a = \frac{\left[H_3O^+\right]\left[A^-\right]}{[HA]} \tag{3}$$

A monoprotic acid, be it strong or weak, undergoes a neutralization reaction when combined with a strong base (BOH):

$$HA(aq) + BOH(aq) \rightarrow H_2O(l) + BA(aq) \tag{4}$$

Given an acid solution of unknown concentration, it is possible to use this neutralization reaction to determine its concentration. The technique for doing this is termed an *acid-base titration*. In an acid-base titration, increments of base of known concentration are added to the acid sample until all the acid has reacted. The base solution usually is added to the acid sample by means of a buret, a calibrated glass dispensing tube. The volume of base necessary to react completely with the acid and its known concentration are used to determine the moles of base consumed in the reaction. The moles of base consumed in the reaction are equal to the moles of acid present initially. The moles of acid divided by the volume of the acid sample yield its concentration.

When performing an acid-base titration, it is necessary to know when the equivalence-point, the point at which the moles of base added exactly equal the moles of acid in the sample, has been reached. Typically, the equivalence-point cannot be observed directly but must be estimated by observing some physical change associated with it.

One common method of equivalence-point detection involves the use of an acid-base indicator. An acid-base indicator is a substance whose color depends on the hydronium ion concentration. The point in the titration at which the indicator's color change is observed is called the *end-point*. A properly chosen indicator will change color at the pH of the acid-base titration equivalence-point. Some common acid-base indicators, their pH ranges of color change, and color changes are listed in Table 27.1.

Table 27.1: Acid-Base Indicators

Indicator	pH Range of Color Change	Color Change
Methyl orange	3.2–4.4	Red to yellow
Bromocresol green	3.8–5.4	Yellow to blue
Methyl red	4.2–6.3	Red to yellow
Bromothymol blue	6.0–7.6	Yellow to blue
Phenolphthalein	8.0–10.0	Colorless to pink

The other common means of detecting the equivalence-point is via monitoring the solution pH throughout the titration. The pH is defined as the negative base-10 logarithm of the hydronium ion concentration:

$$pH = -\log[H_3O^+] \tag{5}$$

The pH is conveniently measured in the laboratory using a pH meter. These data are used to construct a plot of pH as a function of the volume of base added, a so-called titration curve. The pH of the solution undergoes a dramatic increase around the equivalence-point. The equivalence-point is marked by a point of inflection in the titration curve, the point where the curvature shifts from concave to convex.

In this experiment, both methods of equivalence-point detection will be used.

At the equivalence-point of an acid-base titration the principal species in solution is the salt BA. Salts formed from the neutralization reaction between strong acids and strong bases yield neutral solutions (i.e., pH = 7.0). However, salts formed from the neutralization reaction between weak acids and strong bases yield basic solutions (pH > 7). Salts formed from the neutralization reaction of strong acids and weak bases yield acidic solutions (pH < 7).

In this experiment, both strong acid-strong base (HCl–NaOH) and weak acid-strong base (CH_3CO_2H–NaOH) titrations will be performed. These titrations will be monitored using both acid-base indicators and a pH meter. Lastly, a sample of household vinegar will be analyzed for its acid content. Although other acids are present in vinegar, its acid content customarily is expressed in terms of acetic acid (CH_3CO_2H), the principal acidic constituent.

ADDITIONAL READING

Read the sections in the Laboratory Techniques chapter at the beginning of this Lab Manual on cleaning glassware, handling chemicals, pH measurements, volumetric flasks, pipets, and burets prior to performing this experiment.

Safety Precautions:

■ Protective eyewear approved by your institution must be worn at all times while you are in the laboratory.

■ Sodium hydroxide is corrosive, especially to the eyes. Thoroughly wash off any sodium hydroxide spilled on your skin. If you get it into your eyes, immediately flush your eyes with water for at least 10 minutes. If you spill it on your clothing, remove the affected clothing before rinsing with water. Notify your lab instructor, who will assist you.

■ Hydrochloric acid is corrosive and must be kept off skin and clothing. If you get it into your eyes, immediately flush your eyes with water for at least 10 minutes. If you spill it on your clothing, remove the affected clothing before rinsing with water. Notify your lab instructor, who will assist you.

■ Acetic acid is corrosive and must be kept off skin and clothing. If you get it into your eyes, immediately flush your eyes with water for at least 10 minutes. If you spill it on your clothing, remove the affected clothing before rinsing with water. Notify your lab instructor, who will assist you.

PROCEDURE

Prior to use, the pH meter must be standardized. The pH meter will be standardized using buffer solutions of known pH.

Obtain about 25 mL each of pH 4 and pH 7 buffer solutions in two small beakers. Follow the directions supplied with the pH meter for standardization.

Part A: Titration of a Strong Acid (HCl)

Set up a ring stand and a buret clamp. Label a clean, dry 400 mL beaker "NaOH."

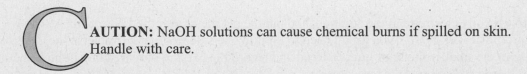

CAUTION: NaOH solutions can cause chemical burns if spilled on skin. Handle with care.

Obtain 80 to 100 mL of the standardized NaOH in this beaker, covering it with a watch glass. Record the exact concentration of the NaOH (from the label on the bottle).

Label a 600 mL beaker "Waste." Obtain a buret. Rinse the buret twice with 5 mL portions of the NaOH solution, draining the rinsings into the waste beaker. Place the buret in the buret clamp, and using a funnel, fill it to a point a little above the zero mark with the NaOH solution. Let the walls drain for about 30 seconds, and then carefully open the stopcock and drain the NaOH solution into the waste beaker until the meniscus is exactly at the zero mark. (Normally, it is a waste of time to exactly zero a buret, but when you prepare a titration curve you must know the exact amount of titrant added at each point in the titration. It would be unnecessarily tedious if you had to subtract an initial reading from each buret reading throughout the titration.)

CAUTION: HCl solutions can cause chemical burns if spilled on skin. Handle with care.

Label a 200 mL beaker "Acid." Obtain about 50 mL of HCl solution in this beaker. The HCl solution is approximately 0.1 *M.* Rinse a volumetric pipet first with distilled water and then with a small portion of the HCl solution, draining the rinsings into the waste beaker. Then carefully measure 10.00 mL of the HCl solution into a 100 mL beaker and add 5 drops of an appropriate acid-base indicator (record the one chosen). Position the beaker under the buret, placing a piece of white paper under the beaker to facilitate observation of the indicator color change (Figure 27.1).

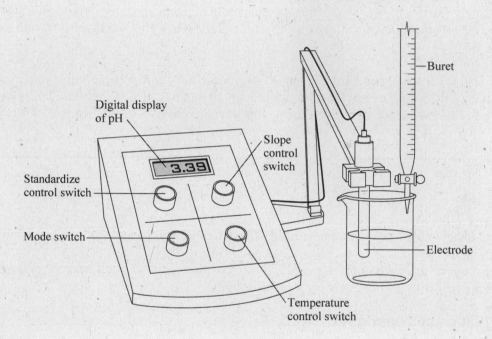

Figure 27.1
pH titration setup.

Immerse the pH electrode in the solution, and put the pH meter into the USE or MEASURE mode. Swirl the beaker gently (you might need to tilt it to one side in order to immerse the electrode sufficiently in the small volume of liquid). Read and record the pH, the buret reading (which should start at 0.00), and the color of the indicator.

Initially, add NaOH from the buret in approximately 1-mL increments. Do not attempt to add exactly 1.00 mL of NaOH; rather, add approximately 1 mL, but be sure to record the actual buret reading. Record the total volume added and the pH after each increment. Once the pH begins to rise, slow the addition of NaOH to 1 or 2 drops at a time. After each addition, swirl the solution gently and record the pH, the buret reading, and the color of the indicator (if there is any change). After the rate of pH increase slows, return to adding the NaOH in 1 mL increments for an additional 4 or 5 mL.

It is essential that you add the NaOH dropwise near the equivalence-point. If you don't, you won't have enough data points to make a good titration curve. If you find that you've added the NaOH too rapidly, repeat the titration.

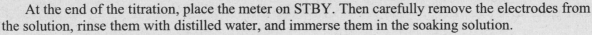

At the end of the titration, place the meter on STBY. Then carefully remove the electrodes from the solution, rinse them with distilled water, and immerse them in the soaking solution.

Place any leftover solutions in the waste beaker.

Part B: Titration of a Weak Acid (CH_3CO_2H)

Refill the buret with the standardized NaOH solution, and adjust the meniscus to 0.00 mL. Rinse the 100 and 200 mL beakers used in Part A thoroughly with distilled water. Dry the 200 mL beaker labeled "Acid."

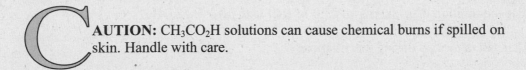

CAUTION: CH_3CO_2H solutions can cause chemical burns if spilled on skin. Handle with care.

Obtain about 50 mL of CH_3CO_2H solution in the beaker labeled "Acid." The CH_3CO_2H solution is about 0.1 M. Rinse the 10.00 mL pipet with distilled water, followed by a small portion of the CH_3CO_2H solution. Then pipet 10.00 mL of CH_3CO_2H solution into the 100 mL beaker. Add 5 drops of the appropriate indicator (record the one chosen). Titrate as in Part A, recording the pH, the buret reading, and the color of the indicator.

Again, at the end of the titration, place the meter on STBY. Then carefully remove the electrodes from the solution, rinse with distilled water, and immerse in the soaking solution.

Place any leftover solutions in the waste beaker.

Part C: Titration of Vinegar

Refill your buret with standard NaOH solution, and adjust the meniscus to 0.00 mL. Rinse the 100 and 200 mL beakers used in Part B thoroughly with distilled water. Dry the 200 mL beaker labeled "Acid." Obtain a 100.0 mL volumetric flask, and rinse it with distilled water.

Obtain about 50 mL of vinegar in the 200 mL beaker labeled "Acid." Record the percent acidity listed on the bottle of vinegar. Rinse the 10.00 mL pipet with distilled water, followed by a small portion of vinegar. Then pipet 10.00 mL of vinegar into the 100.0 mL volumetric flask. Fill the flask exactly to the mark with distilled water (use an eyedropper to add the final drops). Stopper the flask, and mix the contents thoroughly by repeated inversion and shaking. (*Note:* Vinegar must be diluted in this experiment in order to bring its acid concentration into a range convenient for titration.)

Titrate as in Part A, recording the pH, the buret reading, and the color of the indicator.

Again, at the end of the titration, place the meter on STBY. Then carefully remove the electrodes from the solution, rinse with distilled water, and immerse in the soaking solution.

Place any leftover solutions in the waste beaker.

Waste Disposal. Dispose of all waste solutions as directed by your instructor.

Name: _____   Date: _____

Lab Instructor: _____   Lab Section: _____

EXPERIMENT 27

pH Titrations of Strong and Weak Acids

PRELABORATORY QUESTIONS

1. Consider the titration of 10.00 mL of 0.10 M HCl with 0.10 M NaOH.

    (a) What salt is formed during this titration?

    (b) Do you expect the salt solution at the equivalence-point to be acidic, basic, or neutral? Explain.

    (c) From the list provided in Table 27.1, what would be the best indicator for this titration?

2. Consider the titration of 10.00 mL of 0.10 M acetic acid (CH_3CO_2H) with 0.10 M NaOH.

    (a) What salt is formed during this titration?

    (b) Do you expect the salt solution at equivalence to be acidic, basic, or neutral? Explain.

    (c) From the list provided in Table 27.1, what would be the best indicator for this titration?

3. What is the difference between an end-point and an equivalence-point?

Name: _____   Date: _____

Lab Instructor: _____   Lab Section: _____

EXPERIMENT 27

pH Titrations of Strong and Weak Acids

RESULTS/OBSERVATIONS

Part A: Titration of a Strong Acid (HCl)

1. Exact concentration of NaOH solution: _____

2. Volume of HCl solution used: _____

3. Acid-base indicator used: _____

4. Titration data:

mL of NaOH Added	pH	Color

Name: _____   Date: _____

Lab Instructor: _____   Lab Section: _____

Report data for a second HCl–NaOH titration, if necessary.

5. Volume of HCl solution used: _____

6. Acid-base indicator used: _____

7. Titration data:

mL of NaOH Added	pH	Color

Name: _____   Date: _____

Lab Instructor: _____   Lab Section: _____

8. Use the acid-base indicator color change to determine the molarity of the HCl solution.

9. Using the graph below, plot pH as a function of milliliters of NaOH added.

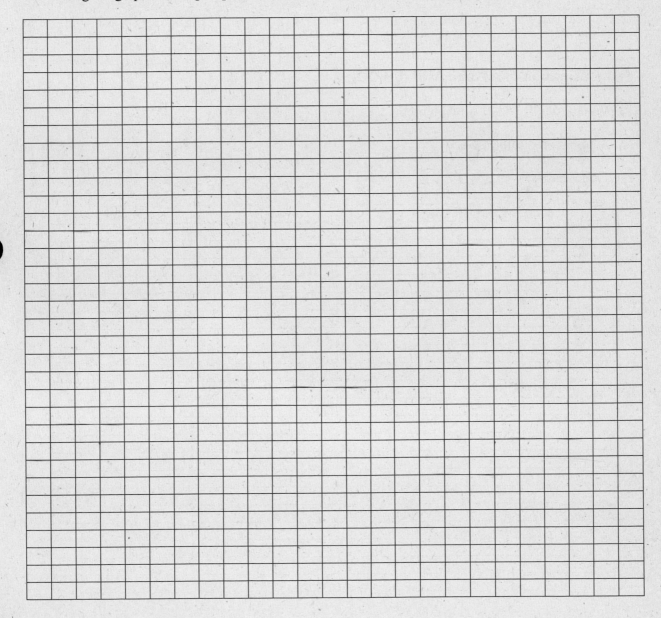

Name: _____   Date: _____

Lab Instructor: _____   Lab Section: _____

10. Use the pH titration curve to determine the molarity of the HCl solution.

11. Are the two determined HCl molarities the same? If not, how close are they?

Part B: Titration of a Weak Acid (CH_3CO_2H)

12. Volume of CH_3CO_2H solution used: _____

13. Acid-base indicator used: _____

14. Titration data:

mL of NaOH Added	pH	Color

Name: _____   Date: _____

Lab Instructor: _____   Lab Section: _____

Report data for a second CH_3CO_2H–NaOH titration, if necessary.

15. Volume of CH_3CO_2H solution used: _____

16. Acid-base indicator used: _____

17. Titration data:

mL of NaOH Added	pH	Color

18. Use the acid-base indicator color change to determine the molarity of the CH_3CO_2H solution.

Name: _____   Date: _____

Lab Instructor: _____   Lab Section: _____

19. Using the graph below, plot pH as a function of milliliters of NaOH added.

Name: _____   Date: _____

Lab Instructor: _____   Lab Section: _____

20. Use the pH titration curve to determine the molarity of the CH_3CO_2H solution.

21. Are the two determined CH_3CO_2H molarities the same? If not, how close are they?

Part C: Titration of Vinegar

22. The percent acidity listed on the bottle of vinegar: _____

23. Volume of undiluted vinegar: _____

24. Final volume the vinegar was diluted to: _____

25. Volume of diluted vinegar used: _____

26. Acid-base indicator used: _____

27. Titration data:

mL of NaOH Added	pH	Color

Name: _____   Date: _____

Lab Instructor: _____   Lab Section: _____

Report data for a second vinegar–NaOH titration, if necessary.

28. Volume of diluted vinegar used: _____

29. Acid-base indicator used: _____

30. Titration data:

mL of NaOH Added	pH	Color

31. Use the acid-base indicator color change to determine the molarity of the CH_3CO_2H in the *undiluted* vinegar solution.

Name: _____ Date: _____

Lab Instructor: _____ Lab Section: _____

32. Using the graph below, plot pH as a function of milliliters of NaOH added.

Name: _____   Date: _____

Lab Instructor: _____   Lab Section: _____

33. Use the pH titration curve to determine the molarity of the CH_3CO_2H in the *undiluted* vinegar solution.

34. Are the two determined CH_3CO_2H molarities in the *undiluted* vinegar solution the same? If not, how close are they?

35. Calculate the percent acid in undiluted vinegar solution (weight/vol %, that is, weight of acid per 100 mL of vinegar):

36. Calculate the percent error in the percent acid determination, assuming the value listed on the bottle to be the accepted value.

Name: _____   Date: _____

Lab Instructor: _____   Lab Section: _____

EXPERIMENT 27

pH Titrations of Strong and Weak Acids

POSTLABORATORY QUESTIONS

1. Calculate the pH value before beginning the titration of 10.00 mL of 0.10 M HCl. Do the same for 10.00 mL of 0.10 M CH_3CO_2H ($K_a = 1.8 \times 10^{-5}$).

2. Calculate the expected equivalence-point pH values for the preceding HCl and CH_3CO_2H titrations when 0.10 M NaOH is used as the base.

Name: _____   Date: _____

Lab Instructor: _____   Lab Section: _____

3. How did your actual starting and equivalence-points compare with those you calculated in questions 1 and 2 above? What might cause any observed differences?

4. Did either HCl or acetic acid have a buffer region (region of stable pH when base was added) in the titration? If so, state which one(s) did and explain why.

5. Compare the titration curves for HCl and CH_3CO_2H. How are they different? How are they similar? Explain why these differences and similarities occur. In particular, discuss the pH at the following key points in the graph: the starting point, the initial addition of NaOH, halfway to the equivalence-point, the equivalence-point, and beyond the equivalence-point.

6. What is the $[H_3O^+]$ in exactly 0.10 M HCl? In exactly 0.10 M CH_3CO_2H? Given equal volumes of each acid, why does it require the same amount of NaOH to reach the equivalence-point?

EXPERIMENT 28

Identification of an Unknown Weak Acid

OBJECTIVE

To determine the identity of an unknown monoprotic weak acid by performing two titration experiments, one using an acid-base indicator and the other using a pH meter. One of the titration experiments should yield the molar mass of the unknown weak acid; the other should determine its pK_a.

INTRODUCTION

A *titration* is a technique for determining the amount of *analyte* (the substance being analyzed) in a sample. In a titration experiment, increments of the *titrant* (a reagent solution of known concentration) are added to the analyte until their reaction is complete. The titrant solution usually is added to the analyte by means of a buret, a calibrated glass dispensing tube. The volume of titrant necessary to complete the reaction and its known concentration are used to determine the moles of titrant consumed in the reaction. The moles of titrant consumed in the reaction are stoichiometrically related to the moles of analyte present initially. The procedure is outlined in the flowchart below:

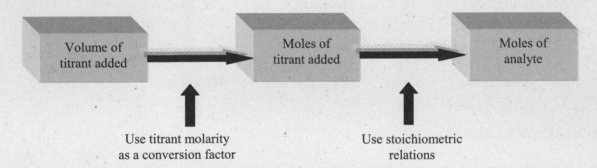

While titrations can be based on any type of chemical reaction, this experiment will be concerned with an acid-base titration. The titrant in this experiment will be a sodium hydroxide (NaOH) solution, and the analyte will be an unknown weak acid.

The progress of an acid-base titration is monitored by either an acid-base indicator or a pH meter. An acid-base indicator is a substance whose color depends on the pH of the solution to which it is added. An acid-base indicator is chosen so that it changes color at the pH of the titration equivalence-point. A pH meter is used to measure analyte solution pH as a function of added titrant, data that usually are expressed as a pH titration curve. The equivalence-point is marked by a sudden change in pH. Each method has its advantages: An acid-base indicator is easier to use. A pH meter allows for the collection of more data. Both types of acid-base titrations will be performed in this experiment.

Your purpose in this experiment is to determine the identity of an unknown monoprotic weak acid. You are to accomplish this by performing two titration experiments, one using an acid-base indicator and the other using a pH meter. One of your titration experiments should yield the molar mass of the unknown weak acid; the other should determine its pK_a. Use the determined molar mass and pK_a to identify your unknown acid from the list of possibilities in Table 28.1.

Table 28.1: Possible Unknown Solid Acids

Name	Formula	Molar Mass (g/mol)	pK_a
Benzoic acid	$C_7H_6O_2$	122.12	4.204
Chloroacetic acid	$C_2H_3ClO_2$	94.50	2.867
p-Chlorobenzoic acid	$C_7H_5ClO_2$	156.57	3.986
Diphenylacetic acid	$C_{14}H_{12}O_2$	212.25	3.939
2-Furoic acid	$C_5H_4O_3$	112.08	3.164
Glycolic acid	$C_2H_4O_3$	76.05	3.831
Hydrocinnamic acid	$C_9H_{10}O_2$	150.18	4.664
Iodoacetic acid	$C_2H_3IO_2$	185.95	3.175
L-Lactic acid	$C_3H_6O_3$	90.08	3.858
DL-Mandelic acid	$C_8H_8O_3$	152.15	3.37
Potassium bitartrate	$C_4H_5KO_6$	188.18	4.36
Potassium hydrogen phthalate	$C_8H_5KO_4$	204.23	5.14
Suberic acid	$C_8H_{14}O_4$	174.20	4.512
Sulfanilic acid	$C_6H_7NO_3S$	173.19	3.227
o-Toluic acid	$C_8H_8O_2$	136.15	3.90
Trimethylacetic acid	$C_5H_{10}O_2$	102.13	5.031

ADDITIONAL READING

Read the sections in the Laboratory Techniques chapter at the beginning of this Lab Manual on cleaning glassware, handling chemicals, weighing, heating liquids and solutions, pH measurements, volumetric flasks, pipets, and burets prior to performing this experiment.

Safety Precautions:

■ Protective eyewear approved by your institution must be worn at all times while you are in the laboratory.

■ Use gloves while handling the solid acids. Several of them are quite caustic in their solid form. If a solid acid comes in contact with your skin, wash with water for 15 minutes. Have a classmate notify your instructor.

■ Sodium hydroxide is corrosive, especially to the eyes. Thoroughly wash off any sodium hydroxide spilled on your skin. If you get it into your eyes, immediately flush your eyes with water for at least 10 minutes. If you spill it on your clothing, remove the affected clothing before rinsing with water. Notify your lab instructor, who will assist you.

■ Hydrochloric acid is corrosive and must be kept off skin and clothing. If you get it into your eyes, immediately flush your eyes with water for at least 10 minutes. If you spill it on your clothing, remove the affected clothing before rinsing with water. Notify your lab instructor, who will assist you.

EXPERIMENT

You are to design and perform two titration experiments in order to determine the identity of an unknown monoprotic weak acid. You should base your identification on measurements of the molar mass and pK_a of your unknown solid acid. Design your experiment around the following restrictions:

1. One of your titration experiments, either to determine the acid molar mass or its pK_a, must use one of the three available indicator dyes (but not a pH meter). The other must use a pH meter (but not an indicator dye).

2. You must test your procedures on a known weak acid (benzoic acid: molar mass of 122 g/mol and pK_a of 4.204) prior to applying them to your unknown.

3. In order to achieve acceptable precision, you must perform at least 3 trials of each titration procedure, a minimum of 12 total titrations: 3 for the molar mass determination of benzoic acid, 3 for the pK_a determination of benzoic acid, 3 for the molar mass determination of the unknown acid, and 3 for the pK_a determination of the unknown acid.

4. You will receive only about 10 g of the unknown weak acid. It will be a pure solid.

5. Some unknown acids are more soluble than others; it might be necessary to heat your solution gently to get the desired amount of acid into solution.

Equipment and Reagents

To perform this experiment, you will have access to all the equipment in your lab drawer as well as the following items:

Standardized base solution: ~0.10 M NaOH (exact concentration listed on label)

Acid-base indicators: methyl orange, methyl yellow, and phenolphthalein (see Table 28.2)

pH meter

100 mL volumetric flasks

50 mL burets

25 mL pipets

Table 28.2

Indicator	pH Range of Color Change	Color Change
Methyl orange	3.2–4.4	Red to yellow
Methyl yellow	1.2–2.4	Red to yellow
Phenolphthalein	8.0–10.0	Colorless to pink

Waste Disposal. Dispose of all chemical waste as directed by your instructor.

Name: _____ Date: _____

Lab Instructor: _____ Lab Section: _____

EXPERIMENT 28

Identification of an Unknown Weak Acid

PRELABORATORY QUESTIONS

1. A benzoic acid solution was prepared by adding 0.488 g benzoic acid to a 100 mL volumetric flask and then adding enough water to bring the total solution volume to 100.00 mL. How many milliliters of 0.0989 M NaOH are required to titrate the entire benzoic acid solution to the equivalence-point? Will the solution be somewhat acidic, neutral, or basic at the equivalence-point?

2. Which of the acid-base indicators available in this experiment (i.e., methyl orange, methyl yellow, or phenolphthalein) should be used when performing the titration in question 1? Explain.

3. If the experiment described in question 1 were repeated using a 50 mL volumetric flask instead of a 100 mL volumetric flask (but the same mass of benzoic acid), how many milliliters of 0.0989 M NaOH would be required to titrate the entire benzoic acid solution to the equivalence-point?

Name: _____   Date: _____

Lab Instructor: _____   Lab Section: _____

4. A benzoic acid solution was prepared as in question 1 (0.488 g benzoic acid dissolved in water to reach a final solution volume of 100.00 mL). By pipet, a 25.00 mL portion of this solution is transferred to an Erlenmeyer flask. How many milliliters of 0.0989 M NaOH are required to titrate the 25.00 mL of benzoic acid solution in the flask to the equivalence-point?

5. The amount of titrant to be used in each titration is determined by the desires to maximize experimental precision and not exceed the capacity of the buret. Only a 50 mL buret is available to you, so significantly less than 50.00 mL of NaOH should be used. Maximizing precision means using at least 10.00 mL; this will give the volume measurement four rather than three significant digits. Given this, what masses of benzoic acid (to be dissolved in water) would make appropriate sample sizes in a titration where 0.10 M NaOH is the titrant?

6. Look up the MSDS (**http://msds.ehs.cornell.edu/**) of benzoic acid, and list any hazards.

Name: _____   Date: _____

Lab Instructor: _____   Lab Section: _____

EXPERIMENT 28

Identification of an Unknown Weak Acid

RESULTS/OBSERVATIONS

1. Unknown acid identification number: _____

    Indicator titration experiment:

2. Name of acid-base indicator used: _____

3. Benzoic acid titrations using the acid-base indicator. Clearly label your data entries in the following table:

Trial Number					
1					
2					
3					

4. Unknown acid titrations using the acid-base indicator. Clearly label your data entries in the following table:

Trial Number					
1					
2					
3					

Name: _____   Date: _____

Lab Instructor: _____   Lab Section: _____

pH meter titration experiment:

5. Benzoic acid titrations using the pH meter:

Trial 1		Trial 2		Trial 3	
Titrant Added (mL)	**pH**	**Titrant Added (mL)**	**pH**	**Titrant Added (mL)**	**pH**

Name: _____   Date: _____

Lab Instructor: _____   Lab Section: _____

6. Unknown acid titrations using the pH meter:

Trial 1		Trial 2		Trial 3	
Titrant Added (mL)	**pH**	**Titrant Added (mL)**	**pH**	**Titrant Added (mL)**	**pH**

7. Write any other pertinent data in the space below. Clearly indicate both the property and the amount.

Name: _____   Date: _____

Lab Instructor: _____   Lab Section: _____

8. Calculation of the molar mass of benzoic acid using the data in trial number _____ (choose a trial):

9. Molar mass of benzoic acid: _____ (trial 1), _____ (trial 2),

   _____ (trial 3)

10. Mean molar mass of benzoic: _____ (average of three trials)

11. Calculation of the molar mass of the unknown acid using the data in trial number _____ (choose a trial):

12. Molar mass of the unknown acid: _____ (trial 1), _____ (trial 2),

    _____ (trial 3)

13. Mean molar mass of the unknown acid: _____ (average of three trials)

Name: _____   Date: _____

Lab Instructor: _____   Lab Section: _____

14. Calculation of the pK_a of benzoic acid using the data in trial number _____ (choose a trial):

15. pK_a of the benzoic acid: _____ (trial 1), _____ (trial 2),

_____ (trial 3)

16. Mean pK_a of the benzoic acid: _____ (average of three trials)

17. Calculation of the pK_a of the unknown acid using the data in trial number _____ (choose a trial):

18. pK_a of the unknown acid: _____ (trial 1), _____ (trial 2),

_____ (trial 3)

19. Mean pK_a of the unknown acid: _____ (average of three trials)

20. The unknown acid is: _____.

Name: _____   Date: _____

Lab Instructor: _____   Lab Section: _____

Plot the pH titration curves on the graph paper provided.

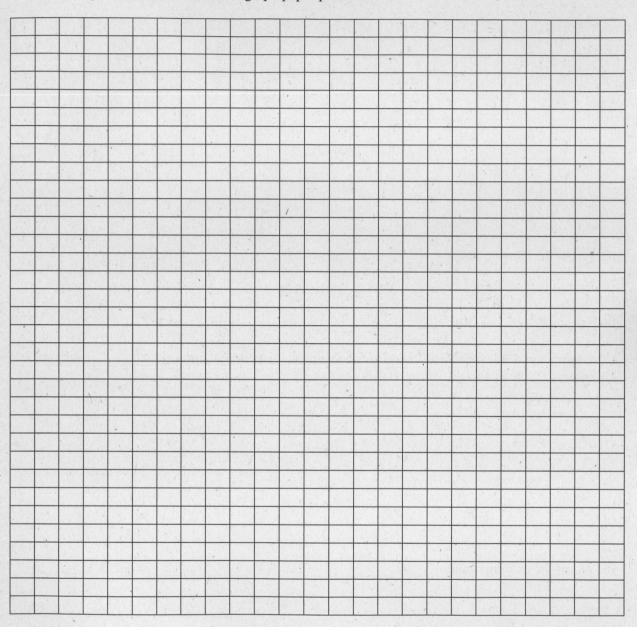

Name: _____   Date: _____

Lab Instructor: _____   Lab Section: _____

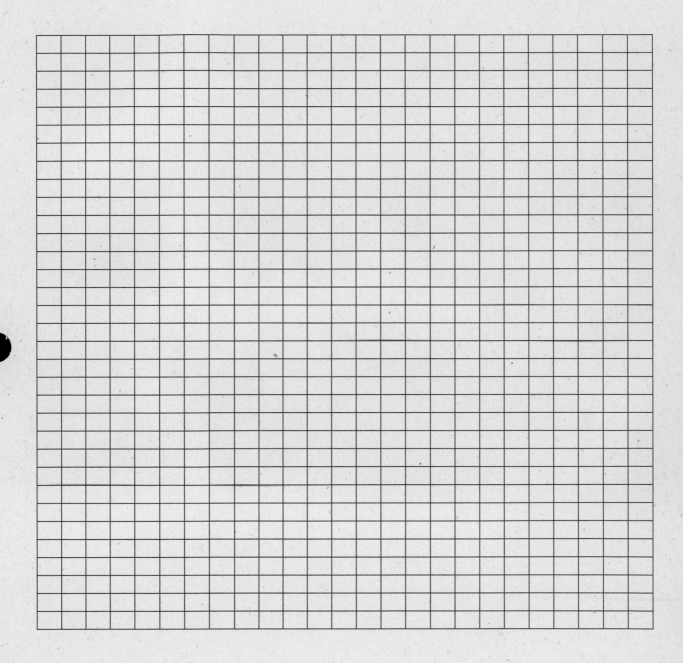

Name: _____   Date: _____

Lab Instructor: _____   Lab Section: _____

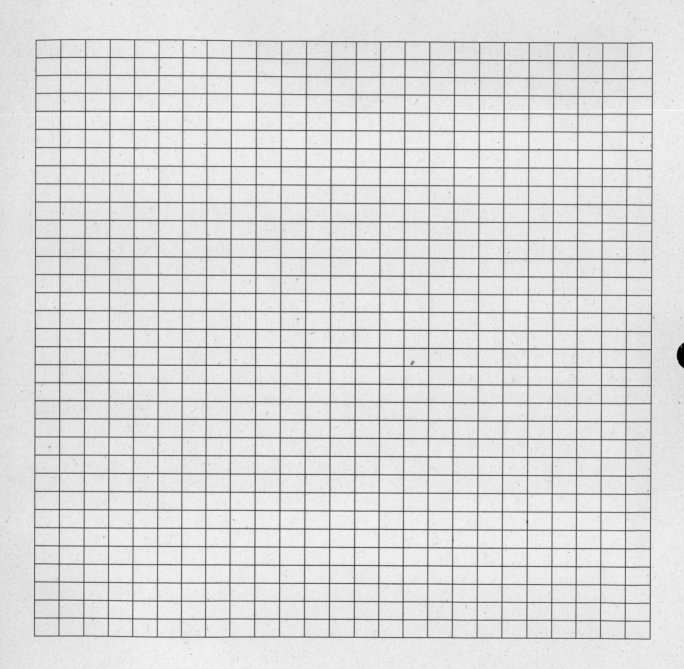

Name: _____   Date: _____

Lab Instructor: _____   Lab Section: _____

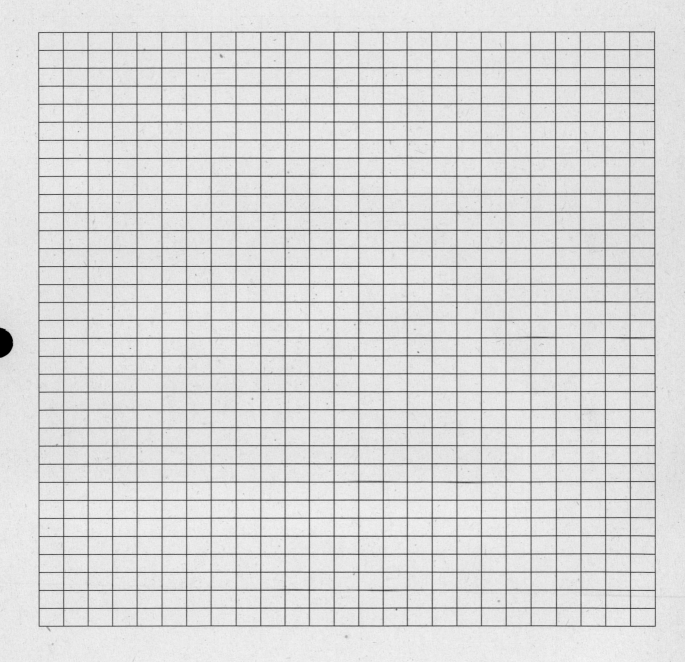

Name: _____   Date: _____

Lab Instructor: _____   Lab Section: _____

Name: _____   Date: _____

Lab Instructor: _____   Lab Section: _____

Name: _____   Date: _____

Lab Instructor: _____   Lab Section: _____

EXPERIMENT 28

Identification of an Unknown Weak Acid

POSTLABORATORY QUESTIONS

1. What are the advantages of performing an acid-base titration using an indicator? What are the advantages of using a pH meter?

2. Why is it necessary to perform each type of titration during this experiment in triplicate?

3. A student performed this experiment with the following results:

Trial Number	Molar Mass (g/mol)	pK_a
1	186.41	4.33
2	184.35	4.38
3	188.37	4.39

What is the identity of this student's unknown weak acid?

Preparation of a Buffer Solution

OBJECTIVE

To prepare approximately 400 mL of a buffer solution with a certain pH. The buffer must have the capacity to absorb 20 mL of either 0.02 M HCl or 0.02 M NaOH and undergo a pH change of no more than ± 0.1.

INTRODUCTION

A buffer solution is composed of a mixture of a weak acid and its conjugate base or a weak base and its conjugate acid. Buffer solutions are so-called because they resist drastic changes in pH; a small amount of acid or base added to a buffer solution has almost no effect on the pH; larger additions of acid or base have only small effects. Buffers are especially important in biologic systems. Blood is the most important practical example of a biologic buffer. Human blood has a pH near 7.4, a value maintained primarily by the buffering ability of the carbonic acid-bicarbonate (H_2CO_3–HCO_3^-) weak acid-conjugate base pair. A blood pH below 7.0 or above 7.8 quickly results in death.

To illustrate the resistance to pH change characteristic of a buffer solution, the effect of adding 0.010 mol NaOH to 1.00 L of 0.10 M $C_6H_5CO_2H$ (an unbuffered solution) will be compared with the addition of an equivalent amount of NaOH to 1.00 L of a solution that is 0.10 M in both $C_6H_5CO_2H$ and $C_6H_5CO_2Na$ (a buffer solution).

Unbuffered Solution

The 0.10 M $C_6H_5CO_2H$ solution has an initial pH of 2.60. Addition of 0.010 mol NaOH raises the pH by consuming some of the acid:

$$OH^-(aq) + C_6H_5CO_2H(aq) \rightarrow C_6H_5CO_2^-(aq) + H_2O(l) \tag{1}$$

As indicated, the reaction between a strong base and a weak acid goes essentially to completion. All the OH^- added will be consumed, and the $[C_6H_5CO_2H]$ will be lowered. Since the stoichiometry of Reaction (1) is 1:1, the amounts of $C_6H_5CO_2H$ consumed and $C_6H_5CO_2^-$ produced are 0.010 mol. The concentrations of $C_6H_5CO_2H$ and $C_6H_5CO_2^-$ after Reaction (1) are

$$[C_6H_5CO_2H] = \frac{0.10 - 0.010 \text{ mol}}{1.00 \text{ L}} = 0.090 \ M \text{ and } \left[C_6H_5CO_2^-\right] = \frac{0.010 \text{ mol}}{1.00 \text{ L}} = 0.010 \ M \tag{2}$$

With these concentrations as the initial conditions, the equilibrium concentration of H_3O^+ and the pH of the solution can be calculated:

$$C_6H_5CO_2H(aq) + H_2O(l) \rightleftharpoons H_3O^+(aq) + C_6H_5CO_2^-(aq)$$

Initial conc. (M)	0.090	0	0.010
Conc. change (M)	$-x$	$+x$	$+x$
Equilibrium conc. (M)	$0.090 - x$	x	$x + 0.010$

$$K_a = 6.5 \times 10^{-5} = \frac{\left[H_3O^+\right]\left[C_6H_5CO_2^-\right]}{\left[C_6H_5CO_2H\right]} = \frac{(x)(x+0.010)}{0.090-x} \approx \frac{x(0.010)}{0.090} \tag{3}$$

$$x = [H_3O^+] = 5.85 \; x \; 10^{-4} \; M \tag{4}$$

$$pH = -\log\left[H_3O^+\right] = -\log\left(5.85 \times 10^{-4}\right) = 3.23 \tag{5}$$

This is a change in pH of 0.63

Buffered Solution

The solution that is 0.10 M in both $C_6H_5CO_2H$ and $C_6H_5CO_2Na$ has an initial pH of 4.19. Addition of 0.010 mol NaOH raises the pH by consuming some of the acid [Reaction (1)]. This changes the concentrations of $C_6H_5CO_2H$ and $C_6H_5CO_2^-$:

$$\left[C_6H_5CO_2H\right] = \frac{0.10 - 0.010 \; \text{mol}}{1.00 \; L} = 0.090 \; M \; \text{ and } \; \left[C_6H_5CO_2^-\right] = \frac{0.10 + 0.010 \; \text{mol}}{1.00 \; L} = 0.11 \; M \tag{6}$$

Using these concentrations as initial conditions, the pH of the solution can be determined:

$$C_6H_5CO_2H(aq) + H_2O(l) \rightleftharpoons H_3O^+(aq) + C_6H_5CO_2^-(aq)$$

Initial conc. (M)	0.090	0	0.11
Conc. change (M)	$-x$	$+x$	$+x$
Equilibrium conc. (M)	$0.090 - x$	x	$x + 0.11$

$$K_a = 6.5 \times 10^{-5} = \frac{\left[H_3O^+\right]\left[C_6H_5CO_2^-\right]}{\left[C_6H_5CO_2H\right]} = \frac{(x)(x+0.11)}{0.090-x} \approx \frac{x(0.11)}{0.090} \tag{7}$$

$$x = [H_3O^+] = 5.32 \times 10^{-5} \; M \tag{8}$$

$$pH = -\log\left[H_3O^+\right] = -\log\left(5.32 \times 10^{-5}\right) = 4.27 \tag{9}$$

This is a change in pH of 0.08.

The pH change of the buffered solution (0.08) is smaller than that of the unbuffered solution (0.63) for addition of the same amount of NaOH.

As can be seen from Equations (3) and (7), the [H$_3$O$^+$], and hence the pH, of a buffer solution depends on the value of K_a and the ratio of [weak acid] to [conjugate base]. It is useful to develop a general equation for use in calculating buffer solution pH. The primary reaction describing a buffer is

$$HA(aq) + H_2O(l) \rightleftharpoons H_3O^+(aq) + A^-(aq) \tag{10}$$

HA is a generic weak acid, and A$^-$ is its conjugate base. The equilibrium concentrations of the species involved in this reaction are determined in the following manner:

	HA(aq) + H$_2$O(l) $\rightleftharpoons$	H$_3$O$^+$(aq) +	A$^-$(aq)
Initial conc. (M)	[HA]	0	[A$^-$]
Conc. change (M)	$-x$	$+x$	$+x$
Equilibrium conc. (M)	[HA] $- x$	x	[A$^-$] $+ x$

Substituting these values into the equilibrium expression, we get

$$K_a = \frac{x([A^-]+x)}{[HA]-x} \approx \frac{x[A^-]}{[HA]} \tag{11}$$

Recognizing that x is the equilibrium concentration of H$_3$O$^+$ and rearranging Equation (11) yields:

$$[H_3O^+] = K_a \frac{[HA]}{[A^-]} \tag{12}$$

Taking the negative logarithm of both sides of Equation (12) and simplifying, we obtain

$$-\log[H_3O^+] = -\log K_a - \log \frac{[HA]}{[A^-]} \tag{13}$$

$$pH = pK_a + \log \frac{[A^-]}{[HA]} \tag{14}$$

This is the Henderson–Hasselbalch equation. The Henderson–Hasselbalch equation shows that the pH of a buffer is quite close to the pK_a of the weak acid, differing from it only by the logarithm of the ratio of the conjugate base to weak acid concentration.

The capacity of a buffer solution is a measure of the amount of acid or base the solution can absorb without a significant change in pH. Buffer capacity depends on the number of moles of weak acid and conjugate base present. For equal concentrations of buffer solution, greater volume leads to greater buffer capacity. For equal volumes of buffer solution, greater concentration leads to greater buffer capacity. To illustrate this latter point, consider the addition of 0.010 mol NaOH to 1.00 L of a buffer solution that is 1.0 M in both C$_6$H$_5$CO$_2$H and

$C_6H_5CO_2Na$. This solution has an initial pH of 4.19, the same initial pH as the buffer solution that was 0.10 M in both $C_6H_5CO_2H$ and $C_6H_5CO_2Na$. Using the Henderson–Hasselbalch equation, the pH of this solution after addition of 0.010 mol NaOH is

$$pH = 4.19 + \log \frac{\left(\dfrac{1.0 + 0.010 \text{ mol}}{1.00 \text{ L}} \right)}{\left(\dfrac{1.0 - 0.010 \text{ mol}}{1.00 \text{ L}} \right)} = 4.20 \tag{15}$$

This is a change in pH of 0.01, a much smaller change than was observed for the equivalent volume of a buffer solution that was 0.10 M in both $C_6H_5CO_2H$ and $C_6H_5CO_2Na$.

Your purpose in this experiment is to prepare approximately 400 mL of a buffer solution with a certain pH. You will be given this pH by your instructor. The buffer must have the capacity to absorb 20 mL of either 0.02 M HCl or 0.02 M NaOH and undergo a pH change of no more than ±0.1. The extent to which you achieve your aim will be tested by another student in your laboratory section, just as you will evaluate another's buffer solution. The criteria used in the testing are given at the end of this experiment.

ADDITIONAL READING

Read the sections in the Laboratory Techniques chapter at the beginning of this Lab Manual on cleaning glassware, handling chemicals, and pH measurements prior to performing this experiment.

 Safety Precautions:

- Protective eyewear approved by your institution must be worn at all times while you are in the laboratory.

- If any of the reagents used in this experiment should spill on your skin, rinse the affected area with water for 15 minutes.

EXPERIMENT

Part I: Preparation of a Buffer Solution

You are to prepare approximately 400 mL of a buffer solution with a certain pH. You will be given this pH by your instructor. The buffer must have the capacity to absorb 20 mL of either 0.02 M HCl or 0.02 M NaOH and undergo a pH change of no more than ±0.1. When preparing your buffer solution, you are restricted to using the items in the Equipment and Reagents section.

Part II: Testing Another's Buffer Solution

After you have designed, prepared, and tested (and redesigned?) your buffer solution, submit it to another student for testing. The testing requires two trials. Provide the other student with two 400 mL samples of your buffer solution.

Just as another student will be testing your buffer solution, you will test the buffer solution of another student. Perform the testing by completing the Buffer Testing Evaluation Sheet.

Equipment and Reagents

To perform this experiment, you will have access to all the equipment in your lab drawer, a pH meter, and the electronic balances. The following solutions will be available:

$0.02\ M$ HCl
$0.02\ M$ NaOH
$0.01\ M$ CH_3CO_2H
$0.01\ M$ $NaCH_3CO_2$
$0.01\ M$ NH_3
$0.01\ M$ NH_4Cl
$0.01\ M$ $C_3H_4N_2$
$0.01\ M$ $C_3H_5N_2Cl$

The following solid reagents also will be available:

$NaCH_3CO_2$
NH_4Cl
$C_3H_5N_2Cl$

Table 29.1: Acid Dissociation Constants at 25°C

Acid	Formula	K_a
Acetic	CH_3CO_2H	1.8×10^{-5}

Base	Formula	K_b
Ammonia	NH_3	1.8×10^{-5}
Imidazole	$C_3H_4N_2$	9.0×10^{-8}

Waste Disposal: Dispose of all chemical waste as directed by your instructor.

Name: _____   Date: _____

Lab Instructor: _____   Lab Section: _____

EXPERIMENT 29

Preparation of a Buffer Solution

PRELABORATORY QUESTIONS

1. Which of the following pairs of compounds could be used to make a buffer solution? Circle those pairs of compounds that could be used to make a buffer solution.

   HCN, NaCN   KOH, KCl HNO_2, $LiNO_2$ HF, NaI   HI, NaI NH_2OH, NH_3OHCl

Substance	K_a (at 25°C)
HCN	4.9×10^{-10}
HNO_2	4.5×10^{-4}
HF	3.5×10^{-4}
NH_3OH^+	1.1×10^{-6}

2. What is the pH change observed on adding 0.010 mol HCl to 1.00 L of 0.10 M $C_6H_5CO_2H$ ($K_a = 6.5 \times 10^{-5}$)?

3. What is the pH change observed on adding 0.010 mol HCl to 1.00 L of a solution that is 0.10 M in both $C_6H_5CO_2H$ and $C_6H_5CO_2Na$? Compare your answer with that of question 2. If they differ from one another, explain why.

Name: _____   Date: _____

Lab Instructor: _____   Lab Section: _____

4. What is the pH change observed on adding 0.010 mol HCl to 1.00 L of a buffer solution that is 1.0 M in both $C_6H_5CO_2H$ and $C_6H_5CO_2Na$? Compare your answer with that of question 3. If they differ from one another, explain why.

5. What is the pH change observed on adding 0.010 mol HCl to 100. mL of a buffer solution that is 1.0 M in both $C_6H_5CO_2H$ and $C_6H_5CO_2Na$? Compare your answer with that of question 4. If they differ from one another, explain why.

Name: _____   Date: _____

Lab Instructor: _____   Lab Section: _____

EXPERIMENT 29

Preparation of a Buffer Solution

RESULTS/OBSERVATIONS

1. Ingredients used to make buffer solution and amount of each:

2. Calculation of the initial pH of the buffer solution:

3. Measured volume of buffer solution: _____ (trial 1)

4. Measured initial pH of buffer solution: _____ (trial 1)

5. Calculation of the pH of the buffer solution after addition of 20 mL of 0.02 M HCl:

Name: _____   Date: _____

Lab Instructor: _____   Lab Section: _____

6. Measured pH of buffer solution after addition of 20 mL of 0.02 M HCl: _____

7. Measured ΔpH of the buffer solution after addition of 20 mL of 0.02 M HCl: _____

8. Measured volume of buffer solution: _____ (trial 2)

9. Measured initial pH of buffer solution: _____ (trial 2)

10. Calculation of the pH of the buffer solution after addition of 20 mL of 0.02 M NaOH:

11. Measured pH of buffer solution after addition of 20 mL of 0.02 M NaOH: _____

12. Measured ΔpH of the buffer solution after addition of 20 mL of 0.02 M NaOH: _____

13. Write any other pertinent data in the space below. Clearly indicate both the property and the amount.

Name: _____   Date: _____

Lab Instructor: _____   Lab Section: _____

BUFFER TESTING EVALUATION SHEET

Name(s) of buffer designer(s): _____ _____

_____ _____

Name(s) of buffer tester(s): _____ _____

_____ _____

Buffer target pH: _____

	Trial 1	Trial 2
Initial buffer solution volume (mL)?		
Initial buffer pH?		
Buffer pH after adding 20 mL of 0.02 M NaOH?		
Buffer pH after adding 20 mL of 0.02 M HCl?		

Place any comments on the buffer solution or the procedure used in testing the buffer solution in this space:

Name: _____   Date: _____

Lab Instructor: _____   Lab Section: _____

Does the buffer solution you are testing do each of the following? (Check the appropriate box.)

	Yes	No
Does the buffer have the correct initial pH?		
Does the pH of the buffer change by ±0.1 or less on addition of 20 mL 0.02 M NaOH?		
Does the pH of the buffer change by ±0.1 or less on addition of 20 mL of 0.02 M HCl?		
Is the volume of the buffer solution 400 mL ± 20?		

Return this completed form to the student who designed this buffer after you have completed it. The designer will attach this form to his or her final report.

Name: _____   Date: _____

Lab Instructor: _____   Lab Section: _____

EXPERIMENT 29

Preparation of a Buffer Solution

POSTLABORATORY QUESTIONS

1. What factors affect the pH of a buffer solution?

2. What factors affect the capacity of a buffer solution?

Thermodynamics of Galvanic Cells

OBJECTIVE

To build a galvanic cell and measure the standard cell potential (E°_{cell}) as a function of temperature for the reaction between aqueous silver ions and Cu metal. These data will be used to calculate the equilibrium constant (K) and the standard Gibbs free-energy change (ΔG°) for the reaction at a variety of temperatures. The standard state enthalpy (ΔH°) and entropy (ΔS°) of the reaction also will be determined.

INTRODUCTION

Galvanic cells rely on a spontaneous chemical reaction to generate an electric current. Batteries are a practical application of the galvanic cell. In principle, it is possible to construct a galvanic cell around any spontaneous oxidation-reduction reaction. For example, immersing a strip of metallic zinc in a beaker containing a solution of lead(II) nitrate [$Pb(NO_3)_2$] will result in the conversion of Pb^{2+} to metallic lead and of metallic zinc to zinc(II). The net ionic equation for this spontaneous reaction is

$$Zn(s) + Pb^{2+}(aq) \rightarrow Zn^{2+}(aq) + Pb(s) \tag{1}$$

In this oxidation-reduction reaction, the Zn is oxidized to Zn^{2+}, and the Pb^{2+} is reduced to Pb. The net oxidation-reduction reaction can be expressed as a sum of two half-reactions, one describing the oxidation and the other describing the reduction:

(Oxidation half-reaction) $Zn(s) \rightarrow Zn^{2+}(aq) + 2\ e^-$ (2)

(Reduction half-reaction) $Pb^{2+}(aq) + 2\ e^- \rightarrow Pb(s)$ (3)

If the reaction is carried out in the manner just described, with the reactants in direct contact with one another, the electrons are transferred directly from Zn to Pb^{2+}, and the enthalpy of reaction is lost to the surroundings as heat. However, if the two half-reactions are physically separated from one another, such as in the manner shown in Figure 30.1, then some of the chemical energy released by the reaction can be extracted and converted to electrical energy, energy that can be used to power some electrically driven instrument. The galvanic cell for this reaction consists of two half-cells, a beaker containing a strip of Zn immersed in an aqueous solution of zinc(II) nitrate [$Zn(NO_3)_2$], and a second beaker in which a strip of Pb is immersed in an aqueous solution of $Pb(NO_3)_2$. The strips of Zn and Pb are termed *electrodes* and are connected by an electrically conducting wire. The two half-cells are also connected by a salt bridge, a U-shaped tube containing a gel permeated with a solution of sodium nitrate ($NaNO_3$), an inert electrolyte.

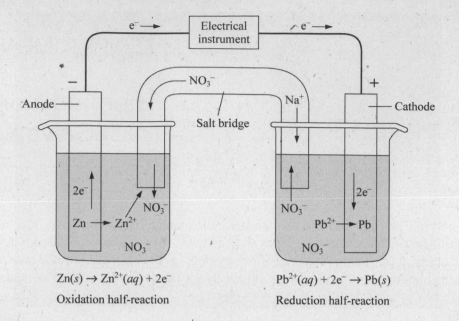

Figure 30.1
A schematic of a galvanic cell that uses oxidation of zinc metal to Zn^{2+}
and reduction of Pb^{2+} to lead metal.

Owing to the physical separation of the two half-reactions, the oxidation-reduction reaction between Zn and Pb^{2+} occurs in the following manner: A tiny amount of Zn from the anode dissolves in solution as Zn^{2+}; electrons are released; the electrons travel through the conducting wire to the lead electrode (electrons are not transferred through the solution because the metal wire is a much more effective conductor of electrons than is water; moreover, free electrons react rapidly with water and therefore are unstable in aqueous solution); electrons react with Pb^{2+} to form Pb at the surface of the lead electrode (cathode); the overall charge balance is maintained by the flow of inert ions through the salt bridge.

The electrodes of the galvanic cell are commonly labeled as positive and negative. Electrons are produced at the negative electrode, the anode, a result of the oxidation reaction that occurs at its surface, and are attracted to the positive electrode, the cathode, where they are consumed in the reduction reaction that occurs at its surface.

Electrons move through the external circuit from the Zn anode to the Pb cathode because they have lower energy when on Pb than on Zn. The driving force that pushes these electrons away from the anode and pulls them toward the cathode is an electrical potential called the *electromotive force* (EMF) or *cell potential* (E_{cell}). Cell potentials depend on both the specific nature of the chemical reactants present and their relative amounts. The former of these two contributions is related by the *standard cell potential* (E°_{cell}), the cell potential when both reactants and products are in their standard states. The standard cell potential of any galvanic cell is the sum of the standard half-cell potentials for the anode and cathode half-reactions:

$$E^{\circ}_{cell} = E^{\circ}_{anode} + E^{\circ}_{cathode} \tag{4}$$

The standard half-cell potentials of many substances, written as reduction potentials, are listed in Appendix C of this Lab Manual.

For the cell described in Figure 30.1, the standard cell potential is

(Anode reaction)	$Zn(s) \rightarrow Zn^{2+}(aq) + 2\ e^-$	E^o_{anode}	$=\ 0.76$ V
(Cathode reaction)	$Pb^{2+}(aq) + 2\ e^- \rightarrow Pb(s)$	$E^o_{cathode}$	$=-0.13$ V
(Net cell reaction)	$Zn(s) + Pb^{2+}(aq) \rightarrow Zn^{2+}(aq) + Pb(s)$	E^o_{cell}	$=\ 0.63$ V

Note that the standard cell potential of the anode half-reaction has a sign opposite that listed in Appendix C. This is because the half-reactions in Appendix C are all written as reductions. To write any of these as an oxidation requires reversing the reaction; when a reaction is reversed, the sign of its standard cell potential is reversed as well.

The dependence of the cell potential on the relative amounts of reactants and products present is expressed as a weighted reaction quotient:

$$\frac{-RT}{nF}\ln Q \tag{5}$$

where R is the universal gas constant (8.314 J/mol $\cdot$ K), T is the absolute temperature, n is the number of electrons transferred in the net cell reaction, F is Faraday's constant (96,485 C/mol of e^-), and Q is the reaction quotient. For the cell described in Figure 30.1, n is 2, and the reaction quotient has the form

$$Q = \frac{\left[Zn^{2+}\right]}{\left[Pb^{2+}\right]} \tag{6}$$

These two contributions to the cell potential are expressed jointly in the Nernst equation:

$$E_{cell} = E^o_{cell} - \frac{RT}{nF}\ln Q \tag{7}$$

As stated previously, the cell potential is the driving force that pushes electrons away from the anode and pulls them toward the cathode. The Gibbs free-energy change (ΔG) is a measure of the driving force of a chemical reaction. Given the similarity of their definitions, it is not surprising that the two are directly related to each other. The specific form of their relationship is given by the equation

$$\Delta G = -nFE_{cell} \tag{8}$$

When 1 M solutions are used, the standard cell potential E^o_{cell} is measured, and the standard Gibbs free-energy change ΔG^o is calculated,

$$\Delta G^o = -nFE^o_{cell} \tag{9}$$

Since the standard Gibbs free-energy is logarithmically related to the equilibrium constant K,

$$\Delta G^o = -RT\ln K \tag{10}$$

so too must be the standard cell potential,

$$E^o_{cell} = \frac{RT}{nF}\ln K \tag{11}$$

Experimentally, it is difficult to measure very small concentrations. Consequently, it is impractical to measure equilibrium constants directly when the equilibrium constant is either very large or very small. Standard cell potentials, however, are quite easy to measure. Many of the tabulated equilibrium constants for reactions with very large or very small equilibrium constants were determined indirectly through the measurement of standard cell potentials.

Once the equilibrium constant has been determined, it is also possible to calculate the standard enthalpy (ΔH°) and entropy (ΔS°) changes for a particular reaction. Recall that the standard Gibbs free-energy change is defined by the equation

$$\Delta G^\circ = \Delta H^\circ - T\Delta S^\circ \qquad (12)$$

Combining Equations (10) and (12) leads to

$$\ln K = -\frac{\Delta H^\circ}{RT} + \frac{\Delta S^\circ}{R} \qquad (13)$$

This equation has the form of a straight line of slope $\Delta H^\circ/R$ and y-intercept $\Delta S^\circ/R$.

In this experiment you will build a galvanic cell and measure the standard cell potential (E°_{cell}) as a function of temperature for the reaction

$$2\,Ag^+(aq) + Cu(s) \rightarrow 2\,Ag(s) + Cu^{2+}(aq) \qquad (14)$$

You will use these data to calculate the equilibrium constant (K) and the standard Gibbs free-energy change (ΔG°) for Reaction (14) at a variety of temperatures. You also will use these data to determine the standard state enthalpy (ΔH°) and entropy (ΔS°) of the reaction.

ADDITIONAL READING

Read the sections in the Laboratory Techniques chapter at the beginning of this Lab Manual on cleaning glassware, handling chemicals, heating liquids and solutions, and digital multimeters prior to performing this experiment.

 Safety Precautions:

- Protective eyewear approved by your institution must be worn at all times while you are in the laboratory.

- If any of the reagents used in this experiment should spill on your skin, rinse the affected area with water for 15 minutes.

PROCEDURE

A water-bath will be used to vary the temperature of the galvanic cell. Prepare the water-bath by filling a 600 mL beaker to which has been added a magnetic stirbar. Place the water-bath on a hotplate-stirrer. Insert a thermometer into a split rubber stopper. Use a ring stand and a clamp to support the thermometer in the water-bath. Heat the water-bath with stirring to about 75°C on a stirrer-hotplate (see Figure 30.2).

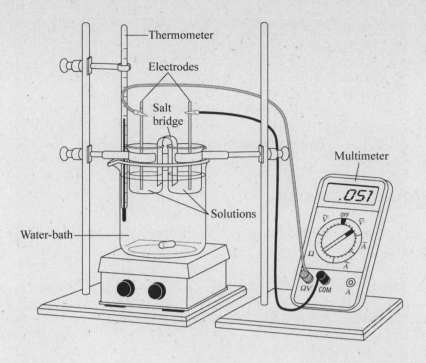

Figure 30.2
Setup for a galvanic cell in a water-bath.

Obtain 1.0 M Cu(NO$_3$)$_2$ and 1.0 M AgNO$_3$ solutions from your instructor. Pour about 20 mL of each solution into separate 30 mL beakers. Use a ringstand and clamp to support each beaker in the water-bath. The height of water in the water-bath should be above that of the solutions in the beakers but below the tops of these beakers. If necessary, add water to the water-bath.

Obtain the copper and silver electrodes (pieces of Cu and Ag wire) from your instructor. Briefly soak these electrodes in a small beaker containing dilute HNO$_3$ to remove surface oxides, which interfere with the experiment. After soaking, rinse the electrodes with distilled water, and place the copper electrode in the Cu(NO$_3$)$_2$ solution and the silver electrode in the AgNO$_3$ solution.

Prepare the salt bridge by soaking a strip cut from the center of a 15 cm piece of filter paper in 0.1 M KNO$_3$. Insert one end of the salt bridge in the Cu(NO$_3$)$_2$ solution and the other in the AgNO$_3$ solution. The salt bridge should not be in direct contact with the electrodes, or erroneous voltages will be measured.

Obtain a digital multimeter. Set the digital multimeter to read a direct current voltage. Connect one electrical lead to the copper electrode and the other to the silver electrode. If the measured voltage is negative, reverse the leads. (By convention, a galvanic cell has a positive cell potential.)

When the temperature in the water-bath reaches approximately 75°C, turn off the hotplate. Allow the temperature to cool to about 70°C. Record the temperature and standard cell potential ($E°_{cell}$). Continue to record the temperature and standard cell potential at 3°C increments until the temperature reaches about 40°C.

Disassemble the experimental setup. Rinse off the electrodes, and return them to your instructor.

Waste Disposal. Dispose of all chemical waste as directed by your instructor.

Name: _____   Date: _____

Lab Instructor: _____   Lab Section: _____

EXPERIMENT 30

Thermodynamics of Galvanic Cells

PRELABORATORY QUESTIONS

1. Consider the galvanic cell to be created in this experiment: 1 M AgNO$_3$, 1 M Cu(NO$_3$)$_2$, Ag and Cu wire for electrodes:

    $$2 \, Ag^+(aq) + Cu(s) \rightarrow 2 \, Ag(s) + Cu^{2+}(aq)$$

    (a) What is the standard cell potential (E°_{cell}) of this galvanic cell?

    (b) Sketch a picture of this galvanic cell, labeling the solutions (with concentrations), salt bridge, direction of electron flow, the electrodes, and the positive and negative terminals of the cell.

Name: _____   .   Date: _____

Lab Instructor: _____   Lab Section: _____

(c) Calculate the equilibrium constant (K) and the standard Gibbs free-energy change ($\Delta G°$) at 25°C for the reaction on which this galvanic cell is based.

(d) Use the standard thermodynamic functions below to calculate $\Delta H°$ and $\Delta S°$ for the reaction on which this galvanic cell is based.

Substance	$\Delta H°_f$ (kJ/mol)	$\Delta G°_f$ (kJ/mol)	$S°$ (J/mol · K)
Cu(s)	0	0	33.1
Cu^{2+}(aq)	64.8	65.5	−99.6
Ag(s)	0	0	42.6
Ag^+(aq)	105.6	77.1	72.7

Name: _____   Date: _____

Lab Instructor: _____   Lab Section: _____

EXPERIMENT 30

Thermodynamics of Galvanic Cells

RESULTS/OBSERVATIONS

1. Standard cell potential measurements as a function of temperature:

Temperature (°C)	Standard Cell Potential, E°_{cell} (V)

2. Calculation of the equilibrium constant K using the data collected in the first temperature-standard cell potential measurement:

3. Calculation of the standard Gibbs free-energy change (ΔG°) using the data collected in the first temperature-standard cell potential measurement:

Name: _____   Date: _____

Lab Instructor: _____   Lab Section: _____

4. Results of calculating the equilibrium constant and standard Gibbs free-energy change as a function of temperature:

Temperature (°C)	Standard Cell Potential, E^o_{cell} (V)	Equilibrium Constant K	Standard Gibbs Free-Energy Change, ΔG^o (kJ/mol)

5. Data necessary for plotting ln K versus $1/T$:

$1/T$ (K^{-1})	ln K

Name: _____   Date: _____

Lab Instructor: _____   Lab Section: _____

6. Plot of ln K as a function of $1/T$:

Name: _____   Date: _____

Lab Instructor: _____   Lab Section: _____

7. Calculation of standard state enthalpy ($\Delta H°$) of the reaction:

8. Percent error in the standard state enthalpy ($\Delta H°$) of the reaction determination. (The accepted value of $\Delta H°$ for the reaction studied in this experiment is −146.4 kJ/mol.)

9. Calculation of standard state entropy ($\Delta S°$) of the reaction:

10. Percent error in the standard state entropy ($\Delta S°$) of the reaction determination. (The accepted value of $\Delta S°$ for the reaction studied in this experiment is −192.9 J/K.)

Name: _____   Date: _____

Lab Instructor: _____   Lab Section: _____

EXPERIMENT 30

Thermodynamics of Galvanic Cells

POSTLABORATORY QUESTIONS

1. Consider a galvanic cell based on the reaction

$$2\,Ag^+(aq) + Pb(s) \rightarrow 2\,Ag(s) + Pb^{2+}(aq)$$

(a) What is the standard cell potential (E°_{cell}) of this galvanic cell?

(b) What are the equilibrium constant (K) and the standard Gibbs free-energy change (ΔG°) at 25°C for the reaction?

(c) What is the cell potential (E_{cell}) of this galvanic cell at 25°C if it is made using 0.10 M $AgNO_3$ and 0.10 $M\,Pb(NO_3)_2$? What is the Gibbs free-energy change (ΔG) at 25°C for the reaction?

Batteries

OBJECTIVE

In the first part of the experiment, the goal is to understand the effect changes in reactant and product concentrations have on the value of the cell potential. A galvanic cell will be constructed using $Cu(NO_3)2$ and $Pb(NO_3)2$ solutions, the concentrations of which will be altered and the resulting cell potentials measured. In the second part of the experiment, the objective is to construct a galvanic cell capable of producing a specific cell potential. The student is responsible for devising and implementing the experimental procedure necessary to construct the galvanic cell in this second part of the experiment.

INTRODUCTION

Batteries are one of the most prevalent forms of energy storage used by our modern society. Their applications in this regard are manifold: They provide the current to start automobiles and to power flashlights, laptop computers, and pacemakers, to list but a few. Batteries are an example of an electrochemical cell, a device that allows for the interconversion of chemical and electrical energy.

Electrochemical cells are divided into two types: galvanic cells and electrolytic cells. Galvanic cells rely on a spontaneous chemical reaction to generate an electric current. Batteries are a practical application of the galvanic cell. Conversely, in electrolytic cells, an electric current drives a nonspontaneous chemical reaction; the production of aluminum metal by the electrolysis of a molten mixture containing aluminum oxide is a particularly important example of an electrolytic process.

In principle, it is possible to construct a galvanic cell around any spontaneous oxidation-reduction reaction. For example, immersing a strip of metallic zinc in a beaker containing a solution of copper(II) nitrate $[Cu(NO_3)_2]$ will result in the conversion of Cu^{2+} to metallic copper and of metallic zinc to zinc(II). The net ionic equation for this spontaneous reaction is

$$Zn(s) + Cu^{2+}(aq) \rightarrow Zn^{2+}(aq) + Cu(s) \tag{1}$$

In this oxidation-reduction reaction, the Zn is oxidized to Zn^{2+}, and the Cu^{2+} is reduced to Cu. The net oxidation-reduction reaction can be expressed as a sum of two half-reactions, one describing the oxidation and the other describing the reduction:

$$\text{(Oxidation half-reaction)} \qquad Zn(s) \rightarrow Zn^{2+}(aq) + 2\text{ e}^- \tag{2}$$

$$\text{(Reduction half-reaction)} \qquad Cu^{2+}(aq) + 2\text{ e}^- \rightarrow Cu(s) \tag{3}$$

If the reaction is carried out in the manner just described, with the reactants in direct contact with one another, the electrons are transferred directly from Zn to Cu^{2+}, and the enthalpy of reaction is lost to the surroundings as heat. However, if the two half-reactions are physically separated from one another, such as in the manner shown in Figure 31.1, then some of the chemical energy released by the reaction can be extracted and converted to electrical energy, energy that can be used to power some electrically driven instrument. The galvanic cell for this reaction consists of two half-cells, a beaker containing a strip of Zn immersed in an aqueous solution of zinc(II) nitrate [$Zn(NO_3)_2$] and a second beaker in which a strip of Cu is immersed in an aqueous solution of $Cu(NO_3)_2$. The strips of Zn and Cu are termed *electrodes* and are connected by an electrically conducting wire. The two half-cells are also connected by a salt bridge, a U-shaped tube containing a gel permeated with a solution of an inert electrolyte, potassium nitrate (KNO_3) in this example.

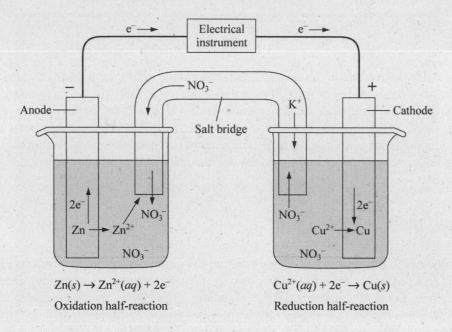

$$Zn(s) \rightarrow Zn^{2+}(aq) + 2e^-$$
Oxidation half-reaction

$$Cu^{2+}(aq) + 2e^- \rightarrow Cu(s)$$
Reduction half-reaction

Figure 31.1
A schematic of a galvanic cell that uses oxidation of zinc metal to Zn^{2+} and reduction of Cu^{2+} to copper metal.

Owing to the physical separation of the two half-reactions, the oxidation-reduction reaction between Zn and Cu^{2+} occurs in the following manner: A tiny amount of Zn dissolves in solution as Zn^{2+}, and electrons are released; the electrons travel through the conducting wire to the copper electrode (electrons are not transferred through solution because the metal wire is a much more effective conductor of electrons than is water; moreover, free electrons react rapidly with water and therefore are unstable in aqueous solution); electrons react with Cu^{2+} to form Cu at the surface of the Cu electrode; the overall charge balance is maintained by the flow of inert ions through the salt bridge.

The electrodes of the galvanic cell are commonly labeled as positive and negative. Electrons are produced at the negative electrode, the anode, a result of the oxidation reaction that occurs at its surface, and are attracted to the positive electrode, the cathode, where they are consumed in the reduction reaction that occurs at its surface.

A common shorthand notation is used to represent electrochemical cells. The shorthand notation for the electrochemical cell shown in Figure 31.1 is

$$Zn(s)|Zn^{2+}(aq)||Cu^{2+}(aq)|Cu(s)$$

In this notation, a vertical line (|) represents a phase boundary, such as that between the solid Zn electrode and the liquid Zn^{2+} solution. Two adjacent vertical lines (||) correspond to the presence of a salt bridge. The half-reaction occurring at the anode is written on the left. The cathode half-reaction is written on the right. The anode electrode is written on the extreme left. The cathode electrode is written on the extreme right. For each half-cell, reactants are written first, products second. Electrons move through the circuit from left to right, from anode to cathode. Unless equal to 1 M, solution concentrations are also included, just after the physical state, for example, $Cu^{2+}(aq, 0.10\ M)$.

Electrons move through the external circuit from the Zn anode to the Cu cathode because they have lower energy when on Cu than when on Zn. The driving force that pushes these electrons away from the anode and pulls them toward the cathode is an electrical potential, called the *electromotive force* (EMF) or *cell potential* (Ecell). Cell potentials depend on both the specific nature of the chemical reactants present and their relative amounts. The former of these two contributions is related by the standard cell potential ($E°$cell), the cell potential when both reactants and products are in their standard states. The standard cell potential of any galvanic cell is the sum of the standard half-cell potentials for the anode and cathode half-reactions:

$$E°_{cell} = E°_{anode} + E°_{cathode} \tag{4}$$

The standard half-cell potentials of many substances, written as reduction potentials, are listed in Appendix C of this Lab Manual.

For the cell described in Figure 31.1 the standard cell potential is

(Anode reaction)	$Zn(s) \rightarrow Zn^{2+}(aq) + 2\ e{-}$	$E°_{anode} = 0.76\ V$
(Cathode reaction)	$Cu^{2+}(aq) + 2\ e^- \rightarrow Cu(s)$	$E°_{cathode} = 0.34\ V$
(Net cell reaction)	$Zn(s) + Cu^{2+}(aq) \rightarrow Zn^{2+}(aq) + Cu(s)$	$E°_{cell} = 1.10\ V$

Note that the standard cell potential of the anode half-reaction has a sign opposite that listed in Appendix C. This is so because the half-reactions in Appendix C are all written as reductions. To write any of these as an oxidation requires reversing the reaction; when a reaction is reversed, the sign of its standard cell potential is reversed as well.

The dependence of the cell potential at 25°C on the relative amounts of reactants and products present is expressed as a weighted reaction quotient:

$$\frac{-0.0592\,V}{n} \log Q \tag{5}$$

where n is the number of electrons transferred in the net cell reaction, and Q is the reaction quotient. For the cell described in Figure 31.1, the weighted reaction quotient has the form

$$\frac{-0.0592V}{2} \log \frac{\left[Zn^{2+} \right]}{\left[Cu^{2+} \right]} \tag{6}$$

The contributions of these two factors to the cell potential are expressed jointly in the Nernst equation:

$$E_{cell} = E°_{cell} - \frac{0.0592\ V}{n} \log Q \tag{7}$$

This experiment involves two parts. In the first part of this experiment, you will construct a galvanic cell using $Cu(NO_3)_2$ and $Pb(NO_3)_2$ solutions. You will become familiar with the way reactant and product concentrations affect the cell potential by varying them according to the stated procedural instructions. You will compare your experimental results with theoretically predicted values, values that you have calculated previously in your prelaboratory assignment through the use of the standard reduction potentials and the Nernst equation. In the second part of this experiment, you will construct a galvanic cell capable of producing a specific cell potential. You will be responsible for devising the experimental procedure necessary to construct this galvanic cell.

ADDITIONAL READING

Read the sections in the Laboratory Techniques chapter at the beginning of this Lab Manual on cleaning glassware, handling chemicals, weighing, pipettes, volumetric flasks, and digital multimeters prior performing this experiment.

 Safety Precautions:

■ Protective eyewear approved by your institution must be worn at all times while you are in the laboratory.

■ If any of the reagents used in this experiment should spill on your skin, rinse the affected area with water for 15 minutes.

EXPERIMENT

Part A: General Galvanic Cell Characteristics

In this part of the experiment you will construct and analyze the general characteristics of a galvanic cell.

Obtain copper and lead electrodes from your instructor. These and all other electrodes used in the experiment should be cleaned by dipping them in a dilute acid solution prior to use. Be sure to rinse the electrodes with distilled water after cleaning in acid so as not to contaminate any half-cell solutions.

Using pipets, transfer 25.00 mL of each 0.0100 M $Pb(NO_3)_2$ and 0.0100 M $Cu(NO_3)_2$ into separate 100 mL beakers. Connect the solutions in the two beakers with a salt bridge. Create the salt bridge by cutting a strip out of the center of a piece of 15 cm filter paper and soaking it in a solution of 0.1 M KNO_3. Place a lead electrode in the $Pb(NO_3)_2$ solution and a copper electrode in the $Cu(NO_3)_2$ (see Figure 31.2).

Obtain a digital multimeter from your instructor. Connect the two electrodes to each other through the digital multimeter. Measure and record the cell potential. Is the result consistent with what you calculated in question 1(a) of the prelaboratory assignment? (*Note:* Slight experimental errors in cell potentials are inherent in this experiment. Ignore variations in measured cell potentials that are within ±10% of the theoretical value.)

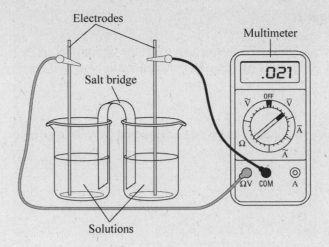

Figure 31.2
Galvanic cell setup.

By pipet, add 25.00 mL of distilled water to both the $Pb(NO_3)_2$ and $Cu(NO_3)_2$ solutions. Is the measured cell potential consistent with your answer to prelaboratory question 1(c)(i)?
Add approximately 0.1000 g of $Na_2SO_4(s)$ to the $Pb(NO_3)_2$ solution. Use glassine weighing paper, not a paper cup, when weighing the solid sample. Is the measured cell potential consistent with your answer to prelaboratory question 1(c)(ii)?

Add about 0.1000 g of $Cu(NO_3)_2 \cdot 5/2H_2O(s)$ to the $Cu(NO_3)_2$ solution. Use glassine weighing paper, not a paper cup, when weighing the solid sample. Is the measured cell potential consistent with your answer to prelaboratory question 1(c)(iii)?

Part B: Battery Design

In this part of the experiment your task is to construct a galvanic cell with a specific cell potential. You will learn the specific chemicals and equipment available to you and your target cell potential when you return the digital multimeter to your instructor after finishing Part A. Perform any necessary calculations, assemble the battery, and call over your lab instructor. Your lab instructor will attach a digital multimeter to your battery and measure the cell potential. If your battery produces the target cell potential (within ±10%), you have completed this part of the experiment successfully. If it does not, you will need to revise your calculations and procedure, assemble the revised battery, and again call over your lab instructor. Your lab instructor will again measure the cell potential with a digital multimeter. If the target cell potential is not achieved in this second trial, you will be allowed a third, and final, attempt at assembling a battery with the given target cell potential.
The information in Appendix C might be useful when performing this portion of the experiment.

Equipment and Reagents

To perform this experiment, you will have access to all the equipment in your lab drawer and those chemicals and pieces of equipment listed on the unknown slip you will receive from your instructor at the start of Part B.

Waste Disposal. Dispose of all chemical waste as directed by your instructor.

Name: _____   Date: _____

Lab Instructor: _____   Lab Section: _____

EXPERIMENT 31

Batteries

PRELABORATORY QUESTIONS

(*Note:* It will be useful to refer to your answers to these prelaboratory questions during the experiment. It is recommended that you photocopy your answers prior to coming to lab.)

1. In the first part of this experiment you will be constructing a battery from 25.00 mL each of 0.0100 M Pb(NO3)2 and 0.0100 M Cu(NO3)2, pieces of lead and copper wire, and a salt bridge.

    (a) What is the cell potential Ecell of this battery?

    (b) Sketch a picture of this battery, labeling the solutions (with concentrations), direction of electron flow, the electrodes, and the positive and negative terminals of the cell. Give the shorthand notation for this electrochemical cell.

Name: _____   Date: _____

Lab Instructor: _____   Lab Section: _____

(c) Assume that the following steps are performed sequentially on the battery of part (b):

   (i) First, 25.00 mL of water is added to both half-cell solutions, Pb(NO3)2 and Cu(NO3)2. Does the cell potential of the operating battery increase, decrease, or remain the same? If the cell potential does change, calculate its value.

   (ii) Next, 0.1000 g of Na2SO4(s) is added to the half-cell containing the Pb(NO3)2 solution. Will the cell potential of the operating battery increase, decrease, or remain the same? (*Hint:* PbSO4(s) $\rightleftharpoons$ $Pb^{2+}(aq) + SO_4^{2-}(aq)$, $K_{sp} = 1.8 \times 10^{-8}$ at 25°C.) If the cell potential does change, calculate its value.

   (iii) And finally, 0.1000 g of Cu(NO3)2 · 5/2H2O(s) is added to the half-cell containing the Cu(NO3)2 solution. Will the cell potential of the operating battery increase, decrease, or remain the same? If the cell potential does change, calculate its value.

Name: _____   Date: _____

Lab Instructor: _____   Lab Section: _____

EXPERIMENT 31

Batteries

RESULTS/OBSERVATIONS

Part A: General Galvanic Cell Characteristics

1. Initial cell potential: _____

2. Calculate the percent error between the measured cell potential and that calculated in question 1(a) of the prelab assignment.

3. Is the result consistent with what you calculated in question 1(a) of the prelab assignment?

    _____

4. Cell potential after addition of 25.00 mL of distilled water to each half-cell: _____

5. Calculate the percent error between the measured cell potential and that calculated in question 1(c)(i) of the prelab assignment.

6. Is the measured cell potential consistent with your answer to prelab question 1(c)(i)?

    _____

7. Mass of Na_2SO_4: _____

8. Cell potential after addition of Na_2SO_4: _____

Name: _____   Date: _____

Lab Instructor: _____   Lab Section: _____

9. Calculate the percent error between the measured cell potential and that calculated in question 1(c)(ii) of the prelab assignment.

10. Is the measured cell potential consistent with your answer to prelab question 1(c)(ii)?

   _____

11. Mass of $Cu(NO_3)_2 \cdot 5/2H_2O$: _____

12. Cell potential after addition of $Cu(NO_3)_2 \cdot 5/2H_2O$: _____

13. Calculate the percent error between the measured cell potential and that calculated in question 1(c)(iii) of the prelab assignment.

14. Is the measured cell potential consistent with your answer to prelab question 1(c)(iii)?

   _____

Part B: Battery Design

15. Target cell potential: _____

16. Calculation of battery cell potential in Part B:

Name: _____   Date: _____

Lab Instructor: _____   Lab Section: _____

16. Calculation of battery cell potential in Part B (continued):

17. Cell potential of battery (trial 1):   _____   Instructor signature: _____

(trial 2): _____   Instructor signature: _____

(trial 3): _____   Instructor signature: _____

18. Battery shorthand notation: _____

Name: _____   Date: _____

Lab Instructor: _____   Lab Section: _____

EXPERIMENT 31

Batteries

POSTLABORATORY QUESTIONS

1. Consider a galvanic cell based on the reaction

   $$2\,Ag^+(aq) + Pb(s) \rightarrow 2\,Ag(s) + Pb^{2+}(aq)$$

   (a) What is the standard cell potential (E°_{cell}) of this galvanic cell?

   (b) What is the cell potential (E°_{cell}) for this galvanic cell if both the Ag^+ and Pb^{2+} concentrations are 0.100 M?

   (c) How would you alter the Ag^+ concentration to increase the cell potential?

   (d) How would you alter the Pb^{2+} concentration to decrease the cell potential?

   (e) What Ag^+ would you use with 0.100 $M\,Pb^{2+}$ to construct a galvanic cell with a cell potential of 0.85 V? Give the shorthand notation of this cell.

EXPERIMENT 32

The Particle-in-a-Box Model

OBJECTIVE

The visible absorbance spectra of three series of polymethine dyes will be measured. The particle-in-a-box model then will be applied to the electronic transitions that lead to the visible absorbances of the examined polymethine dyes.

INTRODUCTION

The visible absorbance bands of polymethine dyes result from electronic transitions involving delocalized π electrons. The wavelengths of these absorbance bands depend on the spacing between the electronic energy levels. The purpose of this experiment will be to test the validity of using the particle-in-a-box model to describe these delocalized π electrons. The model will be considered successful if it can predict the locations of the visible absorbance bands for a series of polymethine dyes.

As an example of why the particle-in-a-box model might well describe electronic transitions that lead to the visible absorbances of polymethine dyes, consider a dilute solution of 1,1'-diethyl-4,4'-carbocyanine iodide:

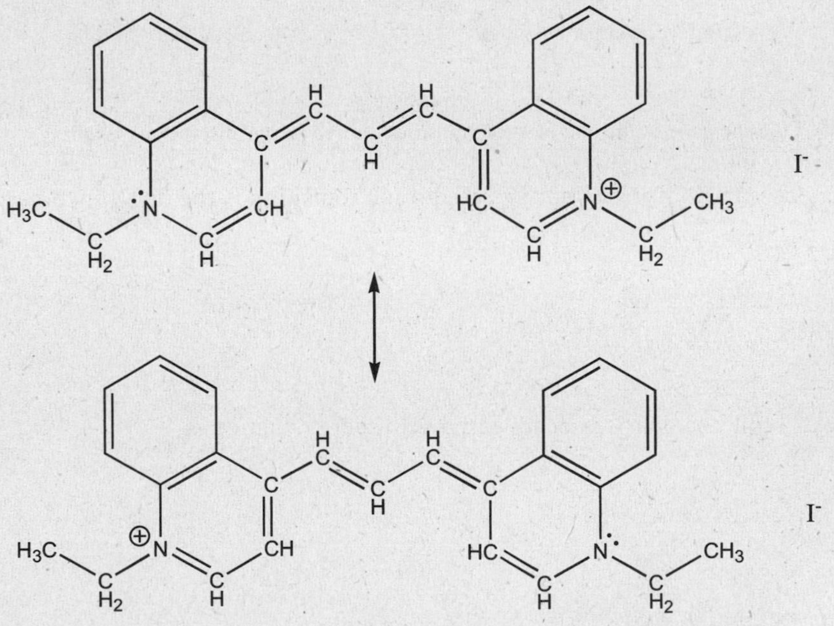

Figure 32.1
1-1' –Diethyl-4,4' –carbocyanine iodide.

This cation is best described as a resonance hybrid of the two resonance structures shown above. The quantum-mechanical interpretation of the resonance hybrid is that the wavefunction for the ion has equal contributions from both states. Consequently, all the bonds along this conjugated chain can be considered equivalent, with a bond order of 1.5 (analogous to the carbon–carbon bonds in benzene). Each carbon atom in the chain and each nitrogen at the end is involved in bonding with three atoms by three localized bonds (σ bonds). The extra valence electrons on the carbon atoms in the chain and the three remaining electrons on the two nitrogens form a delocalized cloud of π electrons along the extent of the chain. Given the delocalization of the π electrons, it is reasonable to assume that the potential energy felt by each is constant along the chain and rises precipitously to infinity at the ends; the π electrons are effectively free electrons moving in a one-dimensional potential box, the length of which is determined by the length of the conjugated system: the particle-in-a-box model. The quantum-mechanical solution for the energy levels (E_n) of this model is

$$E_n = \frac{h^2 n^2}{8 m_e L^2} \qquad n = 1, 2, 3, \ldots \tag{1}$$

where m_e is the electron mass, h is Planck's constant, and L is the box length.

The number of electrons in any given energy level is limited to two by the Pauli exclusion principle. The ground state of a molecule with N π electrons will have the $N/2$ lowest levels filled (if N is even) and all higher levels empty. Absorption of light by the molecule (or ion in this case) is associated with a one-electron transition from the highest filled energy level ($n_1 = N/2$) to the lowest unfilled energy level ($n_2 = N/2 + 1$). See Figure 32.2 for a schematic of this process.

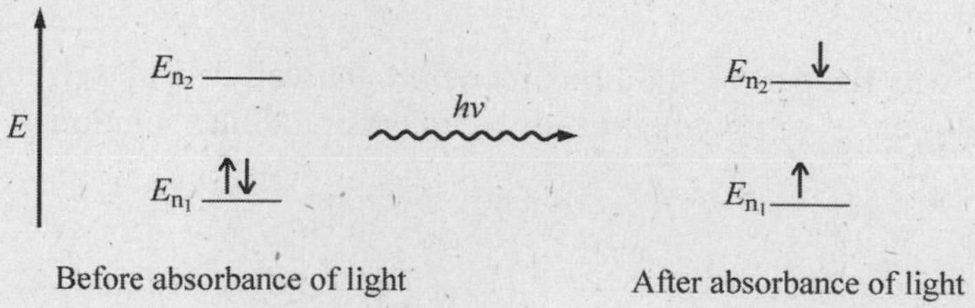

Figure 32.2
Schematic of the one-electron transition caused by the absorption of light.

The energy necessary to bring about this electronic transition is the difference between the two energy levels:

$$\Delta E = \frac{h^2}{8 m_e L^2}\left(n_2^2 - n_1^2\right) = \frac{h^2}{8 m_e L^2}(N+1) \tag{2}$$

The wavelength of light necessary to bring about this energy change is

$$\lambda = \frac{8 m_e c}{h}\frac{L^2}{N+1} \tag{3}$$

where it has been recalled that $\Delta E = h\nu = hc/\lambda$.

It is possible to reexpress this equation in terms of a single variable, the number of carbon atoms in the conjugated chain (p). Each carbon atom in the conjugated chain contributes one π electron, and the two nitrogen atoms contribute three π electrons; thus

$$N = p + 3 \tag{4}$$

Assuming that the length of the potential box within which the π electrons reside is the length of the conjugated chain between nitrogen atoms plus one bond distance on each side,

$$L = (p + 3)l \tag{5}$$

where l is the bond length between atoms along the conjugated chain. Substituting these expressions into Equation (3) yields

$$\lambda = \frac{8m_{e}cl^2}{h}\frac{(p+3)^2}{p+4} \tag{6}$$

Taking l to be equal to 0.139 nm (the bond length in benzene, a molecule with similar bonding) and inserting the numerical values of m_e, c, and h, Equation (6) can be simplified further:

$$\lambda = 63.7\frac{(p+3)^2}{p+4} \tag{7}$$

where λ is in units of nanometers.

The length of the potential box, $(p + 3)l$, was based on two assumptions, that the length of the conjugated chain is between nitrogen atoms plus one bond distance on each side and that the bond length between atoms in the conjugated chain is equivalent to that of benzene. While reasonable, these assumptions are only expected to be approximately correct. To account for the approximate nature of these assumptions, the length of the potential box will be expressed as $(p + 3 + \alpha)$, where α is an adjustable parameter to be determined by experiment:

$$\lambda = 63.7\frac{(p+3+\alpha)^2}{p+4} \tag{8}$$

where λ is in units of nanometers. The value of α should be constant for a series of dye molecules that differ only in the length of their conjugated chains (see Figure 32.4 for examples of such series of dyes), and it should be small, not much larger than 1; an α value significantly larger than 1 implies that the assumed box length is grossly in error. If such a series of dyes is studied experimentally, this empirical parameter may be adjusted to achieve the best fit to the data.

In this experiment, the visible absorbance spectra of three series of polymethine dyes (shown in Figure 32.3) will be measured using a spectrometer. Since the dyes within each series differ from one another only in the length of their conjugated π systems, the wavelength of the visible absorbance for each polymethine dye should be amenable to description by Equation (8). You are to determine the α value that allows for the best fit of the data for each series of dyes. Using the results of these fits, you will decide if the particle-in-a-box model provides an adequate description of the electronic transitions that lead to the visible absorbances of the examined polymethine dyes.

Series 1: 1,1'-diethyl-4,4'-cyanine; 1,1'-diethyl-4,4'-carbocyanine; and 1,1'-diethyl-4,4'-dicarbocyanine

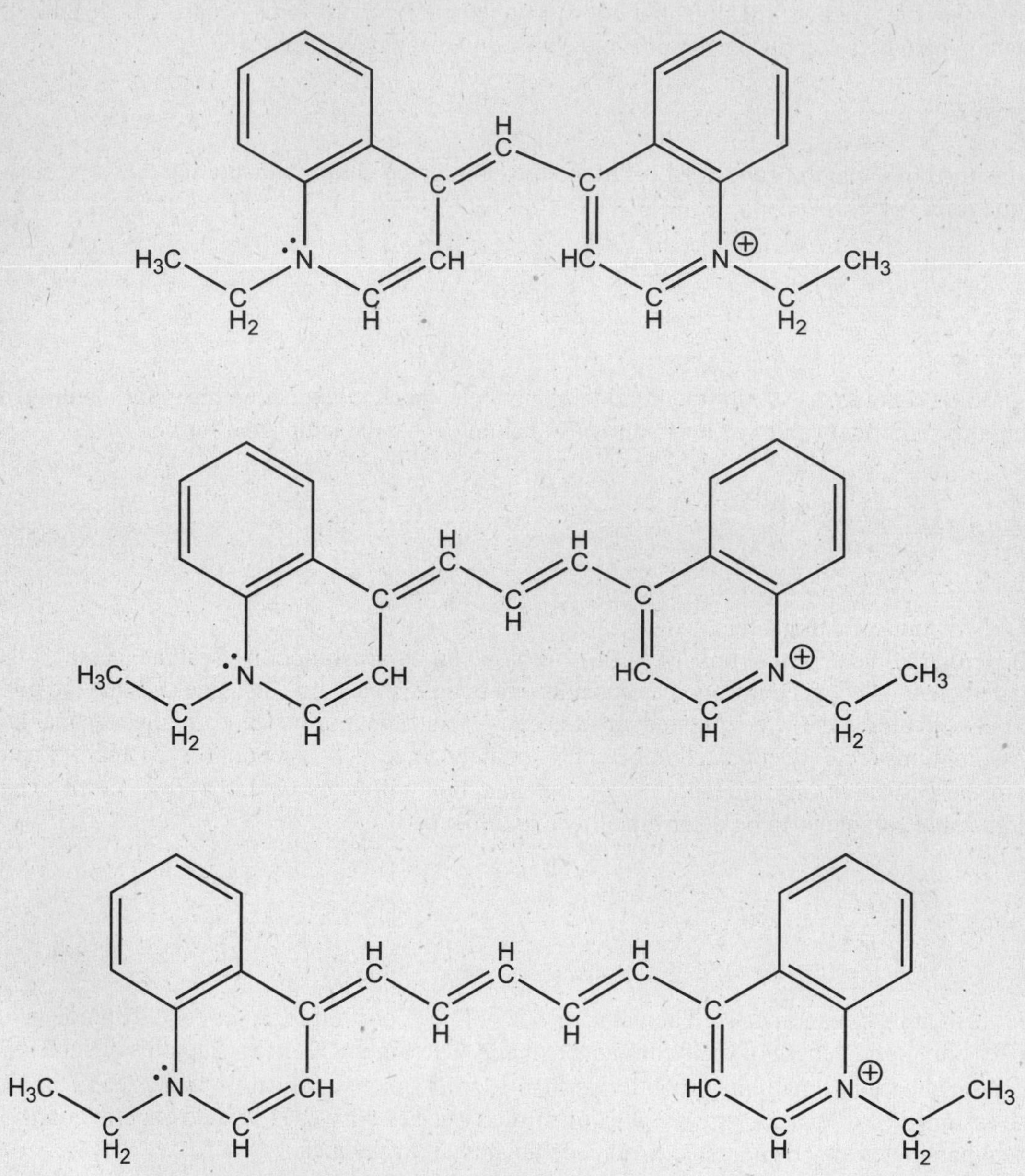

Figure 32.3
The Polymethine Dyes.

Series 2: 1,1'-diethyl-2,2'-cyanine; 1,1'-diethyl-2,2'-carbocyanine; 1,1'-diethyl-2,2'dicarbocyanine

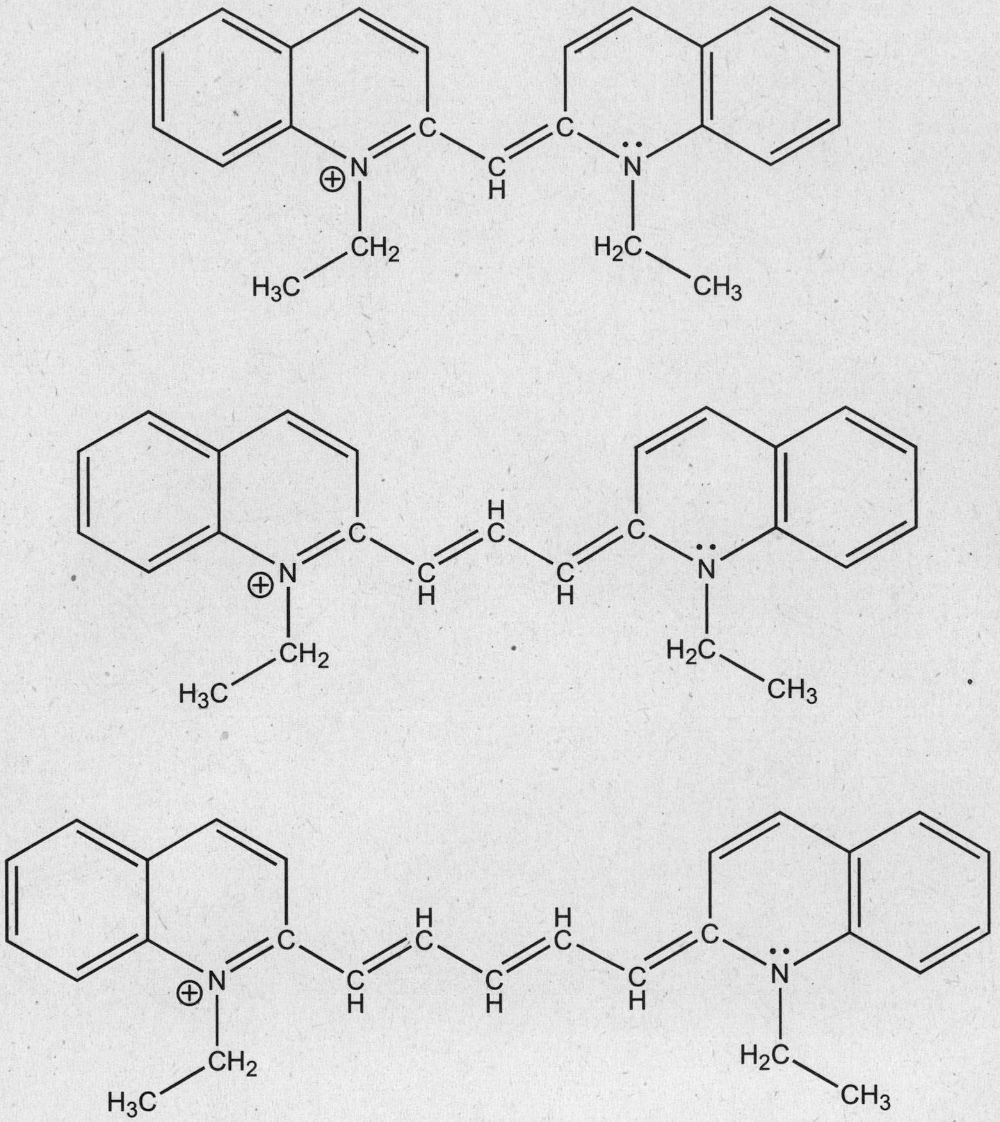

Figure 32.3
The Polymethine Dyes. (Continued)

Series 3: 3,3'-diethyl-2,2'-thiacyanine; 3,3'-diethyl-2,2'-thiacarbocyanine; 3,3'-diethyl-2,2'-thiadicarbocyanine

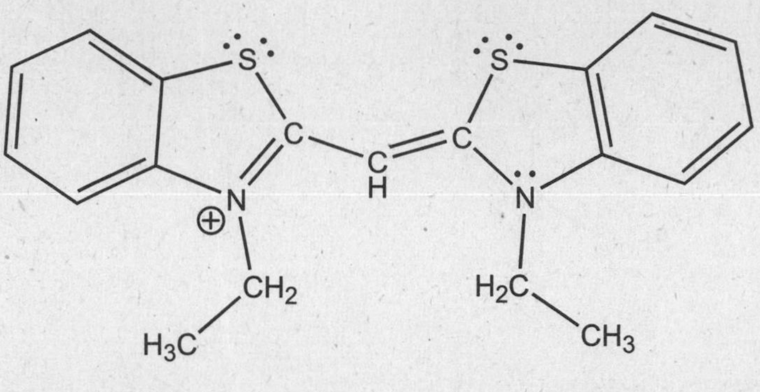

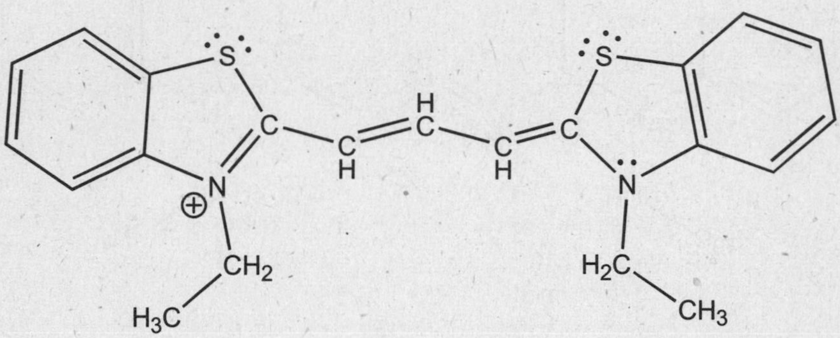

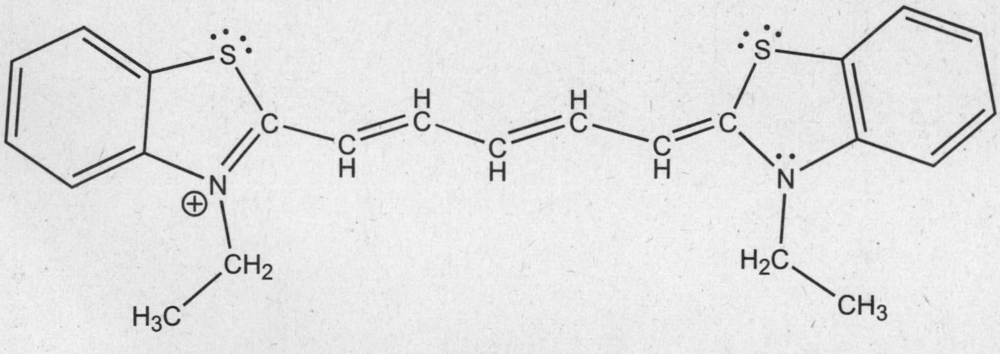

Figure 32.3
The Polymethine Dyes. (Continued)

ADDITIONAL READING

Read the sections in the Laboratory Techniques chapter at the beginning of this Lab Manual on handling chemicals and spectrometers prior to performing this experiment.

Safety Precautions:

■ Protective eyewear approved by your institution must be worn at all times while you are in the laboratory.

■ Gloves should be worn while handling the methanol used in this experiment to lessen the risk of methanol being absorbed through the skin.

■ If either methanol or one of the dye solutions spills on your skin, the affected area should be rinsed with water for 15 minutes.

PROCEDURE

Turn on the spectrometer, and allow it to warm up for 10 minutes prior to use.

This experiment involves determining the wavelengths at which various polymethine dyes absorb light in the visible spectrum, absorbances owing to electronic transitions involving delocalized π electrons. The visible absorbance for each of the dyes occurs over a range of wavelengths; however, the equation by which the validity of the particle-in-a-box model will be evaluated [Equation (8)] contains only a single wavelength value. In this experiment, a single wavelength within the absorbance range will be chosen as representative of the dye absorbance. The wavelength value chosen will be the one that corresponds to the absorbance maximum.

Finding the wavelength of maximum absorbance (λ_{max}) can be expedited by making measurements in the correct area of the visible spectrum. For example, if a solution appears red in white light (normal room light), the solution must be selectively transmitting red light. This implies that the solution is absorbing the complementary color, green. On a visible spectrometer, this solution would be expected to have a maximum absorbance in the green region of the visible spectrum, which corresponds to 490–560 nm. Figure 32.4 summarizes the relationship between the color of light, its wavelength, and its complement.

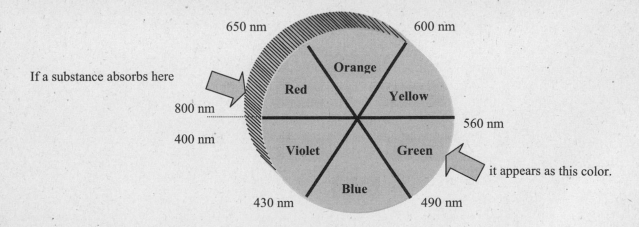

Figure 32.4
A color wheel summarizing the relationship between the color of light, its wavelength, and its complement.

CAUTION: Gloves should be worn while handling the methanol.

Obtain a cuvet and rinse it with several portions of the blank solution, methanol in this case. Fill it three-quarters full with the blank solution. Holding the cuvet by the lip, wipe the outside of the cuvet with a clean Kimwipe to remove any fingerprints or stains that might reduce the intensity of light transmitted to the detector. Place the cuvet in the sample compartment, pushing it fully down. Align the mark on the cuvet with that on the sample holder, and close the sample compartment. Set the wavelength selector to the smallest value within the wavelength range where you expect the first dye in the first dye series to absorb; that is, if the first dye in the first dye series appears red, set the wavelength to 490 nm, the smallest value in the wavelength range where the dye is expected to absorb. Adjust the calibration dial so that the spectrometer reads zero absorbance.

Remove the cuvet from the spectrometer, and empty it into a waste beaker. Rinse the cuvet several times with the first dye in the first dye series. Fill the cuvet three-quarters full with the dye. Holding the cuvet by the lip, wipe the outside of the cuvet clean with a Kimwipe, place the cuvet in the sample compartment, align the mark on the cuvet with that on the sample holder, and close the sample compartment. Read and record the absorbance value.

Increase the wavelength selector by 10 nm. Repeat the blanking of the spectrometer. (Each time the spectrometer wavelength selector is changed, the blanking needs to be repeated.) Again measure and record the absorbance of the dye. If the absorbance increases, increase the wavelength selector by 10 nm, and repeat the blanking and absorbance measurement steps. A decrease in the absorbance implies that λ_{max} is between the current and previous wavelengths. Systematically sample wavelengths in this range, reblanking the spectrometer each time the wavelength is changed, until λ_{max} is determined.

Repeat this procedure for the other dyes in the series. Afterward, apply this procedure to the dyes in the other two series.

Waste Disposal. Dispose of chemical waste as directed by your instructor.

Name: _____   Date: _____

Lab Instructor: _____   Lab Section: _____

EXPERIMENT 32

The Particle-in-a-Box Model

PRELABORATORY QUESTIONS

1. Within the particle-in-a-box model, what general trend occurs in the spacing between energy levels as box length is increased? Explain your answer.

2. Within the particle-in-a-box model, what general trend occurs in the wavelength values as box length is increased? Explain your answer.

3. Examine the three series of polymethine dyes shown in Figure 32.3. When the number of carbon atoms in the conjugated chain increases, what effect will be observed in the wavelength of absorbance? Explain your answer.

Name: _____   Date: _____

Lab Instructor: _____   Lab Section: _____

4. Calculate λ for each dye in each of the three polymethine dye series shown below under the assumption that the particle-in-a-box model as described in the introduction applies perfectly, that is, Equation (8) with α equal to zero.

Series 1:

Dye Molecule	p	λ (nm)
1,1'-Diethyl-4,4'-cyanine		
1,1'-Diethyl-4,4'-carbocyanine		
1,1'-Diethyl-4,4'-dicarbocyanine		

Series 2:

Dye Molecule	p	λ (nm)
1,1'-Diethyl-2,2'-cyanine		
1,1'-Diethyl-2,2'-carbocyanine		
1,1'-Diethyl-2,2'dicarbocyanine		

Series 3:

Dye Molecule	p	λ (nm)
3,3'-Diethyl-2,2'-thiacyanine		
3,3'-Diethyl-2,2'-thiacarbocyanine		
3,3'-Diethyl-2,2'-thiadicarbocyanine		

Name: _____   Date: _____

Lab Instructor: _____   Lab Section: _____

EXPERIMENT 32

The Particle-in-a-Box Model

RESULTS/OBSERVATIONS

1. Absorbance measurements:

Series 1:

Dye Molecule	p	λ_{max} (nm)
1,1'-Diethyl-4,4'-cyanine		
1,1'-Diethyl-4,4'-carbocyanine		
1,1'-Diethyl-4,4'-dicarbocyanine		

Series 2:

Dye Molecule	p	λ_{max} (nm)
1,1'-Diethyl-2,2'-cyanine		
1,1'-Diethyl-2,2'-carbocyanine		
1,1'-Diethyl-2,2'dicarbocyanine		

Series 3:

Dye Molecule	p	λ_{max} (nm)
3,3'-Diethyl-2,2'-thiacyanine		
3,3'-Diethyl-2,2'-thiacarbocyanine		
3,3'-Diethyl-2,2'-thiadicarbocyanine		

Name: _____   Date: _____

Lab Instructor: _____   Lab Section: _____

2. Plot λ_{max} as a function of p for series 1 on the attached sheet of graph paper. On this same graph, plot Equation (8) for $\alpha = 0$. (*Note:* You determined the necessary values when you solved prelaboratory question 4.) Be sure to distinguish the curves from one another; that is, make one dashed.

    Visually inspect the two curves. If they are not close, choose a value of α that you believe will improve agreement between the two. Calculate λ values according to the particle-in-a-box model [Equation (8)] for α equal to this new value. Plot these new data.

    Repeat this procedure with different α values until agreement between experiment and the model has been maximized.

3. Value of α that maximizes agreement between experiment and the model for series 1: _____

4. Does the particle-in-a-box model provide an adequate description of the electronic transitions that lead to the visible absorbances of the series 1 dyes? _____ Explain:

5. Apply the graphic analysis described above to series 2.

6. Value of α that maximizes agreement between experiment and the model for series 2: _____

7. Does the particle-in-a-box model provide an adequate description of the electronic transitions that lead to the visible absorbances of the series 2 dyes? _____ Explain:

8. Apply the graphic analysis to series 3.

9. Value of α that maximizes agreement between experiment and the model for series 3: _____

10. Does the particle-in-a-box model provide an adequate description of the electronic transitions that lead to the visible absorbances of the series 3 dyes? _____ Explain:

Name: _____   Date: _____

Lab Instructor: _____   Lab Section: _____

Series 1:

Name: _____   Date: _____

Lab Instructor: _____   Lab Section: _____

Series 2:

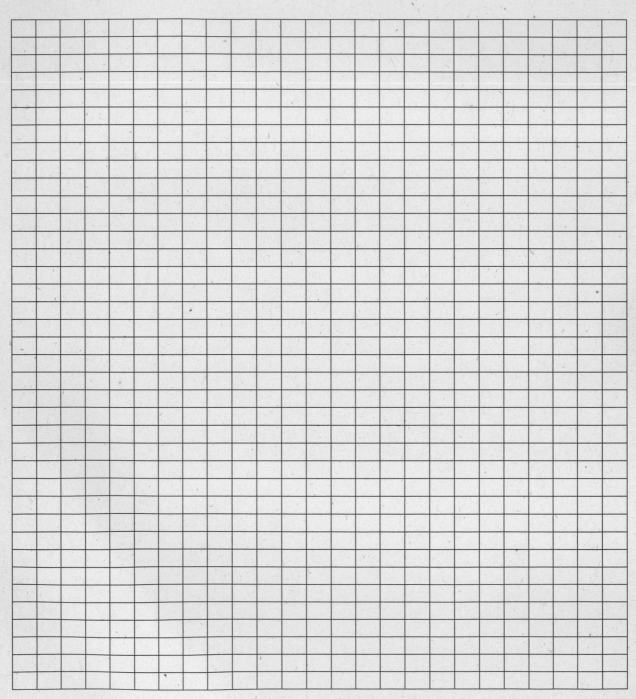

Name: _____ Date: _____

Lab Instructor: _____ Lab Section: _____

Series 3:

Name: _____   Date: _____

Lab Instructor: _____   Lab Section: _____

EXPERIMENT 32

The Particle-in-a-Box Model

POSTLABORATORY QUESTIONS

1.   What assumptions were made in this experiment regarding the length of the potential box?

2.   The α parameter was included to account for the approximate nature of the assumptions made regarding the length of the potential box. Give a physical interpretation of an α value greater than zero.

3.   Give a physical interpretation of an α value less than zero.

EXPERIMENT 33

Optical
Spectroscopy

OBJECTIVE

To observe emission spectra from a variety of sources using a spectrometer. To design and implement a procedure capable of determining the composition of a solution that contains two or more ionic salts. To construct a partial energy-level diagram for hydrogen.

INTRODUCTION

Optical spectroscopy involves the measurement and analysis of electromagnetic radiation (a.k.a. light). Many of the properties of light are conveniently described by means of a classical wave model. Within this model, light waves are characterized by such variables as frequency and wavelength. The frequency (ν), which describes the number of wave crests passing a given point per second for the light wave, is inversely proportional to the wavelength (λ), the distance between successive wave crests. The light we can see, visible light, corresponds to a very small portion of the electromagnetic spectrum, from about 400 nm (violet) to 800 nm (red). The light wave frequency and wavelength are related to one another by the equation

$$c = \lambda \nu \tag{1}$$

where c is the speed of the light wave. The speed of light in a vacuum is 3.00×10^8 m/s.

For phenomena where the classical wave model of light proves insufficient, a particle model is invoked. In the particle model, light is composed of a stream of discrete particles called *photons*. The energy (E) of each photon is directly proportional to its frequency and inversely proportional to its wavelength:

$$E = h\nu = \frac{hc}{\lambda} \tag{2}$$

where the proportionality constant h is Planck's constant (6.626×10^{-34} J $\cdot$ s).

Thus electromagnetic radiation can be described by either the wave or particle model. The model applied is the one that accurately describes the phenomenon being investigated.

In spectroscopy, light is used as a means of probing matter. One means of probing matter with light uses the phenomenon of absorption. When an atom, molecule, or ion absorbs a photon, its energy increases. The energy change of the atom must be equivalent to the energy of the photon. Thus the absorbed wavelengths of light reveal the differences between energy levels in the atom (see Figure 33.1).

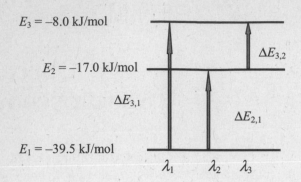

$E_3 = -8.0$ kJ/mol

$E_2 = -17.0$ kJ/mol

$E_1 = -39.5$ kJ/mol

$\lambda_1 \quad \lambda_2 \quad \lambda_3$

$E_{3,1} = E_3 - E_1 = 31.5$ kJ/mol
$\lambda_1 = hcN_A/\Delta E_{3,1} = 3800$ nm

$\Delta E_{2,1} = E_2 - E_1 = 22.5$ kJ/mol
$\lambda_2 = hcN_A/\Delta E_{2,1} = 5320$ nm

$\Delta E_{3,2} = E_3 - E_2 = 9.0$ kJ/mol
$\lambda_3 = hcN_A/\Delta E_{3,2} = 13301$ nm

Figure 33.1

For an atom containing solely the three energy states shown here, there are only three wavelengths of light that can be absorbed, each of which is illustrated by an arrow pointing up. The values of the three absorbed wavelengths of light, as calculated from the energy-level differences, are shown to the right of the energy-level diagram. (Avogadro's number is included in the equation for the wavelength because the energies are given per mole.)

The converse of absorption is emission. When an atom emits a photon of light, its energy decreases. The energy of the emitted photon must be equivalent to the energy change of the atom. The emitted wavelengths of light correspond to the differences between energy levels in the atom (see Figure 33.2).

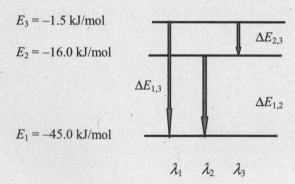

$E_3 = -1.5$ kJ/mol

$E_2 = -16.0$ kJ/mol

$E_1 = -45.0$ kJ/mol

$\lambda_1 \quad \lambda_2 \quad \lambda_3$

$\Delta E_{1,3} = E_1 - E_3 = -43.5$ kJ/mol
$\lambda_1 = hcN_A/\Delta E_{1,3} = 2752$ nm

$\Delta E_{1,2} = E_1 - E_2 = -29.0$ kJ/mol
$\lambda_2 = hcN_A/\Delta E_{1,2} = 4128$ nm

$\Delta E_{2,3} = E_2 - E_3 = -14.5$ kJ/mol
$\lambda_3 = hcN_A/\Delta E_{2,3} = 8256$ nm

Figure 33.2

For an atom containing solely the three energy states shown here, there are only three wavelengths of light that can be emitted, each of which is illustrated by an arrow pointing down. The values of the three emitted wavelengths of light, as calculated from the energy-level differences, are shown to the right of the energy-level diagram. (The negative energies arise from the convention that energy lost by an atom is negative, whereas that gained by an atom is negative, whereas that gained by an atom is positive. Avogadro's number is included in the equation for the wavelength because the energies are given per mole.)

In many cases, the pattern of wavelengths absorbed or emitted by a pure substance is characteristic of that substance. Thus the pattern of emitted or absorbed wavelengths can be used as a means of identifying a substance, a sort of "fingerprint" in light.

The patterns of wavelengths absorbed or emitted by atoms, molecules, or ions are known as *spectra*. Spectra may be classified as emission spectra or absorption spectra. In an emission experiment, the source, which could be an ordinary incandescent or fluorescent light bulb, a salt in a flame, or an electrically excited gas in a tube, emits the light (see Figure 33.3).

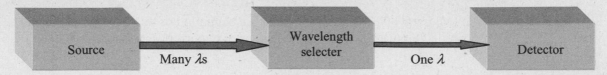

Figure 33.3
Schematic of an emission experiment.

   The emitted light is passed through a wavelength selector (a prism or diffraction grating) to select one wavelength. The intensity of light at this wavelength is measured at the detector. Adjusting the wavelength selector changes the wavelength of light whose intensity is measured at the detector. The collection of these measurements over a range of wavelengths makes up the spectrum. If only a few characteristic wavelengths are emitted, the result is a bright-line emission spectrum. If emission occurs at all wavelengths within a given range, the result is a continuous emission spectrum. Continuous emission spectra result when the number of available energy levels is very large and the spacing between them approaches the infinitesimal.

   In an absorption experiment, the light from a source passes through an absorbing medium, such as a gas sample or a solution, the effect of which is to remove certain wavelengths. The wavelengths of unabsorbed light then are passed through a wavelength selector and onto a detector (see Figure 33.4).

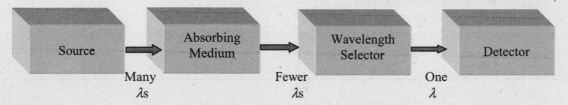

Figure 33.4
Schematic of an absorption experiment.

   The result is a dark-line spectrum. If a band of wavelengths is absorbed, the result will be a dark area in that part of the spectrum.

   In this experiment you will use a spectrometer to acquire emission spectra from a variety of sources. The emission spectra of a variety of materials will be observed: a fluorescent lightbulb, an incandescent lightbulb, helium gas, and a variety of salt solutions. These materials will provide examples of both continuous and line spectra. Your purpose in this experiment is twofold. One goal is to determine the composition of a solution that contains two or more ionic salts. The second goal is to construct a partial energy-level diagram for hydrogen (see Figures 33.1 and 33.2 for examples of this). In Part A of this experiment, you make a number of observations of emission spectra. In Part B, you will use your initial observations to design procedures for determining the composition of a solution that contains two or more ionic salts and for constructing a partial energy-level diagram for hydrogen.

ADDITIONAL READING

Read the sections in the Laboratory Techniques chapter at the beginning of this Lab Manual on handling chemicals, cleaning glassware, and spectrometers prior to performing this experiment.

 Safety Precautions:

- Protective eyewear approved by your institution must be worn at all times while you are in the laboratory.

- In addition to visible light, the discharge tubes emit ultraviolet radiation, which is damaging to the eyes. While safety goggles will absorb most of this radiation, it is recommended that you look at the radiation source for only short periods of time.

- The power supply to the discharge tubes develops a voltage of several thousand volts. Do not touch any portion of the power supply, wire leads, or discharge tubes unless the power supply is unplugged from the electrical outlet.

- Always unplug the power supply from the electrical outlet prior to adjusting the position of the discharge tubes or any other part of the apparatus.

EXPERIMENT

When recording spectral data, include the relative intensity of each line. If your spectrometer lacks an intensity scale, estimate the intensities by eye using a scale of 1 to 10, where 10 means very bright and 1 means you can barely see the line. For continuous spectra, record wavelength ranges corresponding to each color that you can distinguish.

Part A

Fluorescent Light Spectrum

Observe the emission spectrum of the fluorescent light with the spectrometer. Some of the lines you are likely to see occur at 405, 436, 546, 577, 579, 615, and 691 nm. Don't be concerned if you can't see all the lines; some are very faint.

Incandescent Lightbulb Spectrum

Observe the emission spectrum of the incandescent lightbulb with the spectrometer. You should observe a continuous spectrum because an incandescent lightbulb gives off white light.

Helium Spectrum

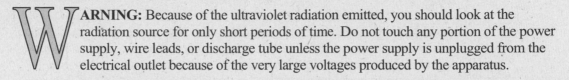

W **ARNING:** Because of the ultraviolet radiation emitted, you should look at the radiation source for only short periods of time. Do not touch any portion of the power supply, wire leads, or discharge tube unless the power supply is unplugged from the electrical outlet because of the very large voltages produced by the apparatus.

There is a device in your lab consisting of a glass tube (discharge tube) filled with helium that is attached to a voltage source used to excite the helium gas, causing it to glow. Turn the discharge tube on, and record the color of the glowing helium gas. Use the spectrometer to observe the helium emission spectrum. Helium lines reportedly occur at 447, 502, 588, 668, and 707 nm, but some are much easier to see than others.

Spectrum of Salt Solutions: NaCl, LiCl, KCl, CaCl₂, and SrCl₂

Half fill five of the smallest test tubes in your equipment drawer with one each of the following salt solutions: sodium chloride, lithium chloride, potassium chloride, calcium chloride, and strontium chloride. Obtain 10 cotton swabs.

Light a Bunsen burner. Adjust the air flow using the knurled knob at the bottom of the burner so as to get two distinct cones of flame. Arrange the spectrometer so that the flame is visible in the slit.

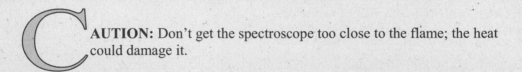

C **AUTION:** Don't get the spectroscope too close to the flame; the heat could damage it.

Use the spectrometer to observe the emission spectrum of each salt solution. To accomplish this, soak one end of a cotton swab in a salt solution. One student should hold the tip of the swab in the hottest portion of the burner flame (just above the inner cone) while another looks through the spectroscope at the flame. It may take a few seconds for the swab to dry out before the intensely colored flame appears. Try not to ignite the swab. If you cannot find the slit image when the color is intense, soak the swab again, and repeat the trial. Repeat this procedure for each of the other salts.

To avoid contamination, be sure to use a new swab when you change salt solutions.

The emission from potassium chloride is so faint that you may have difficulty seeing any lines at all. If you cannot record the wavelength of any potassium lines, just record the color.

Part B

Identification of Unknown Salts in Solution

You are to design and carry out a procedure that will allow for the identification of the unknown salts in a solution. Your unknown solution will contain two or more of the following salts: NaCl, LiCl, KCl, CaCl₂, and SrCl₂. Your instructor will provide your group with its unknown. Record the unknown number in your laboratory notebook.

Prior to examining its unknown sample, your group is required to test your procedure on a mixture of known composition. Use the salt solutions available in lab—NaCl, LiCl, KCl, CaCl₂, and SrCl₂—to create a mixture of known composition. Record the results of testing your procedure. Use these data to verify (or improve) the efficacy of your procedure.

Partial Energy-Level Diagram for Hydrogen

W**ARNING:** Because of the ultraviolet radiation emitted, you should look at the radiation source for only short periods of time. Do not touch any portion of the power supply, wire leads, or discharge tube unless the power supply is unplugged from the electrical outlet because of the very large voltages produced by the apparatus.

There is also a discharge tube filled with hydrogen in the lab. Your group is to design and implement a procedure for collecting data from this discharge tube that can be used to generate a partial energy-level diagram for the electronic states of hydrogen.

For the partial energy-level diagram of hydrogen, assume that all the observed transitions terminate at the same state; for example, if you observe two transitions, they are from state A → state X and state B → state X. Also, set the value of the highest energy state to 0.0 kJ/mol. This will cause the other energy levels to have negative values.

Equipment and Reagents

To perform Part B of this experiment, you will have access to all the equipment in your lab drawer, cotton swabs, a spectrometer, and a hydrogen discharge tube. Aqueous solutions of NaCl, LiCl, KCl, $CaCl_2$, and $SrCl_2$ also will be available.

Waste Disposal. Dispose of all chemical waste as directed by your instructor.

Name: _____   Date: _____

Lab Instructor: _____   Lab Section: _____

EXPERIMENT 33

Optical Spectroscopy

PRELABORATORY QUESTIONS

1. Explain the difference between continuous and line spectra.

2. Explain the difference between absorption and emission spectra.

3. For an atom with the energy levels below, what wavelength of light (in nanometers) will be emitted in a transition between E_2 and E_1 (indicated by the down arrow below)? What wavelength of light must be absorbed to cause a transition between E_3 and E_5 (indicated by the up arrow below)?

$E_5 = -8.0$ kJ/mol

$E_4 = -17.0$ kJ/mol

$E_3 = -39.5$ kJ/mol

$E_2 = -200.0$ kJ/mol

$E_1 = -1100.0$ kJ/mol

Name: _____   Date: _____

Lab Instructor: _____   Lab Section: _____

4.   Can the atom of question 3 absorb or emit light with a wavelength of 555 nm? Can it absorb or emit light with a wavelength of 5320 nm? If so, state which energy levels that the transition occurs between.

5.   Consider the emission spectra of the two hypothetical elements X and Z.

Emission spectrum of X:

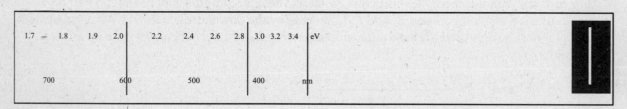

Emission spectrum of Z:

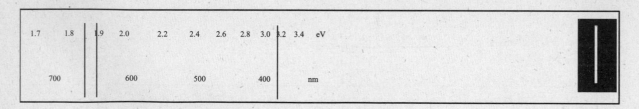

Draw a picture of the emission spectra expected from a sample containing a mixture of X and Z on the spectrum blank below.

Emission spectrum of a mixture of X and Z:

| 1.7 | 1.8 | 1.9 | 2.0 | 2.2 | 2.4 | 2.6 | 2.8 | 3.0 | 3.2 | 3.4 | eV |

| 700 | | 600 | | 500 | | 400 | | nm |

Name: _____   Date: _____

Lab Instructor: _____   Lab Section: _____

EXPERIMENT 33

Optical Spectroscopy

RESULTS/OBSERVATIONS

Part A

1. Observations of emission spectra.

Fluorescent Light Spectrum

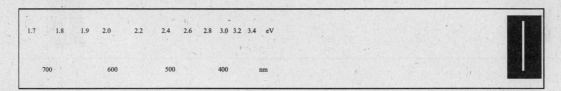

Incandescent Lightbulb Spectrum

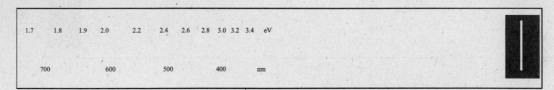

Helium Spectrum

Name: _____   Date: _____

Lab Instructor: _____   Lab Section: _____

Sodium Chloride Spectrum

| 1.7 | 1.8 | 1.9 | 2.0 | 2.2 | 2.4 | 2.6 | 2.8 | 3.0 | 3.2 | 3.4 | eV |

| 700 | | 600 | | 500 | | 400 | | nm |

Lithium Chloride Spectrum

| 1.7 | 1.8 | 1.9 | 2.0 | 2.2 | 2.4 | 2.6 | 2.8 | 3.0 | 3.2 | 3.4 | eV |

| 700 | | 600 | | 500 | | 400 | | nm |

Potassium Chloride Spectrum

| 1.7 | 1.8 | 1.9 | 2.0 | 2.2 | 2.4 | 2.6 | 2.8 | 3.0 | 3.2 | 3.4 | eV |

| 700 | | 600 | | 500 | | 400 | | nm |

Calcium Chloride Spectrum

| 1.7 | 1.8 | 1.9 | 2.0 | 2.2 | 2.4 | 2.6 | 2.8 | 3.0 | 3.2 | 3.4 | eV |

| 700 | | 600 | | 500 | | 400 | | nm |

Strontium Chloride Spectrum

| 1.7 | 1.8 | 1.9 | 2.0 | 2.2 | 2.4 | 2.6 | 2.8 | 3.0 | 3.2 | 3.4 | eV |

| 700 | | 600 | | 500 | | 400 | | nm |

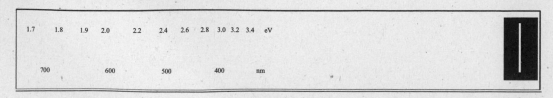

Name: _____   Date: _____

Lab Instructor: _____   Lab Section: _____

Part B:

Test of Procedure for Determining the Composition of an Unknown Salt Solution:

Test mixture contains: _____

1.7   1.8   1.9   2.0   2.2   2.4   2.6   2.8   3.0 3.2 3.4   eV
700   600   500   400   nm

Unknown Salt Solution (Unknown Identification Number   )

1.7   1.8   1.9   2.0   2.2   2.4   2.6   2.8   3.0 3.2 3.4   eV
700   600   500   400   nm

The unknown salt solution contains: _____

Hydrogen Spectrum

1.7   1.8   1.9   2.0   2.2   2.4   2.6   2.8   3.0 3.2 3.4   eV
700   600   500   400   nm

2. Use the observed wavelengths from the hydrogen emission spectrum to calculate the differences between hydrogen energy levels.

Name: _____   Date: _____

Lab Instructor: _____   Lab Section: _____

3. Draw a partial energy-level diagram for hydrogen. Assume that all the observed transitions
    terminate at the same state; for example, if you observe two transitions, they are from state A →
    state X and state → state X. Also, set the value of the highest energy state to 0.0 kJ/mol. This will
    cause the other energy levels to have negative values.

Name: _____   Date: _____

Lab Instructor: _____   Lab Section: _____

EXPERIMENT 33

Optical Spectroscopy

POSTLABORATORY QUESTIONS

1. Consider the emission spectra of the hypothetical elements α, β, and δ.

    Emission spectrum of *α*:

    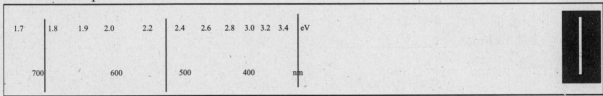

    Emission spectrum of *β*:

    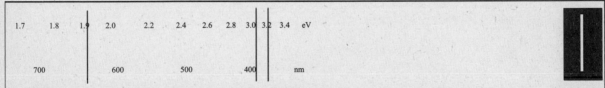

    Emission spectrum of *δ*:

    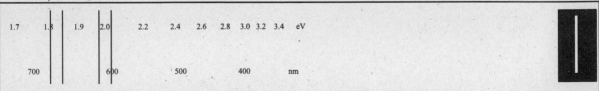

    Which of these three elements are present in a mixture of unknown composition, the emission spectrum of which is below?

    Emission spectrum of mixture:

    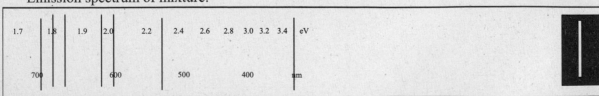

Name: _____ Date: _____

Lab Instructor: _____ Lab Section: _____

2. For an atom with the energy levels below, what wavelengths of light (in nanometers) can be emitted?

$E_4 = 0.0$ kJ/mol _____

$E_3 = -22.8$ kJ/mol _____

$E_2 = -278.3$ kJ/mol _____

$E_1 = -1234.5$ kJ/mol _____

Experiment 34

Analysis of an Acetone-Water Mixture

OBJECTIVE

To design and implement an experimental procedure using Reichardt's dye to determine the composition of an acetone-water mixture.

INTRODUCTION

When an atom, molecule, or ion absorbs light, its energy increases. The energy change of the molecule must be equivalent to the energy of the photon absorbed (see Figure 34.1).

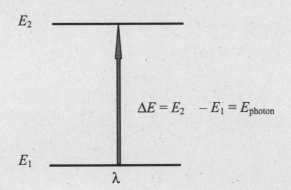

$$\Delta E = E_2 - E_1 = E_{photon}$$

Figure 34.1
The energy change of the molecule must be equivalent to the energy of the photon absorbed.

The energy of a photon (E_{photon}) is directly proportional to its frequency and inversely proportional to its wavelength:

$$E_{photon} = h\nu = \frac{hc}{\lambda} \tag{1}$$

Thus the absorbed wavelength of light is a measure of the difference between energy levels for the molecule: The greater the wavelength, the smaller the energy difference.

The spacing between energy levels is determined by the forces each molecule experiences. In the gas phase, where the average spacing between molecules is very large, the forces are all internal. In the solid, liquid, or solution phases, however, external forces also have an effect. A particularly dramatic example of how external forces affect energy spacing is provided by *solvatochromism:* the change in wavelength of an electronic absorption with a change in solvent polarity. Reichardt's dye exhibits a large solvatochromic effect. When Reichardt's dye is dissolved in solvents of increasing polarity, the wavelength of absorbed light decreases (see Table 34.1).

Table 34.1: Absorbance of Reichardt's Dye in Solvents of Varying Polarity

Increasing Solvent Polarity	Solvent	Color Solution Appears	Color Solution Absorbs	Decreasing Wavelength of Maximum Absorbance (λ_{max})
↓	Acetone	Green	Red	↓
	Ethanol	Violet	Yellow	
	Methanol	Red	Green	

The pronounced solvatochromism of Reichardt's dye is attributable to the large dipole moment exhibited by the molecule in its ground electronic state (see Figure 34.2).

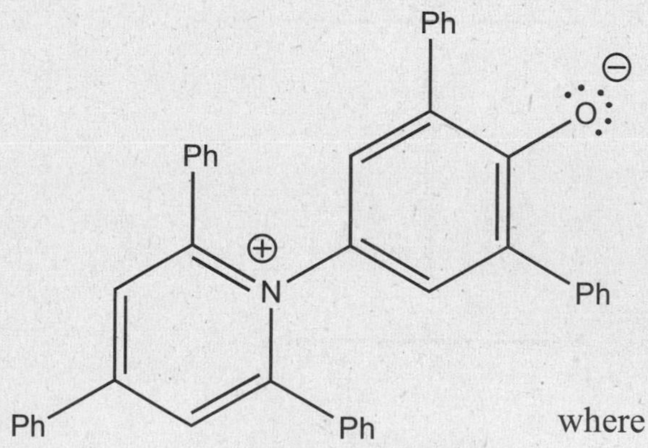

Figure 34.2
Reichardt's dye.

where Ph is C_6H_5

Polar solvents interact favorably with this huge dipole moment, lowering the energy of the ground electronic state relative to that of the significantly less polar excited electronic state. The more polar the solvent, the greater the energy separation between the ground and excited electronic states.

Your purpose in this experiment is to use Reichardt's dye to determine the composition of an acetone-water mixture.

ADDITIONAL READING

Read the sections in the Laboratory Techniques chapter at the beginning of this Lab Manual on cleaning glassware, handling chemicals, pipets, burets, and spectrometers prior to performing this experiment.

 Safety Precautions:

■ Protective eyewear approved by your institution must be worn at all times while you are in the laboratory.

■ Acetone is flammable. No open flames should be used during this experiment. In the event of a fire, alert your instructor. A fire in a small container can be smothered by covering the vessel with a watchglass. Afterwards, the container will be hot, so let it cool before touching it. Use a fire extinguisher to eliminate a larger fire. If the fire is burning over too large an area to be extinguished easily, evacuate the area and activate the fire alarm. If your clothing should catch on fire, use the safety shower or a fire blanket to extinguish the flames. Be ready to assist your neighbors should they need it.

EXPERIMENT

You are to design and perform an experiment using Reichardt's dye to determine the composition of an acetone-water mixture. Report the composition of your mixture as a volume of water to volume of acetone ratio, that is, the volume of water that was mixed with a fixed volume of acetone in order to prepare your mixture. *Note:* For best results, the measured absorbances should less than 1.2.

Equipment and Reagents

To perform this experiment, you will have access to all the equipment in your lab drawer as well as the following items:

> Reichardt's dye
> Acetone
> Distilled water
> Visible spectrophotometer
> Volumetric pipets
> 50 mL buret

Waste Disposal. Dispose of all chemical waste as directed by your instructor.

Name: _____   Date: _____

Lab Instructor: _____   Lab Section: _____

EXPERIMENT 34

Analysis of an Acetone-Water Mixture

PRELABORATORY QUESTIONS

1. What color light would you expect Reichardt's dye to absorb when dissolved in a solvent whose polarity is between that of acetone and ethanol? Explain.

2. Acetone-water mixtures of different composition will exhibit different wavelengths of maximum absorbance (λ_{max}). Why is this?

3. What is the danger associated with acetone? How can this danger be avoided?

Name: _____   Date: _____

Lab Instructor: _____   Lab Section: _____

EXPERIMENT 34

Analysis of an Acetone-Water Mixture

RESULTS/OBSERVATIONS

1. Record any pertinent data in the space below. Clearly indicate both the property and the amount.

2. Perform any relevant calculations in the space below. Perform any plots on the graph paper provided.

Name: _____   Date: _____

Lab Instructor: _____   Lab Section: _____

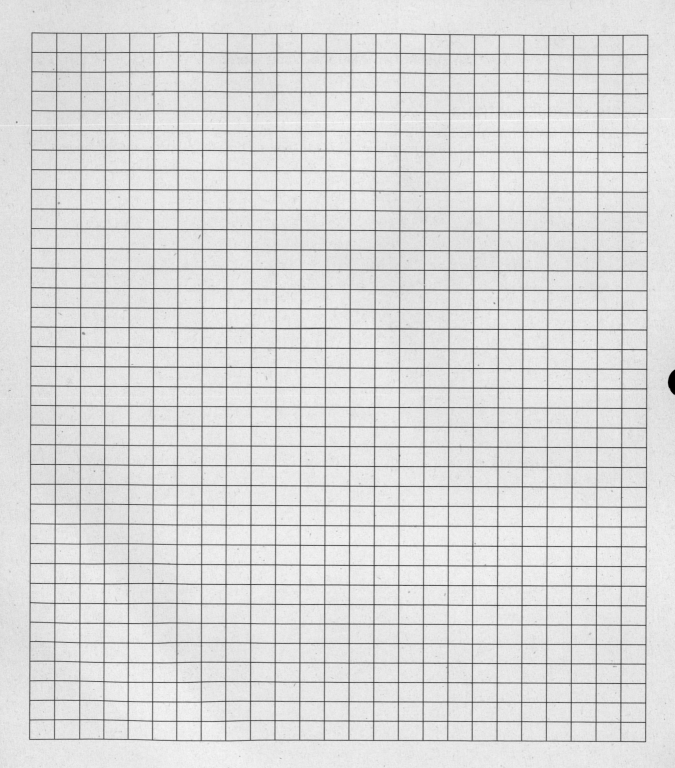

Name: _____   Date: _____

Lab Instructor: _____   Lab Section: _____

EXPERIMENT 34

Analysis of an Acetone-Water Mixture

POSTLABORATORY QUESTIONS

1. Which would absorb light of a greater wavelength, a 5 mL H_2O–10 mL acetone solution or a 5 mL H_2O–5 mL acetone solution? Explain.

2. Should the amount of Reichardt's dye used affect the value of λ_{max}? Explain.

Polymer Syntheses

OBJECTIVE

To synthesize both an addition polymer (polystyrene) and a condensation polymer (nylon) and examine their physical properties.

INTRODUCTION

Polymers are compounds in which chains or networks of small repeating units form very large molecules. Polymers are composed of small repeating units because of the manner in which they are formed. A polymerization reaction involves the repetitive bonding together of many smaller molecules, called *monomers*. The two major types of polymerization reactions are addition polymerization and condensation polymerization.

In addition polymerization, monomers react to form a polymer strand without net loss of atoms. The overall process of addition polymerization entails three steps: initiation, propagation, and termination. The most common examples of addition polymerization involve free-radical chain reactions of monomers containing carbon–carbon double bonds. The polymerization of ethylene ($H_2C = CH_2$) is a convenient example. The initiators of the free-radical chain reaction are produced from a small concentration of molecules with bonds weak enough to be broken by heating or the absorption of light and so produce free radicals. Organic peroxides, such as benzoyl peroxide $[(C_6H_5CO)_2O_2]$, are often used for this purpose. The weak O—O bond is broken, and two free radicals are formed:

$$\text{Initiation:} \quad (C_6H_5CO)_2O_2 \xrightarrow{\text{heat}} 2\ (C_6H_5CO)O\cdot \tag{1}$$

On account of their electron deficiency, the free radicals are highly reactive. The π bond of ethylene provides an immediate source of electron density, but the resulting reaction produces another, different free radical:

Propagation:

$$\tag{2}$$

This free radical can react with another ethylene molecule to continue the chain reaction:

Propagation:

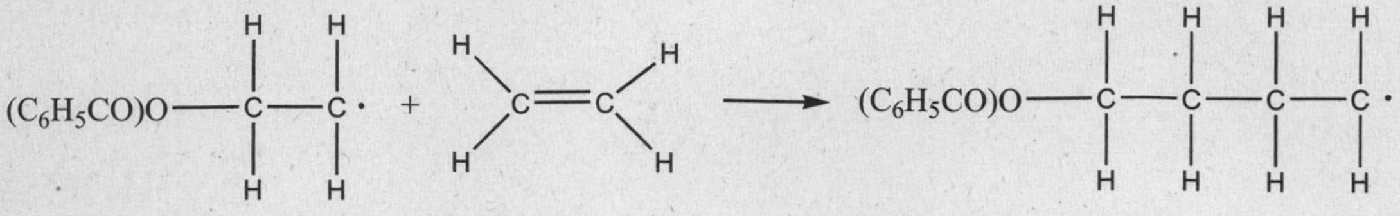

$$(C_6H_5CO)O—CH_2CH_2· \quad + \quad CH_2{=}CH_2 \quad \longrightarrow \quad (C_6H_5CO)O—CH_2CH_2CH_2CH_2· \quad (3)$$

The propagation steps will continue several hundred to several thousand more times, producing polyethylene, a polymer with the repeating unit shown in Figure 35.1.

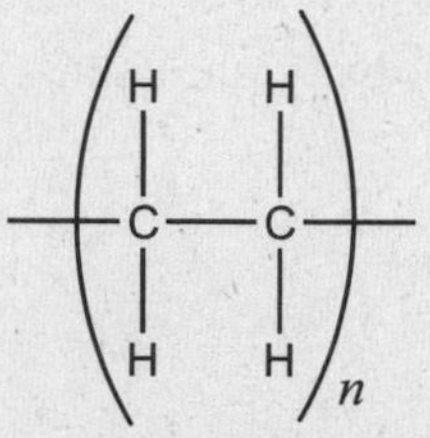

$$\left(\!\!\begin{array}{c} CH_2{-}CH_2 \end{array}\!\!\right)_n$$

Polyethylene

Figure 35.1

Termination occurs when two free radicals react to form a nonradical species:

Termination:

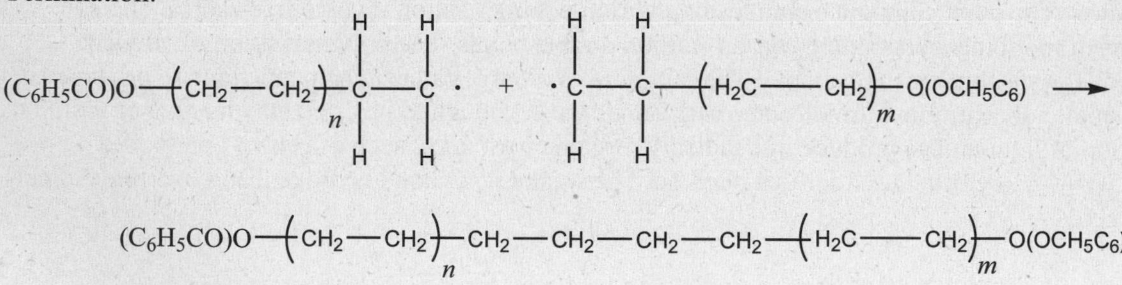

$$(C_6H_5CO)O{-}\!\!\left(CH_2{-}CH_2\right)_n\!\!{-}CH_2{-}CH_2· \ + \ ·CH_2{-}CH_2{-}\!\!\left(H_2C{-}CH_2\right)_m\!\!{-}O(OCH_5C_6) \longrightarrow$$

(4)

$$(C_6H_5CO)O{-}\!\!\left(CH_2{-}CH_2\right)_n\!\!{-}CH_2{-}CH_2{-}CH_2{-}CH_2{-}\!\!\left(H_2C{-}CH_2\right)_m\!\!{-}O(OCH_5C_6)$$

Polystyrene (Figure 35.2) will be prepared in this experiment by addition polymerization.

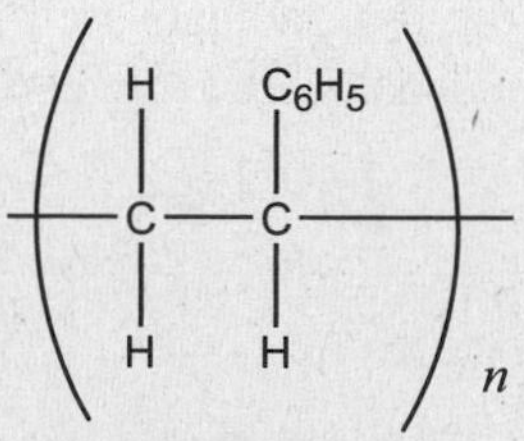

$$\left(\!\!\begin{array}{c} CH{-}CH(C_6H_5) \end{array}\!\!\right)_n$$

Polystyrene

Figure 35.2

Polystyrene will be synthesized from the monomer unit styrene (Figure 35.3) using the initiator 2-butanone peroxide [$CH_3C(OOH)_2CH_2CH_3$].

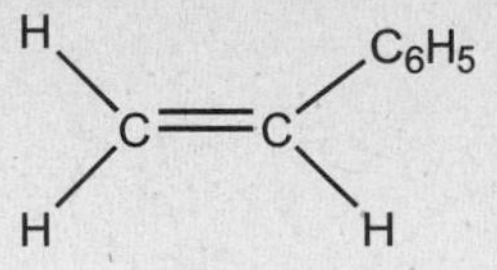

Styrene

Figure 35.3

Condensation polymerization involves the loss of a small molecule, often water, as each monomer unit is attached to the polymer strand. For example, Dacron, a polymer used widely in fibers for the manufacture of clothing, is formed by condensation polymerization of ethylene glycol and p-terephthalic acid:

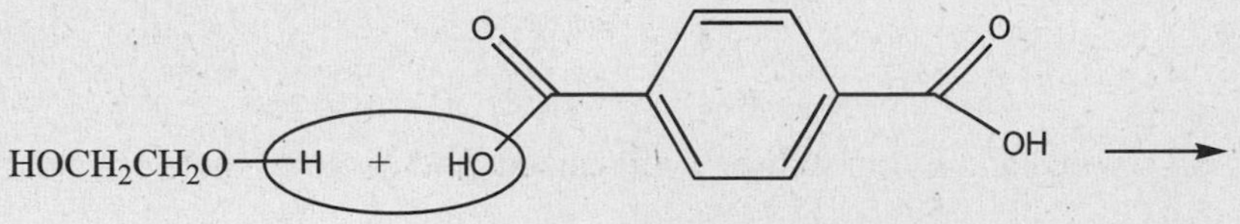

Ethylene glycol   p-Terephthalic acid   (5)

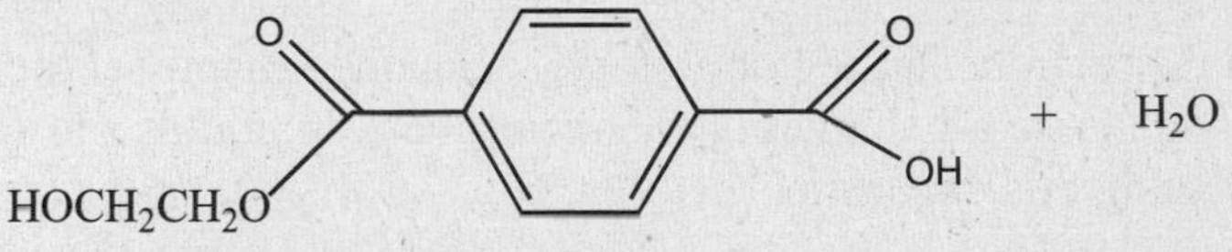

where the circled atoms are those that became water. This product molecule is a dimer (a combination of two molecules). Notice that this dimer contains the same functional groups that were involved in the reaction between ethylene glycol and p-terephthalic acid, so it also can undergo reaction, but now at both ends. The addition of monomers (and elimination of water molecules) repeats to create a long-chain molecule with the repeating unit shown in Figure 35.4.

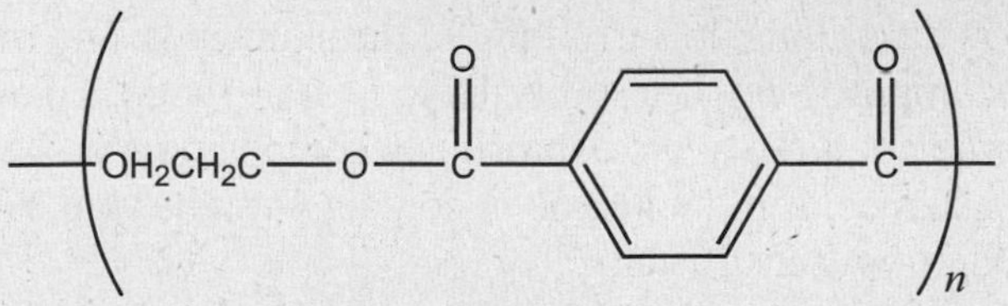

Dacron

Figure 35.4

Nylon, one of the polymers that will be synthesized in this experiment, is an example of condensation polymerization. Nylon can be prepared by reacting hexamethylenediamine with adipoyl chloride. For each reaction between the two starting materials, one HCl molecule is split out, and an amide linkage (—NH—CO—) is formed.

Hexamethylenediamine   Adipoyl chloride

(6)

Nylon

In this experiment, both polystyrene and nylon will be synthesized, and some of their physical properties will be examined.

ADDITIONAL READING

Read the sections in the Laboratory Techniques chapter at the beginning of this Lab Manual on cleaning glassware, handling chemicals, heating liquids and solutions, and decantation prior to performing this experiment.

Safety Precautions:

■ Protective eyewear approved by your institution must be worn at all times while you are in the laboratory.

■ 2-Butanone peroxide is hazardous. It is irritating to the skin, eyes, and mucous membranes. It can undergo explosive decomposition when subjected to heat or shock. To avoid inhalation, it should be used only in the fume hood. To avoid explosive decomposition, it should be kept away from heat, sparks, and open flames. If it does come in contact with the skin, rinse the area for 15 minutes with water. Alert your instructor.

■ Styrene is hazardous. It is irritating to the skin, eyes, and mucous membranes. It is flammable. To avoid inhalation, it should be used only in the fume hood. To avoid fires, no open flames should be used during this experiment. If it does come in contact with the skin, rinse the area for 15 minutes with water. Alert your instructor.

■ Both hexamethylenediamine and cyclohexane are flammable. Use these chemicals only in the fume hood. No open flames should be used during this experiment. In the event of a fire, alert your instructor. A fire in a small container can be smothered by covering the vessel with a watchglass. After the fire is extinguished, the glass will be hot, so avoid touching it until it cools. Use a fire extinguisher to eliminate a larger fire. If the fire is burning over too large an area to be extinguished easily, evacuate the area and activate the fire alarm. If your clothing should catch on fire, use the safety shower or a fire blanket to extinguish the flames. Be ready to assist your neighbors.

PROCEDURE

Part A: Preparation of an Addition Polymer (Polystyrene)

CAUTION: Because of the hazardous nature of the reagents, all work in this part of the experiment must be performed in the fume hood.

Commercially obtained styrene contains 4-*tert*-butylcatechol, a polymerization inhibitor. The inhibitor can be removed by passing the styrene through a chromatographic column containing alumina. Prepare a microscale chromatographic column by placing a small wad of glass wool in the bottom of the upper stem of a Pasteur pipet. Add dry alumina on top of the glass wool until the pipet is about half full. In a fume hood, use a clamp and a ringstand to support the chromatographic column. Place a 10 × 75 mm testtube in a small beaker for support. Position the testtube so that the tip of the chromatographic column is inside it. Still working in the fume hood, obtain about 2 mL of styrene in another 10 × 75 mm testtube. Transfer the styrene to the top of the chromatographic column using a Pasteur pipet. The purified styrene will collect in the original 10 × 75 mm testtube.

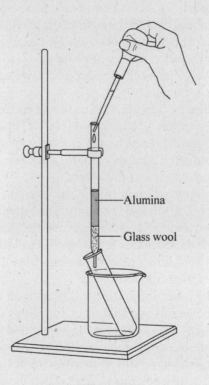

Figure 35.5
A microscale chromatographic column.

In the fume hood, prepare a hot-water bath by partially filling a 400 mL beaker to which has been added a magnetic stirbar. Place the water bath on a hotplate-stirrer, turn on the stirrer, and heat to boiling.

Using a Pasteur pipet, add 20 drops of 2-butanone peroxide, the polymerization initiator, to the purified styrene in the testtube. Slowly stir the mixture with a glass stirring rod so as not to introduce any air bubbles. Use a ring stand and a clamp to support the testtube containing the reaction mixture in the hot-water bath. Heat the testtube contents in the hot-water bath for about 2 hours. Check the volume of water in the hot-water bath frequently, and add water if necessary to offset any losses to vaporization. (While the reaction mixture is heating in the hot-water bath, proceed to Part B.)

After 2 hours of heating, turn off the heat on the stirrer-hotplate, and remove the testtube from the hot-water bath by loosening the clamp on the ringstand and lifting the testtube out, holding it by the clamp. Retighten the clamp on the ringstand, and let the testtube cool. Once the testtube has cooled, examine the appearance and hardness of the polystyrene. To examine its hardness, insert a glass stirring rod into the testtube, and press against the polystyrene. Record your observations.

Part B: Preparation of a Condensation Polymer (Nylon)

CAUTION: Because of the hazardous nature of the reagents, work in this part of the experiment must be performed in the fume hood. Do not work at your regular bench space until the nylon has been formed.

The hexamethylenediamine necessary for this experiment is dissolved in water. The adipoyl chloride is dissolved in cyclohexane. Water and cyclohexane do not mix when poured together. Cyclohexane, being less dense, floats atop water. The reaction between the hexamethylenediamine and adipoyl chloride occurs at the interface of the water and cyclohexane liquid layers. An indicator dye (methyl red, methyl orange, or bromocresol green) will be added to the water solution to make the interface between these two liquid layers more distinct.

Take a 100 mL beaker, a graduated cylinder, and a stirring rod to one of the fume hoods. Pour 25 mL of hexamethylenediamine solution (4% in 0.75 M NaOH) into the 100 mL beaker, using the graduations on the beaker to measure the quantity. Do *not* use your graduated cylinder to measure this solution; it must be dry for measuring the adipoyl chloride solution. Add about 1 mL (20 drops) of indicator dye to the solution in the beaker, and stir to mix it.

Measure 25 mL of adipoyl chloride solution (4% in cyclohexane) in your graduated cylinder.

Grasping the beaker in one hand, tilt it, and taking the graduated cylinder in your other hand, slowly and carefully pour the adipoyl chloride solution down the side of the tilted beaker so that it ends up floating atop the hexamethylenediamine solution with as little mixing as possible. Mixing causes the nylon to congeal into a blob rather than remaining threadlike. Nylon immediately should begin to form at the interface of the two solutions.

Being careful not to disturb the solutions in the beaker, set it on a pad of paper towels. Using your stirring rod, free the walls of the beaker of any strands or film of nylon, and push the material toward the center. Then reach into the beaker with your testtube holder and grasp the center of the nylon film. Lift a rope of nylon out of the solution slowly and smoothly; if you pull it too fast, the nylon thread will break. Wind the strand onto a wooden splint or stirring rod. If the thread breaks, it can be picked up again at the interface between the two solutions. Try to keep the fresh nylon off the benchtop by keeping it above the paper towel.

Transfer the nylon to a beaker of water for washing. Allow it stand in the water for several minutes. Pour the solutions remaining in the reaction beaker into the waste bottle provided, not down the sink. (The remaining solutions must not be poured down the drain because they could combine to form nylon in the drain pipes, and nylon is extremely difficult to remove.)

Take the beaker of nylon to your lab bench. Decant off the water. Wash the nylon with about 5 mL of acetone in a beaker. Decant the used acetone into the sink, and wash it down with water. Place the nylon on a paper towel to dry.

Once the nylon is dry, examine its color, hardness, and strength when pulled. Record your observations.

Waste Disposal. Dispose of all chemical waste as directed by your instructor.

Name: _____   Date: _____

Lab Instructor: _____   Lab Section: _____

EXPERIMENT 35

Polymer Syntheses

PRELABORATORY QUESTIONS

1. Identify the monomer(s) used to make poly(vinyl chloride) by addition polymerization.

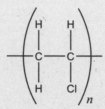

Poly(vinyl chloride)

Name: _____   Date: _____

Lab Instructor: _____   Lab Section: _____

2. Identify the monomers used to make Kevlar by condensation polymerization. (The small
    molecule split off in this condensation polymerization is HCl.)

Kevlar

Name: _____   Date: _____

Lab Instructor: _____   Lab Section: _____

3. Write out the structure for a polystyrene strand containing a total of 24 carbon atoms. (Don't include the carbons of the initiator in the count)."

4. Write out the structure for a nylon strand containing a total of four nitrogen atoms.

5. In the nylon preparation, the reaction will occur at the interface between two solutions. What is contained in each solution? Which solution forms the upper layer? Which the lower?

6. After you have prepared some nylon, why is it important not to the dispose of the remaining reaction solutions by pouring them down the sink drain?

Name: _____   Date: _____

Lab Instructor: _____   Lab Section: _____

EXPERIMENT 35

Polymer Syntheses

RESULTS/OBSERVATIONS

1. Color of polystyrene: _____

2. Hardness of polystyrene: _____

3. Color of nylon: _____

4. Hardness of nylon: _____

5. Strength of nylon when pulled: _____

Name: _____ Date: _____

Lab Instructor: _____ Lab Section: _____

EXPERIMENT 35

Polymer Syntheses

POSTLABORATORY QUESTIONS

1. Nylon also can be formed by the reaction of hexamethylenediamine with sebacoyl chloride:

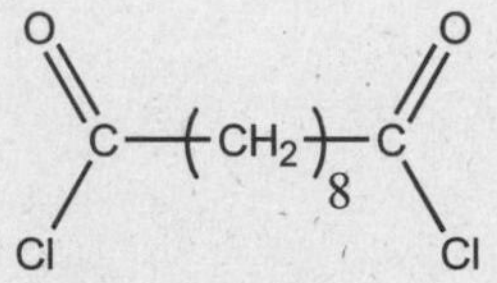

Sebacoyl chloride

What is the structure of nylon produced from this reaction?

2. Proteins are an example of a biologic polymer. The monomers of proteins are amino acids. Twenty different amino acids are commonly found in proteins. The simplest of these amino acids is glycine:

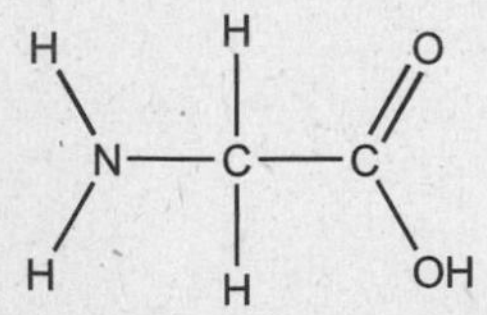

Glycine

What is the structure of the protein that would result from a condensation polymerization of glycine? What small molecule is also produced during this reaction? Compare this protein with nylon. How are the two polymers similar? How are they different?

EXPERIMENT 36

Polymer Cross-Linking and Viscosity

OBJECTIVE

To devise and perform a procedure capable of determining the effect cross-linking has on polymer solution viscosity. Borate ion $[B(OH)_4^-]$ will be used to cross-link strands of the polymer poly(vinyl alcohol).

INTRODUCTION

Polymers are compounds in which chains or networks of small repeating units form very large molecules. Polymers are composed of small repeating units owing to the manner in which they are formed. A polymerization reaction involves the repetitive bonding together of many smaller molecules, called *monomers*. For example, polyethylene, one of the simplest polymers, is formed from ethylene monomers:

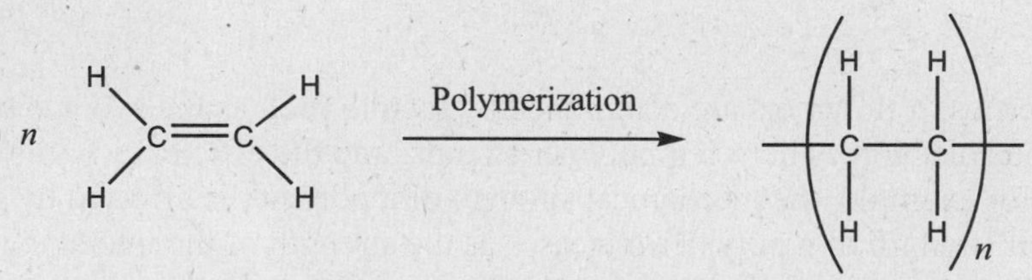

(1)

Ethylene　　　　Polyethylene

where n represents a very large number (usually several thousand).

The polymer to be examined in this experiment is poly(vinyl alcohol), which is often referred to by its initials, *PVA*. Interestingly, PVA is not made by polymerizing vinyl alcohol

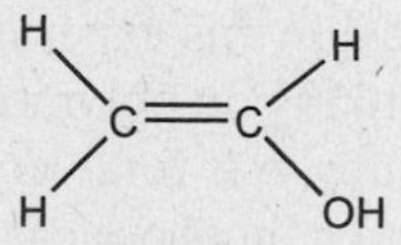

Vinyl alcohol

Figure 36.1

because vinyl alcohol is unstable. Rather, PVA is made by first polymerizing vinyl acetate, yielding poly(vinyl acetate):

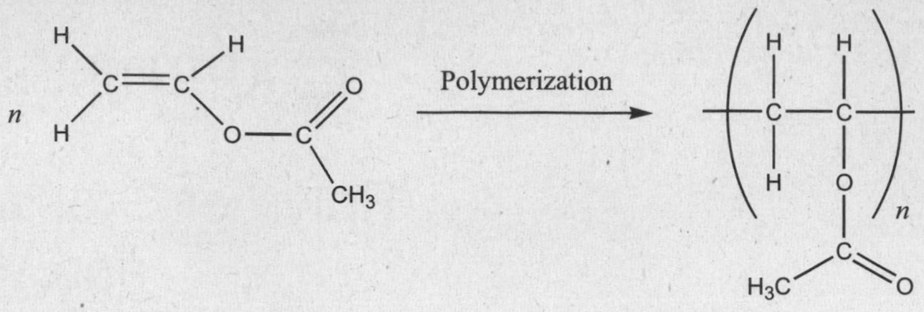

Vinyl acetate   Poly(vinyl acetate)

(2)

Poly(vinyl acetate) then is hydrolyzed in basic solution to give acetate ion ($CH_3CO_2^-$) and PVA, which has the structure shown in Figure 36.2.

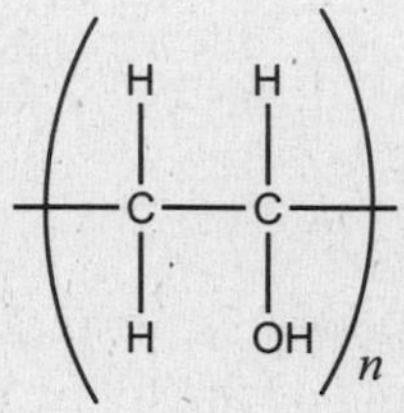

Poly(vinyl alcohol) (PVA)

Figure 36.2

The physical properties of polymers are determined by factors such as the average chain length, the strength of the intermolecular forces between polymer strands, and the efficiency with which polymer chains pack together. For example, the mechanical strength of a polymer is affected by all three of these factors. The mechanical strength of a polymer increases as the strength of the interactions between chains increases. Increasing the average chain length increases the accumulated intermolecular forces between chains, resulting in greater polymer strength. For polymers of the same average chain length, greater intermolecular forces lead to greater mechanical strength. Polymer chains that interact via dipole–dipole forces, for example, are stronger than those which interact by London dispersion forces. Lastly, chain packing arrangements that maximize intermolecular contact also maximize intermolecular forces, resulting in greater strength. Unbranched polymer chains better maximize intermolecular contact than do branched polymer chains.

Another manner of affecting the physical properties of a polymer is by cross-linking, the introduction of chemical linkages between polymer strands. In this experiment, the extent of cross-linking in a PVA solution will be varied. The effect of cross-linking on polymer solution viscosity will be observed.

PVA contains a large number of OH groups, one for every two carbon atoms in the polymer chain. These OH groups are potential hydrogen bond donors or acceptors. Consequently, PVA is quite soluble in water. PVA also can form hydrogen bonds with other substances, such as borate ion [$B(OH)_4^-$]. In this experiment, a solid sample of PVA first will be dissolved in water. A solution of borate ion then will be added to the PVA solution. The solution of borate ion is formed by dissolving borax ($Na_2B_4O_7 \cdot 10H_2O$) in water:

$$Na_2B_4O_7 \cdot 10H_2O(s) \xrightarrow{H_2O} 2\,B(OH)_3(aq) + 2\,Na^+(aq) + 2\,B(OH)_4^-(aq) + 3\,H_2O(l)$$

(3)

The borate ion has the tetrahedral structure shown in Figure 36.3.

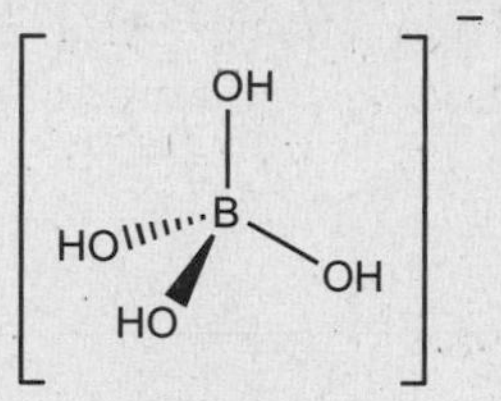

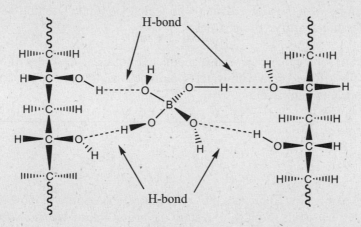

Borate ion

Figure 36.3

The borate ions will cross-link with the PVA chains by forming hydrogen bonds between strands of the polymer, producing a mixture consisting of a three-dimensional network of polymer strands that are hydrogen-bonded to borate ions and entrapped water molecules (see Figure 36.4).

Figure 36.4
Borate ion cross-links PVA chains through hydrogen bonds.

The nascent cross-linking then will be removed by addition of sulfuric acid (H_2SO_4), which reacts with the borate ion according to the reaction

$$2\ B(OH)_4^-(aq) + H_2SO_4(aq) \rightarrow 2\ B(OH)_3(aq) + SO_4^{2-}(aq) + 2\ H_2O(l) \tag{4}$$

Lastly, addition of base (NaOH) will be used to neutralize the just-added acid, restoring the cross-linking.

Your purpose in this experiment is to determine the effect cross-linking has on the viscosity of the polymer solution. Does cross-linking increase, decrease, or have no effect on the viscosity of the polymer solution? If there is an observable effect, is it proportional to the extent of cross-linking? Does the removal of cross-linking by the addition of sulfuric acid bring about a complete restoration of the viscosity of the pre-cross-linked polymer solution? Does neutralization of the acid lead to a complete restoration of the viscosity of the cross-linked polymer solution?

ADDITIONAL READING

Read the sections in the Laboratory Techniques chapter at the beginning of this Lab Manual on cleaning glassware, handling chemicals, weighing, and heating liquids and solutions prior to performing this experiment.

 Safety Precautions:

- Protective eyewear approved by your institution must be worn at all times while you are in the laboratory.

- Chemical burns can result when 3 M H_2SO_4 comes in contact with skin. If you spill 3 M H_2SO_4 on your skin, immediately wash the affected area with water. Continue washing with water for 15 minutes. Have a classmate notify your instructor.

- Handle the 1 M NaOH with care; it can cause chemical burns. If you spill 1 M NaOH on your skin, immediately wash the affected area with water. Continue washing with water for 15 minutes. Have a classmate notify your instructor.

EXPERIMENT

The procedure below explains how to prepare the cross-linked PVA solution, how to remove the cross-linking by addition of sulfuric acid, and how to restore the cross-linking by neutralizing the acid through the addition of base. What it does not provide are instructions concerning when or how to measure viscosity changes. This task is left to you. Will your measurements be quantitative, qualitative, or a mixture of both? If the former, how will you quantify the viscosity? Answering the prelaboratory questions will help you to answer these questions. The reagents and equipment available to you are listed in the Equipment and Reagents section of this experiment.

PROCEDURE

Part A: Cross-Linking PVA

Use a hotplate to heat 50 mL of tap water in a 200 mL beaker to boiling. Weigh out about 2 g of PVA [poly(vinyl alcohol)]. Turn off the heat when the water in the beaker starts to boil. Allow the water to cool to about 90°C. Sprinkle a bit of powdered PVA onto the surface of the hot water, and stir to dissolve the solid. Heat the solution gently, trying to keep the solution just below the boiling point. Once the first portion of PVA has dissolved, sprinkle a bit more onto the surface of the water, and heat with stirring until dissolution occurs. Repeat this until you have dissolved the entire 2 g sample of PVA. Attempting to dissolve too much of the powdered polymer at once will result in the formation of a gooey, unusable mess. Overheating will produce an insoluble scum on the surface. The final solution should be clear and nearly colorless. When dissolution of the powdered polymer is complete, discontinue heating, and allow the beaker to cool.

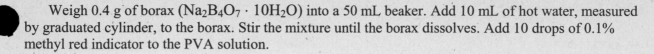

Weigh 0.4 g of borax ($Na_2B_4O_7 \cdot 10H_2O$) into a 50 mL beaker. Add 10 mL of hot water, measured by graduated cylinder, to the borax. Stir the mixture until the borax dissolves. Add 10 drops of 0.1% methyl red indicator to the PVA solution.

Note: Assuming that the PVA solution is not too hot, it is safe to handle it with your bare hands from this point onward in Part A.

Once the PVA solution has cooled sufficiently to allow safe handling of the beaker, add 1.0 mL of borax solution with continuous stirring. Use a calibrated transfer pipet to measure this volume. Continue stirring until a homogeneous mixture forms. The mixture may separate into two phases; with time and sufficient stirring, a homogeneous mixture will be achieved. The resulting mixture has a pH of about 9, causing the methyl red indicator to appear yellow.

Add, with stirring, a second 1.0 mL portion of the borax solution to the mixture. Stir until the mixture is homogeneous.

Stir in a third 1.0 mL portion of the borax solution until the mixture is homogeneous.

Part B: Removal and Resurrection of Cross-Linking

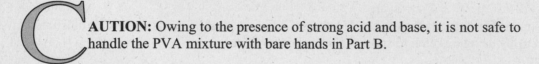

CAUTION: Owing to the presence of strong acid and base, it is not safe to handle the PVA mixture with bare hands in Part B.

Add 3.0 M H_2SO_4, dropwise, to the mixture with stirring until it is red; the color change indicates that the mixture now has an acidic pH. It will require approximately the amount you calculated in prelaboratory question 5. During the intermediate stages of mixing, before the mixture is homogeneous, you may see lumps of yellow in a red mixture. If this happens, keep stirring and mixing until the solution is homogeneous.

Add 2.0 mL of 1.0 M NaOH to the mixture. This is more than enough base to neutralize the added H_2SO_4. Stir the mixture again. The methyl red indicator again should be yellow.

EQUIPMENT AND REAGENTS

To perform this experiment, you will have access to all the equipment in your lab drawer, 3.0 M H_2SO_4, 1.0 M NaOH, a calibrated transfer pipet, popsicle sticks, marbles, ball bearings, and the electronic balances.

Waste Disposal. Dispose of all chemical waste as directed by your instructor.

Name: _____   Date: _____

Lab Instructor: _____   Lab Section: _____

EXPERIMENT 36

Polymer Cross-Linking and Viscosity

PRELABORATORY QUESTIONS

1. Write out the structure for a PVA strand containing a total of six carbon atoms.

2. PVA strands can form hydrogen bonds between different parts of the same strand. Illustrate how this can occur by using a sketch to show the hydrogen bonding.

3. Define viscosity. In what units is viscosity typically reported? Give one example of an experimental technique used to measure viscosity.

Name: _____   Date: _____

Lab Instructor: _____   Lab Section: _____

4. State four places within the procedure where you will make measurements of viscosity.

5. Calculate the amount of 3.0 M H_2SO_4 necessary to react with 3 mL of the solution of borate ions created in the procedure [0.4 g of borax ($Na_2B_4O_7 \cdot 10H_2O$) in 10 mL H_2O].

Name: _____   Date: _____

Lab Instructor: _____   Lab Section: _____

EXPERIMENT 36

Polymer Cross-Linking and Viscosity

RESULTS/OBSERVATIONS

1. Polymer viscosity measurements (state the stage in the procedure at which the polymer solution viscosity was measured, the technique used for measuring the viscosity, and the viscosity):

2. Does cross-linking increase, decrease, or have no effect on polymer solution viscosity?

3. If there is an observable effect, is it proportional to the extent of cross-linking?

4. Does the removal of cross-linking by the addition of sulfuric acid bring about a complete restoration of the pre-cross-linked polymer solution viscosity?

5. Does neutralization of the acid lead to a complete restoration of the cross-linked polymer solution viscosity?

Name: _____   Date: _____

Lab Instructor: _____   Lab Section: _____

EXPERIMENT 36

Polymer Cross-Linking and Viscosity

POSTLABORATORY QUESTIONS

1. Which would be expected to have a greater viscosity, a polymer sample with strands cross-linked by hydrogen bonds or a polymer sample with strands cross-linked by covalent bonds, all other factors being equal? Explain.

2. What is happening at the molecular level when the polymer sticks to your fingers? (Recall that proteins, such as skin tissue, contain many side groups that contain nitrogen and oxygen atoms.)

APPENDIX A

Handling Data

Part A-1: Recording Data

There is a degree of uncertainty in every measurement: Nothing can be measured exactly. An experimenter must know the degree of uncertainty associated with each measurement he or she makes and record each measurement in such a manner that the degree of uncertainty is unambiguous to a reader.

The accepted manner for recording a measured value is to state all the digits that are certain, followed by a single last digit that has been estimated. The last digit indicates the uncertainty of the measurement. The total number of digits in a measurement is called the number of significant figures.

For a device with a digital display, such as an electronic balance, recording the correct number of significant figures is easy. Simply record all the displayed digits, even if the last one or more digits are zeros. Neglecting to record trailing zeros suggests that the measurement was less precise than it actually was.

For a device with a scale, such as a graduated cylinder or thermometer, record all the certain digits, followed by one last estimated digit. For example, Figure A.1 shows portions of two graduated cylinders. The volume of liquid in the graduated cylinder on the left is definitely between 8 and 9 mL. This volume would be recorded as 8.4 or 8.5 mL. The first digit is certain. The last digit is uncertain; its value was estimated. Recording the volume as 8.45 mL would be misleading because the use of three significant figures suggests to the reader that the measured value is known with a greater certainty than it actually is. The volume of liquid in the graduated cylinder on the right would be recorded as 8.47 or 8.48 mL. In this case, recording the volume as 8.5 mL would undervalue the amount of certainty in the measurement.

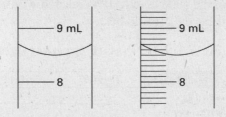

Figure A.1
Portions of two graduated cylinders of different precision. The curved line represents the meniscus of the liquid.

Some instruments, such as pipets and volumetric flasks, bear a single marking instead of a series of graduations. For these instruments, the manufacturer has indicated on the glassware the uncertainty in the measurement when it is filled to the marking.

Figure A.2
A volumetric flask.

For the volumetric flask shown in Figure A.2, the uncertainty lies in the hundredths place, so the volume should be recorded to four significant figures, 25.00 ± 0.05 mL at 20°C in this case. As was demonstrated in the preceding examples, all recorded measurements must include descriptive units (g, mL, °C, etc.) to be meaningful.

Part A-2: Organizing Data

Organization is essential if the frequently large amount of data and observations collected during an experiment is to be understandable to a reader. When a number of repetitive observations or measurements are made in an experiment, the best way to organize these data is in tabular form. A well-constructed table is easy to read and facilitates interpretation of the data. The table should be given a descriptive title, as should each column (or row). The descriptive column headings should include the units of the quantity, if appropriate. For example, in a certain acid-base titration experiment, a measured amount of the solid acid KHP is reacted with a measured volume of NaOH solution. A table appropriate for these data would be:

Table A.1: KHP-NaOH Titration Data

	Trial 1	Trial 2	Trial 3	Trial 4
Mass of 125 mL Erlenmeyer flask (g)				
Combined mass of 125 mL Erlenmeyer flask and KHP (g)				
Mass of KHP (g)				
Initial NaOH volume in buret (mL)				
Final NaOH volume in buret (mL)				
Volume of NaOH (mL)				

Part A-3: Graphing Data

Graphs are invaluable at presenting data. Graphs make it is easy to decipher the changing relationship between two variables. The slope and y-intercept of many graphs have important physical meaning, thus making it possible to determine some quantities without resorting to separate experiments. Many graphs can be extrapolated to data regions difficult to access experimentally.

Plotting Variables

A line graph consists of two axes perpendicular to one another. The horizontal axis is called the *x*-axis, or abscissa. The vertical axis is termed the *y*-axis, or ordinate. The *x*-axis represents the independent variable, the variable deliberately altered by the experimenter. The *y*-axis represents the dependent variable, the variable whose value changes in response to changes in the independent variable. Take as an example an experimenter who desires to determine the relationship between gas pressure and temperature. The experimenter systematically changes the gas temperature, noting the corresponding pressure change. Since the experimenter purposely altered the temperature of the gas, temperature is the independent variable and should be represented by the *x*-axis. Pressure is the dependent variable, represented by the *y*-axis. See Figure A.3 for an example.

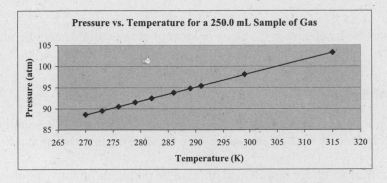

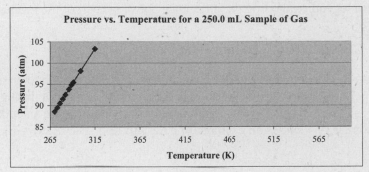

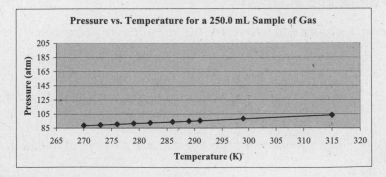

Figure A.3

The three graphs represent the same data using different scales for the axes. The top graph uses axis scales that spread the data over almost the entirety of the graph. This is the best choice of scales because it represents the trend between the variables most clearly. The middle graph uses an x-axis scale that is too large. The bottom graph uses a y-axis scale that is too large. All three graphs bear a descriptive title, and the variables plotted on each axis are clearly labeled, including the units.

Selecting Axis Scales

The scale of each axis should be chosen so that the data span nearly the entirety of the graph, not just one small region. Figure A.3 provides examples of both well and poorly constructed graphs.

Choosing a Title

Each graph should be given a title or caption that accurately describes the variables plotted, the chemicals used, the quantities to be determined, and any special conditions pertinent to the experiment. See Figure A.3 for an example.

Labeling Axes

Each of the axes must be labeled with the name or symbol of the variable and the units. See Figure A.3 for an example.

Drawing Data Points

Represent the data points as dots, squares, circles, triangles, or diamonds. The choice of symbol is arbitrary. All that is necessary is to clearly convey the location of each data point. Figure A.3 contains an example of this.

   If data from multiple trials are represented on a single graph, different symbols should be used for each data set. A figure legend explaining the difference between each of the trials should be included.

Drawing Lines or Curves Through Data

When connecting data points, avoid dot-to-dot connections. Instead, use a line or smooth curve. If the data do not all fit on a line or curve, draw the so-called best-fit line or curve through the data. The best-fit line or curve is the one for which an equal number of data points lie an equal distance above and below the line. Figure A.4 shows an example of a best-fit line. The best-fit line helps to eliminate random error.

Determining the Slope and Intercept

For data that are linear, the slope and intercept of the line often are of importance. The slope is calculated from the equation

$$\text{Slope} = \frac{y_2 - y_1}{x_2 - x_1} \tag{1}$$

where (x_1, y_1) and (x_2, y_2) are two points on the line. For example, the slope of the line in Figure A.4 is

$$\text{Slope} = \frac{172\,\text{atm} - 142\,\text{atm}}{350\,\text{K} - 290\,\text{K}} = 0.500\,\text{atm/K} \tag{2}$$

   Once the slope has been determined, either intercept of the line can be determined from Equation (1). To determine the y-intercept, set x_1 equal to zero, let (x_2, y_2) be any point on the line, and solve for y_1. For example, to determine the y-intercept of the line in Figure A.4,

$$0.500 \text{ atm/K} = \frac{172 \text{ atm} - y_1}{350 \text{ K} - 0 \text{ K}} \tag{3}$$

$$y_1 = -3 \text{ atm}$$

Determine the x-intercept by setting y_1 equal to zero and solving for x_1.

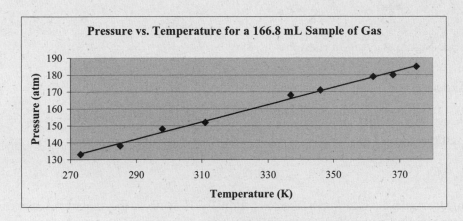

Figure A.4
An example of a best-fit line through experimental data.

Computer-Generated Graphs

Software programs capable of generating plots are seemingly ubiquitous, easy to use, and ever more popular with both students and instructors. In large part, the popularity of these software programs is due to their ability to automatically perform such tasks as determining best-fit lines, slope and intercept calculations. They also can be used to determine the correlation between the data points and the best-fit line. The correlation or R^2 value is expressed as a number between 0 and 1. The closer the R^2 value is to 1, the greater the correlation between the data points and the best-fit line, that is, the more reliable is the best-fit line. Moreover, such software programs produce plots that are aesthetically pleasing. If your instructor allows it and you are so inclined, generate your graphs using such a software program. All the graphing instructions given in this appendix are still applicable. But be warned: The default manipulations made by the software program might not lead to the best presentation of your data. Examine the resulting graph with a critical eye, and edit it as necessary to best illustrate your data.

Part A-4: Evaluating Data

There is a degree of uncertainty in every measurement: Nothing—be it the mass of a solid sample, the volume of an aqueous solution, or any other physical observable—can be measured exactly. As an experimenter, it is necessary to know the degree of uncertainty associated with each measurement and the relationship between these uncertainties and the reliability of the final result.

Precision and accuracy are the terms used commonly to describe the degree of uncertainty in a measurement. Precision refers to the degree of refinement in the performance of an operation or the degree of perfection in the instruments and methods used to obtain a result; it is an indication of the reproducibility of a result. Accuracy refers to how close a measured value is to the accepted or "true" value. Precision relates to the quality of an operation by which a result is obtained. Accuracy relates to the quality of a result. To illustrate the difference between precision and accuracy, consider the analogy of a marksman for which the "truth" is represented by a bullseye. In Figure A.5, the marksman achieved reproducibility, but the results differ from the "true" value. Thus the results can be described as precise but inaccurate. In Figure A.6, the results are not reproducible and must be considered imprecise. However, the results are clustered around the "true" value, making them accurate. The results in Figure A.7 are both precise and accurate. Those in Figure A.8 are neither precise nor accurate.

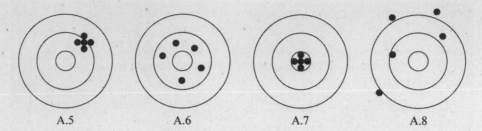

A.5   A.6   A.7   A.8

Figures A.5-A.8
Targets illustrating the results of four marksmen provide examples of the difference between precision and accuracy.

Precision

In the preceding example, the data were presented pictorially. However, a pictorial representation of data isn't always possible. And even when it is possible, it is often an inconvenient way of presenting data. When repetitive measurements are made in an experiment, it is typical to express the data as a single representative number. This number is called the mean $(\bar{x})$, or average. The mean is defined by the equation

$$\bar{x} = \frac{\sum_i x_i}{N} \tag{4}$$

where each x_i is the result of an individual measurement. The symbol Σ denotes summation:

$$\sum_i x_i = x_1 + x_2 + x_3 + \cdots + x_N \tag{5}$$

The mean is the sum of the measured values divided by the total number of values (N).

Precision is frequently expressed quantitatively in terms of deviation from the mean $(\bar{x} - x_i)$. This is the difference between an individual measured value and the mean value. The average deviation $(\Delta\bar{x})$,

$$\Delta\bar{x} = \frac{\sum_i |\bar{x} - x_i|}{N} \tag{6}$$

is used to express precision when the data set is small (less than five repetitive measurements). A large average deviation indicates imprecision. An average deviation that is small relative to the mean indicates a high degree of precision.

The size of the average deviation compared with the mean can be conveniently expressed as the relative average deviation:

$$\text{Relative average deviation} = \frac{\Delta\bar{x}}{\bar{x}} 100 \tag{7}$$

The larger the relative average deviation, the lower the degree of precision.

For larger data sets (five or more repetitive measurements), the standard deviation (s) is used to express the precision:

$$s = \sqrt{\frac{\sum_i (\bar{x} - x_i)^2}{N-1}} \tag{8}$$

The relative standard deviation, which expresses how the standard deviation compares with the mean, is defined as

$$\text{Relative standard deviation} = \frac{s}{\bar{x}} 100 \tag{9}$$

An example will demonstrate the calculation of these various quantities. Suppose that an experiment to determine the concentration of a potassium hydroxide (KOH) solution was performed. To enable the precision of the method to be determined, the experiment was repeated four times. The results are shown in Table A.2. Using these results, the mean, average deviation, relative average deviation, standard deviation, and relative standard deviation will be calculated.

Table A.2: KOH Concentration Determination

Trial	KOH Concentration (M)
1	1.483
2	1.508
3	1.588
4	1.427

$$\text{Mean concentration: } \bar{M} = \frac{1.483\ M + 1.508\ M + 1.588\ M + 1.427\ M}{4} = 1.502\ M \tag{10}$$

Average deviation:

$$\Delta\bar{M} = \frac{|1.502 - 1.483\ M| + |1.502 - 1.508\ M| + |1.502 - 1.588\ M| + |1.502 - 1.427\ M|}{4} \tag{11}$$

$$\Delta\bar{M} = 0.047\ M \tag{12}$$

$$\text{Relative average deviation: } \frac{0.047\ M}{1.502\ M} 100 = 3.1\% \tag{13}$$

Standard deviation:

$$s = \sqrt{\frac{(1.502 - 1.483\ M)^2 + (1.502 - 1.508\ M)^2 + (1.502 - 1.588\ M)^2 + (1.502 - 1.427\ M)^2}{4-1}} \tag{14}$$

$$s = 0.067\ M \tag{15}$$

$$\text{Relative standard deviation: } \frac{0.067\ M}{1.502\ M} 100 = 4.5\% \tag{16}$$

Note the different values of the average deviation, which is appropriate for a small set of data such as was present in this example, and the standard deviation, which is more aptly applied to a larger set of data.

Accuracy

In cases where the "true" or accepted value is known, the accuracy of a result can be quantified by calculating the percent error:

$$\text{Percent error} = \left|\frac{\text{accepted value} - \text{mean value}}{\text{accepted value}}\right| \times 100 \tag{17}$$

The two vertical lines indicate absolute value. The smaller the percent error, the greater the accuracy.

For example, if the true value of the KOH concentration from the preceding example is 1.500 M, the percent error is

$$\text{Percent error} = \left|\frac{1.500\ M - 1.502\ M}{1.500\ M}\right| \times 100 = 0.133\% \tag{18}$$

Dealing with Bad Data

Occasionally a measured value appears to be inconsistent with the remaining data; that is, it is much higher or lower than the mean. The Q test is used to help decide whether to keep such an outlying value or discard it as unreliable.

The quantity Q is the absolute difference between the questionable measurement (x_q) and the next closest measurement (x_n) divided by the range (ω) of the entire data set:

$$Q = \frac{|x_q - x_n|}{\omega} \tag{19}$$

If the calculated value of Q is greater than the value listed in Table A.3, the measurement should be discarded.

Table A.3: Values of Q for Rejecting Data

Q (90% confidence)	3	4	5	6	7	8	9	10
Number of observations	0.94	0.76	0.64	0.56	0.51	0.47	0.44	0.41

An example will better illustrate the use of the Q test. Suppose that the experiment of the preceding example to determine KOH concentration was repeated a fifth time with a result of 1.836 M. This value is significantly larger than the other measured values (see Table A.2). It might be that it would be best to discard this measurement as unreliable. The Q value for this measurement is

$$Q = \frac{|1.836\ M - 1.588\ M|}{1.836\ M - 1.427\ M} = 0.606 \tag{20}$$

Since Q is less than the value given in Table A.3, it should be retained. Put another way, there is more than a 10% chance that the value 1.836 M is reliable.

Vapor Pressure of Water as a Function of Temperature

Temperature (°C)	Pressure (mmHg)
16	13.63
17	14.53
18	15.48
19	16.48
20	17.54
21	18.65
22	19.83
23	21.07
24	22.38
25	23.76
26	25.21
27	26.74
28	28.35
29	30.04
30	31.82
31	33.70
32	35.66
33	37.73
34	39.90
35	42.18
40	55.32
45	71.88
50	92.51

Standard Reduction Potentials at 25°C

Half-Reaction	E^o (V)
$Co^{3+}(aq) + e^- \rightarrow Co^{2+}(aq)$	1.81
$Ce^{4+}(aq) + e^- \rightarrow Ce^{3+}(aq)$	1.72
$Hg^{2+}(aq) + 2\,e^- \rightarrow Hg(l)$	0.85
$Ag^+(aq) + e^- \rightarrow Ag(s)$	0.80
$Hg_2^{2+}(aq) + 2\,e^- \rightarrow 2\,Hg(l)$	0.80
$Fe^{3+}(aq) + e^- \rightarrow Fe^{2+}(aq)$	0.77
$Cu^+(aq) + e^- \rightarrow Cu(s)$	0.52
$Cu^{2+}(aq) + 2\,e^- \rightarrow Cu(s)$	0.34
$Cu^{2+}(aq) + e^- \rightarrow Cu^+(aq)$	0.15
$Sn^{4+}(aq) + 2\,e^- \rightarrow Sn^{2+}(aq)$	0.15
$2\,H^+(aq) + 2\,e^- \rightarrow H_2(g)$	0
$Fe^{3+}(aq) + 3\,e^- \rightarrow Fe(s)$	−0.04
$Pb^{2+}(aq) + 2\,e^- \rightarrow Pb(s)$	−0.13
$Sn^{2+}(aq) + 2\,e^- \rightarrow Sn(s)$	−0.14
$Cr^{3+}(aq) + e^- \rightarrow Cr^{2+}(aq)$	−0.41
$Fe^{2+}(aq) + 2\,e^- \rightarrow Fe(s)$	−0.45
$Cr^{3+}(aq) + 3\,e^- \rightarrow Cr(s)$	−0.74
$Zn^{2+}(aq) + 2\,e^- \rightarrow Zn(s)$	−0.76
$Cr^{2+}(aq) + 2\,e^- \rightarrow Cr(s)$	−0.91
$Mg^{2+}(aq) + 2\,e^- \rightarrow Mg(s)$	−2.37

The Laboratory Notebook

The purpose of a laboratory notebook is to keep a permanent, accurate record of the work done in the laboratory. This record should include, among other things, procedural steps, data, observations, the results of preparatory research, and calculations. All observations, data, notes, and calculations should be recorded in the laboratory notebook concurrent with performing the experiment lest these important items be noted incorrectly later or forgotten. Data never should be recorded on scrap pieces of paper because they are too easily misplaced.

The following are guidelines for keeping a lab notebook:

Type of Notebook

Many different types of lab notebooks are available. Your instructor might specify the type you are to use. If your instructor doesn't, use the following general guidelines to direct your choice. The notebook should be bound. Notebooks with loose or spiral pages are discouraged because pages can be added or removed. The pages should be square ruled in order to allow graphing. A notebook containing carbonless duplicating sets is useful. Your instructor might not allow you to remove your notebook from the laboratory, or your instructor sometimes might collect your notebook for grading. Carbonless duplicate pages allow both you and your instructor to retain a copy of the lab notebook. Alternatively, carbon paper itself could be used to make duplicate copies on plain paper.

Numbering Pages

The pages of the notebook must be numbered in sequence. If the notebook you purchase does not come with prenumbered pages, number them yourself.

Table of Contents

The first few pages of the notebook should be reserved for a table of contents. The table of contents should be updated as each experiment is begun.

Title, Purpose, and Physical Constants

Typically, only the right-hand pages of the notebook are used for recording information. Begin each experiment on a fresh page. At the top of the page, write the title of the experiment, the date, the names of any lab partners, and a brief statement of purpose. List any pertinent physical constants: molecular weights, boiling points, melting points, solubilities, toxicities, etc. Provide any necessary prelaboratory calculations.

Procedure

The required level of written detail concerning the procedure will vary from one instructor to another. It might be that your instructor allows you simply to reference the appropriate experimental procedure in this Lab Manual. At the other end of the spectrum, you might be asked to provide a detailed explanation of all your procedural steps. Or you might be asked to provide something in between. If you are asked to write an experimental procedure in your lab notebook, it should consist of a detailed running account of what you have done in the lab, specifying the reagents, quantities, concentrations, and equipment used. When describing what you did, you should write, "I added 24.56 mL of 0.9988 M NaOH . . . ," for example, not, "Add 25 mL of ~1 M standardized NaOH. . . ." Proper English grammar and punctuation should be used. Brevity is prized. You should sketch and label pictures of all experimental apparatus.

Data and Observations

Observations include the appearances of chemical reagents, intermediates, and products and temperature changes, color changes, and changes of phase, for example, formation of gases or precipitates from a liquid solution. Use descriptive terms when noting observations: "slow gas formation," "bright-yellow precipitate," or "opaque blue solution." Data entries must be unambiguous. When recording numerical data, use units and significant digits to express the type and sensitivity of the measurement. Tabulate data wherever possible.

Calculations and Graphs

Show a complete example of each type of calculation, using proper units. A sample calculation usually includes a general relationship or formula, a substitution of data into the equation, and an answer. It is not necessary to show the algebraic manipulation of an equation. Repetitive calculations should not be shown, but the results should be tabulated. Results for each of two or more samples should be calculated separately, and the final results should be averaged, if appropriate. Don't average the raw data and then perform one calculation. Carry extra digits through your calculations, and round the final answer to the correct number of significant figures.

Approximate calculations and notes often are placed on the left-hand pages of the lab notebook.

Graphs should be drawn large enough and axes scaled such that data are represented clearly. Each graph should bear a descriptive title and labeled axes, including units. Curves drawn through data points should be smooth lines. (See Appendix A, Part A-3: Graphing Data.) If a software program is used to generate a graph instead of hand drawing, a copy of the graph should be pasted into the lab notebook.

Discussion and Error Analysis

Briefly discuss the results. Evaluate the quality of the results, discussing such issues as precision and accuracy, if possible (see Appendix A, Part A-4: Evaluating Data).

Conclusions

Summarize any conclusions you have drawn, such as the result of testing a hypothesis, the relationship between two variables, or a value for a constant that you have determined.

Format

Do not use pencil. The lab notebook is a *permanent* record of the work you have done in the lab. Accordingly, all lab notebook entries must be made in ink. Never obliterate data or comments if you decide they are erroneous. Draw a single line through the apparent ~~mistake~~, leaving it legible. You might later discover that the item that first appeared to be wrong actually was correct. Remember that the lab notebook is a "work in progress"; it will be messy, perhaps even stained; it is not expected to resemble a finished manuscript.

A well-kept lab notebook is one that answers the following question in the affirmative: Using only the notebook as a guide, could another student understand what problem was being investigated, perform the same experiment, and acquire the same data or make the same observations?